Springer Water

The book series Springer Water comprises a broad portfolio of multi- and interdisciplinary scientific books, aiming at researchers, students, and everyone interested in water-related science. The series includes peer-reviewed monographs, edited volumes, textbooks, and conference proceedings. Its volumes combine all kinds of water-related research areas, such as: the movement, distribution and quality of freshwater; water resources; the quality and pollution of water and its influence on health; the water industry including drinking water, wastewater, and desalination services and technologies; water history; as well as water management and the governmental, political, developmental, and ethical aspects of water.

Manoj Chandra Garg
Editor

The AI Cleanse: Transforming Wastewater Treatment Through Artificial Intelligence

Harnessing Data-Driven Solutions

Springer

Editor
Manoj Chandra Garg
Amity Institute of Environmental Sciences
Amity University
Noida, Uttar Pradesh, India

ISSN 2364-6934 ISSN 2364-8198 (electronic)
Springer Water
ISBN 978-3-031-67236-1 ISBN 978-3-031-67237-8 (eBook)
https://doi.org/10.1007/978-3-031-67237-8

This Springer imprint is published by the registered company Springer Nature Switzerland AG
The registered company address is: Gewerbestrasse 11, 6330 Cham, Switzerland

Contents

Chapter 1
The Integration of Artificial Intelligence in Advanced Wastewater Treatment Systems

Manoj Chandra Garg, Sheetal Kumari, and Smriti Agarwal

Abstract The integration of Artificial Intelligence (AI) into advanced wastewater treatment systems signifies a transformative advancement in environmental engineering, addressing the intricate challenges of modern wastewater management. This manuscript explores the necessity and impact of incorporating AI into these systems, emphasizing how intelligent technologies enhance the efficacy and adaptability of wastewater treatment processes. Modern wastewater management is increasingly complex due to urbanization, industrial expansion, and population growth, which traditional treatment methods struggle to address effectively. Historical wastewater treatment methods, evolving from basic pollutant containment to sophisticated processes, now face limitations in adapting to diverse and dynamic pollutants. AI emerges as a critical solution to these challenges. With capabilities in learning, adaptation, and data-driven decision-making, AI can analyze vast datasets, identify patterns, and optimize treatment processes in real time. This makes AI an ideal tool for navigating the complexities of contemporary wastewater, marking a paradigm shift towards sustainable, efficient, and adaptive treatment solutions. This manuscript delves into the fundamentals of AI and its applications in wastewater treatment, aiming to bridge the gap between theoretical knowledge and practical implementation. Advanced wastewater treatment systems have evolved to meet stringent environmental standards, characterized by their ability to address complex contaminant matrices. These systems, which emphasize resource recovery and sustainability, transform wastewater into a reservoir of recoverable resources. Key features such as membrane filtration, advanced oxidation processes, and nutrient removal

M. C. Garg (✉) · S. Kumari
Amity Institute of Environmental Science (AIES), Amity University Uttar Pradesh, Sector-125, Noida, Gautam Budh Nagar 201313, India
e-mail: manoj28280@gmail.com

S. Kumari
e-mail: Sheetalluhania99@gmail.com

S. Agarwal
Department of Electronics and Communication Engineering, Motilal Nehru National Institute of Technology Allahabad, Prayagraj, U.P, India

M. C. Garg (ed.), *The AI Cleanse: Transforming Wastewater Treatment Through Artificial Intelligence*, Springer Water, https://doi.org/10.1007/978-3-031-67237-8_1

strategies distinguish advanced systems by their precision, efficiency, and adaptability. The manuscript highlights the multifaceted nature of contemporary wastewater, influenced by urbanization, industrialization, and societal changes, and underscores the importance of adaptable and resilient treatment solutions. AI serves as a catalyst in overcoming these challenges. Its intelligent integration optimizes treatment processes, enhances efficiency, reduces energy consumption, and addresses operational intricacies. By leveraging AI's adaptive capabilities and real-time data processing, advanced wastewater treatment systems can achieve greater sustainability and efficiency. This manuscript provides a comprehensive view of the evolution, characteristics, and challenges of advanced wastewater treatment, setting the stage for exploring AI's transformative role in this dynamic landscape. It underscores the transformative potential of integrating AI, inspiring a deeper appreciation for the synergy between intelligent technologies and environmental sustainability.

Keywords Artificial intelligence (AI) · Wastewater treatment · Environmental engineering · Resource recovery · Sustainable management · Machine learning

1.1 Introduction

The integration of AI into advanced wastewater treatment systems marks a pivotal moment in the realm of environmental engineering, ushering in an era where intelligent technologies converge with the complexities of water management. The contemporary landscape of wastewater management is characterized by unprecedented challenges, propelled by the evolving composition of wastewater in the face of urbanization, industrial expansion, and population growth [54]. Traditional wastewater treatment methods, once stalwarts in addressing environmental concerns, now find themselves grappling with the limitations imposed by the complex and diverse nature of modern pollutants. The journey of wastewater treatment unfolds against the backdrop of historical methodologies that, over time, have transitioned from rudimentary practices to the sophisticated systems in place today. Early forms of wastewater treatment were primarily concerned with basic containment and removal of visible pollutants [50]. As societies progressed, so did the methodologies, incorporating filtration and sedimentation techniques to enhance treatment efficacy. The rationale for exploring AI in wastewater treatment stems from the inherent limitations of conventional methods in coping with this evolving scenario [54]. Conventional treatment, often designed for specific pollutants and standardized conditions, struggles to adapt to the dynamic and multifaceted nature of modern wastewater. The need for a more nuanced, adaptive, and intelligent approach becomes imperative to address the diversity and intensity of pollutants effectively.

Imperative Need for AI Integration

The imperative need for AI integration arises as a response to this shifting paradigm. AI, with its capacity for learning, adaptation, and data-driven decision-making, emerges as a promising solution to navigate the complexities of modern wastewater [47]. Unlike rigid rule-based systems, AI can analyze vast datasets, identify patterns, and optimize treatment processes in real time [6]. Its ability to evolve and learn from ongoing processes aligns seamlessly with the dynamic nature of contemporary wastewater challenges.

Setting the Stage for Transformation

As the historical narrative transitions from simplistic approaches to the intricate challenges of today, the integration of AI emerges as a beacon of innovation. The stage is now set for a paradigm shift, where AI, equipped with its cognitive capabilities, becomes an integral ally in the quest for sustainable, efficient, and adaptive wastewater treatment solutions. The chapter's scope transcends the ordinary by encompassing a holistic view of AI's role in wastewater treatment. Beyond a surface-level exploration, it seeks to delve into the fundamentals of AI, unravel its adaptability, and elucidate how it optimizes a spectrum of diverse treatment processes [6]. This expansive approach positions the chapter as a roadmap for readers, guiding them through a nuanced exploration that spans from foundational AI knowledge to its intricate applications in the wastewater treatment landscape. The chapter's objectives manifest in its intent to seamlessly guide readers through a spectrum of knowledge—from foundational insights about AI to its pragmatic applications in advanced wastewater treatment. By doing so, it seeks to bridge the gap between theoretical understanding and practical implementation, fostering a comprehensive comprehension of the symbiotic relationship between AI and the evolving landscape of wastewater treatment.

Beyond mere informative content, the objectives of the chapter extend to instigating a profound understanding. It endeavours to go beyond surface-level knowledge dissemination, aiming to cultivate a deeper appreciation for the symbiosis between AI and advanced wastewater treatment. By the chapter's conclusion, readers are not only informed but inspired to recognize the transformative potential that lies in the synergy of AI and the intricate processes of wastewater treatment.

1.2 Overview of Advanced Wastewater Treatment Systems

In the pursuit of sustainable water management, traditional wastewater treatment systems are evolving into advanced frameworks, reflecting the urgency to meet stringent environmental standards. Delving into the landscape of advanced wastewater treatment systems provides a comprehensive overview that sets the backdrop for understanding the integration of AI [68].

The historical narrative of wastewater treatment systems unfolds as a tapestry woven with threads of innovation and adaptation. From the humble origins of septic tank designs to the sophistication introduced by activated sludge processes, this journey epitomizes a relentless pursuit of more effective and environmentally conscious methodologies [68]. Each phase in the evolution becomes a chapter in the story, echoing the industry's determination to progress beyond the status quo. Central to this evolution is the unyielding quest for effectiveness and environmental consciousness. The storyline encapsulates the industry's dedication to refining and redefining methods that not only treat wastewater but also align with ecological sensitivities [70]. The shift from rudimentary designs to advanced processes mirrors a paradigmatic transformation, where sustainability and efficiency emerge as guiding principles. Central to this exploration is the emphasis on the factors propelling the need for wastewater treatment advancements. Population growth amplifies the wastewater volume, industrialization introduces complex pollutants, and the emergence of novel contaminants further complicates the treatment landscape [37]. Understanding these propelling factors becomes pivotal in comprehending why conventional systems, while groundbreaking in their time, reach a threshold demanding the integration of cutting-edge technologies.

The evolution of wastewater treatment becomes a story not just of chronological progress but also of strategic adaptation to meet contemporary challenges. As the chapter unfolds, the groundwork is laid for the subsequent exploration of how AI intertwines with this dynamic landscape, offering solutions and innovations for the challenges that have marked the historical evolution of wastewater treatment systems.

1.2.1 Characteristics of Advanced Wastewater Treatment Systems

Foundations of Complexity

Delving into the intricate realm of advanced wastewater treatment systems requires a foundational understanding of the characteristics that define their complexity. These systems transcend the conventional paradigm of mere pollutant removal; they evolve to address complex matrices of contaminants [28]. The essence of advanced wastewater treatment lies in its departure from a singular focus on pollutant removal. Instead, these systems are crafted to navigate through intricate matrices that include pharmaceuticals, personal care products, and persistent organic pollutants. This nuanced approach aligns with the contemporary reality of wastewater, reflecting the ever-expanding spectrum of challenges posed by diverse and persistent contaminants [28]. One of the defining characteristics is the emphasis on resource recovery. Advanced wastewater treatment systems recognize that within the complexity of wastewater, there lies untapped potential and value. Beyond purification, these systems are designed to extract and reclaim resources embedded in the wastewater matrix. This paradigm shift aligns with the broader concept of sustainability, transforming wastewater from a mere waste to a reservoir of recoverable resources.

Distinguishing Features: Membrane Filtration, Advanced Oxidation Processes, and Nutrient Removal

Unravelling the distinguishing features that set advanced wastewater treatment systems apart, membrane filtration, with its precision in separating contaminants, becomes a cornerstone. Advanced oxidation processes, leveraging cutting-edge chemical reactions, add another layer of sophistication [28]. Nutrient removal strategies address the intricate dynamics of nutrient imbalances, contributing to the overall efficiency of these systems [68].

Multifaceted Nature of Contemporary Wastewater

The discourse underscores the multifaceted nature of contemporary wastewater. As urbanization, industrialization, and societal changes influence wastewater composition, the treatment landscape evolves in response [77]. The exploration emphasizes that these systems are not just purifiers, they are adaptable and resilient, navigating through the evolving complexity of contaminants.

Efficiency, Resource Reclamation, and Environmental Impact

Purification efficiency is intertwined with resource reclamation, forming a dual objective that defines the success of advanced wastewater treatment systems. The chapter brings attention to how these systems not only meet the demands of pollutant removal but also contribute to minimizing environmental impact [77]. The synergy between efficiency, resource reclamation, and environmental sustainability becomes a focal point, setting the stage for the subsequent exploration of how Artificial Intelligence integrates into this intricate landscape [21].

1.2.2 Challenges in Advanced Wastewater Treatment

Navigating Complexity

The journey towards advanced wastewater treatment is a progressive ascent but not devoid of intricate challenges. Unravelling the multifaceted hurdles that characterize the implementation of sophisticated treatment systems, as the treatment landscape evolves to address contemporary wastewater complexities, challenges such as energy consumption, operational intricacies, and the demand for tailored solutions emerge [72]. One of the primary challenges highlighted is the significant energy consumption associated with advanced wastewater treatment [72]. The intricacies involved in processes like membrane filtration and advanced oxidation demand substantial energy inputs. Navigating the delicate balance between treatment efficiency and energy sustainability becomes a pivotal aspect [38]. This challenge necessitates innovative strategies to optimize energy consumption without compromising the efficacy of treatment. The operational intricacies of advanced treatment systems add another layer of complexity [38]. As these systems become more sophisticated to handle diverse contaminants, operational challenges arise. Maintaining seamless operations amidst intricate processes, adapting to variable influent compositions, and ensuring consistent performance become formidable tasks [39].

Tailored Solutions Requirement

Unlike conventional treatment methods, advanced wastewater treatment demands tailored solutions. The diverse nature of contaminants and the dynamic composition of wastewater require customized approaches [38]. This challenge emphasizes the necessity for adaptive systems that can tailor treatment strategies based on real-time conditions. The call for flexibility in approach becomes apparent, setting the stage for the subsequent exploration of AI as a transformative solution.

AI as the Catalyst

Setting the stage for understanding the pivotal role of AI in overcoming these challenges, the intelligent integration of AI is not merely about optimizing treatment processes; it becomes a catalyst for mitigating inherent challenges [32]. The exploration dives into how AI, with its adaptive capabilities and real-time data processing, offers solutions to enhance efficiency, reduce energy consumption, and address the operational intricacies of advanced wastewater treatment [52]. It lays the groundwork for the subsequent exploration of AI's transformative role in this dynamic landscape, emphasizing the synergy between technological innovation and sustainable, efficient wastewater treatment practices [32]. In essence, this subsection provides a panoramic view of advanced wastewater treatment, encapsulating its evolution, intrinsic characteristics, and the hurdles that necessitate a paradigm shift [22]. It lays the groundwork for the subsequent exploration of AI's transformative role in this dynamic landscape, emphasizing the synergy between technological innovation and sustainable, efficient wastewater treatment practices.

1.3 Fundamentals of Artificial Intelligence (AI)

AI stands as a transformative force reshaping the landscape of various industries, and its integration into wastewater treatment heralds a new era in efficient, adaptive, and data-driven environmental management [22]. AI's prowess lies in its versatility in problem-solving. Whether it's image recognition, natural language processing, or strategic decision-making, AI extends its reach across diverse domains. The narrative concludes by accentuating the contemporary paradigm of human–machine symbiosis [69]. AI is not positioned as a replacement for human intelligence but rather as a collaborative partner, augmenting human capabilities.

1.3.1 *Introduction to AI in Environmental Sciences*

In the realm of wastewater treatment, deciphering the core principles of AI becomes imperative. At the heart of AI lies the dynamic landscape of machine learning. This subsection unravels the intricate world of machine learning algorithms, the driving force behind the AI landscape (Fig. 1.1). Ranging from supervised algorithms, where models are trained on labelled data, to unsupervised algorithms that explore patterns autonomously, machine learning forms the bedrock of AI's problem-solving capabilities [69]. The emphasis here is on how these algorithms empower systems to learn, predict, and adapt, fostering continuous evolution in response to the intricacies of wastewater treatment scenarios. Delving into the realm of neural networks, this part provides comprehensive insights into the architecture inspired by the human brain. Neural networks, particularly in the context of deep learning, contribute significantly to AI's prowess [79].

Natural Language Processing (NLP)

Another facet explored is Natural Language Processing (NLP), a domain within AI that enables systems to understand and respond to human language. In wastewater treatment, where effective communication is crucial, NLP plays a pivotal role [11]. This aspect of AI contributes to enhancing the usability and accessibility of AI applications in the wastewater treatment landscape. In essence, this exploration underscores the dynamic adaptability of AI, where machine learning, neural networks, and NLP converge to create intelligent systems capable of learning, reasoning, and interacting [45]. As these core principles unfold, the subsequent sections will delve into

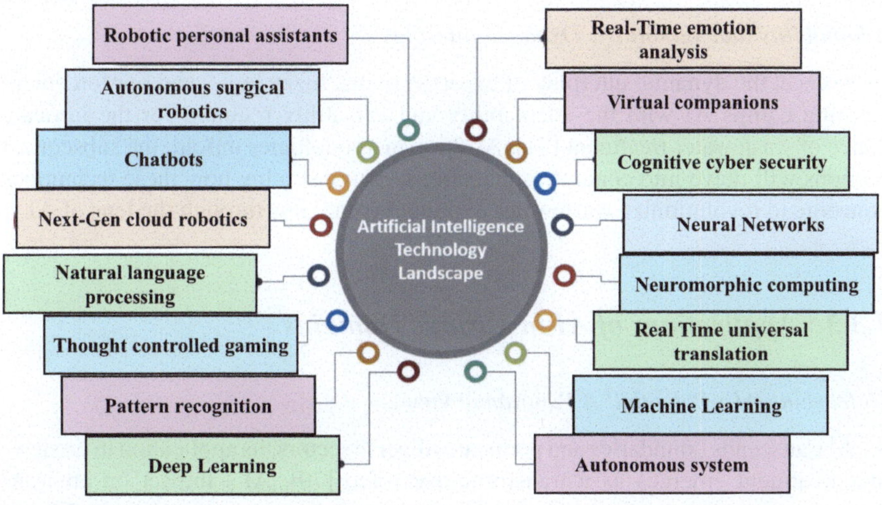

Fig. 1.1 Multinational AI landscape

how these principles translate into tangible applications, revolutionizing wastewater treatment paradigms through the lens of AI [75].

1.3.2 AI Methodologies and Techniques

In the landscape of AI, a diverse array of methodologies and techniques stand as the tools for crafting solutions tailored to specific problem domains. Delving into the nuanced world of AI methodologies, the exploration unveils their unique attributes and applications in the context of wastewater treatment challenges.

Harnessing Human Expertise: Expert Systems

The journey begins with expert systems, a methodology that harnesses human expertise to create rule-based decision support systems. In wastewater treatment, where the amalgamation of human knowledge and AI is pivotal, expert systems serve as beacons of informed decision-making [45]. Fuzzy logic emerges as a crucial player in accommodating uncertainty, mirroring human decision-making processes in complex and ambiguous environments. In wastewater treatment scenarios fraught with dynamic variables and uncertainties, fuzzy logic becomes a valuable tool [82]. A pivotal paradigm within AI, reinforcement learning takes centre stage as a methodology that allows systems to learn through trial and error. The dynamic interplay of receiving feedback and refining strategies over time empowers AI systems with a learning mechanism akin to human adaptation. In the realm of wastewater treatment, where real-time adjustments are crucial, reinforcement learning emerges as a dynamic force [82].

Adaptability and Versatility: Dynamic Interplay

In essence, the dynamic interplay of expert systems, fuzzy logic, and reinforcement learning equips AI with the adaptability and versatility required for the intricate dance of wastewater treatment [81]. As these methodologies unfold, the subsequent sections will delve into concrete applications, demonstrating how these techniques converge to revolutionize wastewater treatment paradigms through the lens of AI.

1.3.3 Applications of AI in Various Domains

Unleashing AI's Potential: A Panoramic View

As AI transcends boundaries and permeates diverse sectors, its application in wastewater treatment emerges as a transformative force [70]. AI's impact on environmental management unfolds through its prowess in optimizing processes. Within the realm of wastewater treatment, where intricacies abound, AI becomes a catalyst for enhancing operational efficiency [19]. The predictive capabilities of AI take centre stage, offering wastewater treatment plants the foresight needed to anticipate challenges. Through advanced algorithms and machine learning models, AI empowers environmental managers to foresee fluctuations in pollutant levels, enabling proactive measures [50]. This exploration unravels the layers of predictive modelling, illustrating how AI becomes a reliable ally in the quest for efficient and anticipatory wastewater treatment. The integration of AI into wastewater treatment heralds a paradigm shift, promising not only enhanced operational efficiency but also resource recovery.

1.4 Applications of AI in Wastewater Treatment

1.4.1 Real-Time Monitoring and Control

In the intricate realm of wastewater treatment, the pivotal role of real-time monitoring and control takes centre stage, guided by the transformative influence of AI.

Foundations of Efficiency: AI and Advanced Sensors

At the heart of real-time monitoring and control lies the integration of advanced sensors and data analytics, orchestrated by the capabilities of AI. These sophisticated sensors, attuned to the nuances of wastewater composition, continuously collect data on crucial parameters. From pH levels and dissolved oxygen concentrations to chemical compositions, the real-time data stream serves as a dynamic canvas reflecting the evolving state of the treatment system [54]. The application of AI in environmental sustainability are shown in Fig. 1.2.

Machine Learning Alchemy: Transforming Data into Insights

The real-time data generated by advanced sensors undergoes a metamorphosis, courtesy of AI's machine learning algorithms. This transformative alchemy unravels insights from the data stream, offering a nuanced understanding of the treatment system's current dynamics. Machine learning becomes the key that unlocks patterns, identifies trends, and unveils subtle deviations, providing a comprehensive perspective on the ever-changing landscape of wastewater treatment [63]. Armed with the insights gleaned from real-time monitoring, AI paves the way for adaptive control strategies. These strategies, finely tuned to the intricacies of wastewater treatment, empower treatment plants to dynamically respond to fluctuations. Whether in influent composition or flow rates, the adaptive control mechanisms ensure that the treatment system operates at optimal efficiency [24]. This not only minimizes operational costs but also contributes to the overarching goal of establishing a sustainable and resilient wastewater treatment infrastructure. The synergy between AI-driven real-time monitoring and adaptive control strikes a delicate balance, orchestrating a symphony of efficiency, cost reduction, and sustainability. By enabling precise interventions in response to real-time data, AI ensures that wastewater treatment processes operate

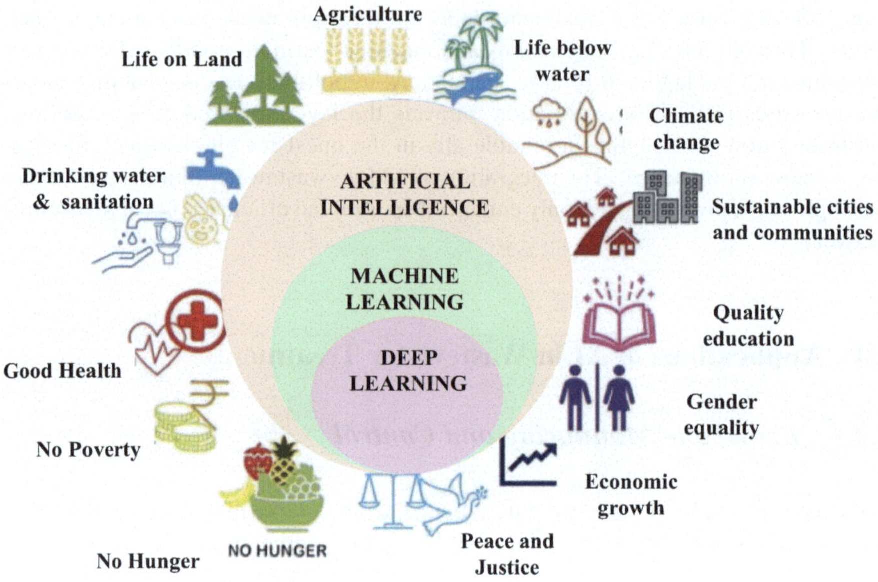

Fig. 1.2 Application of artificial intelligence in environmental sustainability and climate change

in harmony with the ever-shifting variables [10]. This not only enhances the overall efficiency of the treatment plant but also establishes a foundation for sustainable practices, reinforcing the resilience of wastewater treatment infrastructure. The integration of advanced sensors, AI algorithms, and adaptive control strategies heralds a new era, where the efficiency, cost-effectiveness, and sustainability of wastewater treatment converge seamlessly [10]. As we navigate this transformative landscape, the promise of AI in crafting the future of wastewater treatment stands resolute, offering a vision where technology becomes the cornerstone of environmental stewardship.

1.4.2 Predictive Modeling for Process Optimization

Unleashing the Power of Anticipation: AI-Infused Predictive Modeling

In the dynamic landscape of wastewater treatment, the integration of AI ushers in a new era marked by predictive modelling—an innovative paradigm that reshapes the contours of process optimization is shown in Fig. 1.3 [10]. Unravelling the transformative impact of AI in creating predictive models, this section elucidates how historical data becomes the catalyst for anticipating and optimizing crucial facets of wastewater treatment [56].

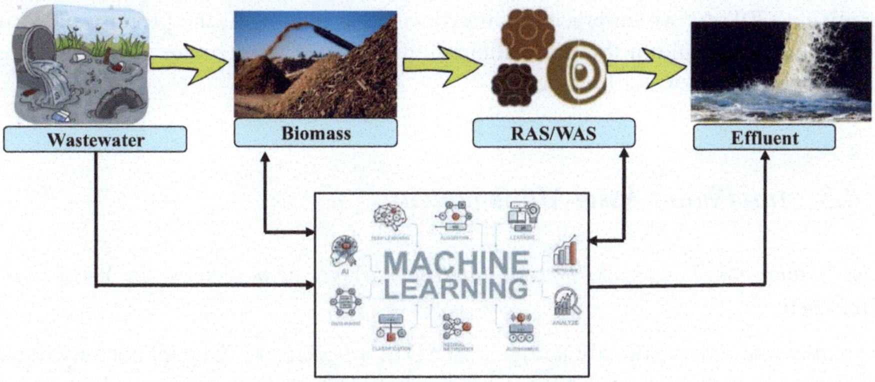

*RAS-Return Activated Sludge, WAS- Waste Activated Sludge

Fig. 1.3 Structure of machine learning ensemble

Anticipating the Unseen: Forecasting Variations and Scenarios

The prowess of predictive modelling shines brightest in its ability to anticipate the unseen. By leveraging historical data, AI-driven models become adept at forecasting variations in influent characteristics and hydraulic loading, among other critical parameters. These forecasts extend beyond mere predictions, enabling treatment plants to envision plausible scenarios that may unfold [51]. This foresight becomes a strategic tool for decision-makers, offering a nuanced understanding of potential changes in the wastewater treatment landscape. The predictive models crafted by AI serve as invaluable companions for decision-makers in the realm of wastewater treatment [42]. Armed with insights derived from historical data and machine learning algorithms, these models contribute to strategic planning endeavours. Decision-makers gain the ability to peer into the future, armed with a comprehensive understanding of potential scenarios [5]. This foresightedness not only aids in strategic planning but also empowers treatment plants to proactively optimize their processes.

Optimization in Advance: Improving Resource Utilization and Efficiency

The true essence of predictive modelling lies in its capacity to optimize processes in advance. By foreseeing variations and potential scenarios, treatment plants can proactively fine-tune their operations. This proactive optimization leads to improved resource utilization, reduced chemical usage, and an overall enhancement in treatment efficiency [49]. The marriage of AI and predictive modelling becomes a catalyst for instilling a proactive, anticipatory approach in the intricate dance of wastewater treatment [65]. The integration of AI-driven predictive modelling transcends traditional paradigms, ushering in an era where wastewater treatment is not a reactive process but a proactive, anticipatory endeavour. The synergy between historical data, machine learning algorithms, and predictive models becomes a powerful force, enabling treatment plants to navigate the complexities of the future with informed precision [30]. As we embrace this transformative paradigm, the promise of AI in redefining optimization through anticipation stands as a testament to the evolving landscape of wastewater treatment.

1.4.3 Intelligent Asset Management

Unleashing the Power: AI-Driven Intelligent Asset Management in Wastewater Treatment

The integration of Artificial Intelligence (AI) in wastewater treatment not only transforms processes but also heralds the era of intelligent asset management. At the heart of AI-driven intelligent asset management lies the paradigm of predictive maintenance [74]. AI algorithms, endowed with the ability to analyze performance data from various treatment plant assets, function as oracles of asset health [62]. By deciphering patterns and anomalies in real time, these algorithms predict potential equipment failures before they manifest as critical issues. One of the tangible benefits of AI-driven intelligent asset management is the minimization of unscheduled downtime. By predicting maintenance needs in advance, treatment facilities can implement timely interventions, avoiding unexpected breakdowns [67]. This resilience factor not only ensures continuous operational efficiency but also contributes to the overall reliability of wastewater treatment processes.

Extending Lifespan: Optimizing Plant Reliability

AI's role in intelligent asset management extends beyond avoiding downtime. It becomes a guardian of the equipment's lifespan. By optimizing maintenance schedules based on real-time data, AI contributes to extending the lifespan of treatment plant assets [59]. This longevity factor aligns with sustainability goals, as it minimizes the need for premature replacements and capitalizes on the full potential of existing infrastructure.

1.4.4 Resource Recovery and Energy Optimization

Harvesting Value: AI-Driven Resource Recovery in Wastewater Treatment

AI's transformative footprint in wastewater treatment extends to resource recovery and energy optimization, reshaping how treatment plants harness value from operational processes. Delving into the profound impact of AI, we uncover opportunities for resource extraction and refine energy-intensive operations [61].

Data-Driven Insights: Unveiling Resource Recovery Opportunities

At the core of AI-driven resource recovery lies the ability to sift through vast datasets and identify untapped opportunities. Machine learning algorithms, fuelled by operational data, analyze the complex matrices of wastewater to pinpoint valuable by-products [8]. This data-driven approach unveils hidden potentials for resource extraction, ranging from nutrients and metals to organic compounds.

Energy Optimization: AI as the Maestro of Efficiency

In parallel, AI plays a pivotal role in optimizing energy-intensive processes within wastewater treatment. By analyzing real-time data, machine learning algorithms fine-tune energy consumption, ensuring that treatment processes operate at the nexus of efficiency and sustainability [23]. This energy optimization not only reduces operational costs but also aligns with global efforts towards greener and more environmentally conscious practices. The symbiotic relationship between resource recovery and energy optimization becomes the hallmark of AI's dual role in wastewater treatment [13]. By simultaneously harvesting value from operational processes and fine-tuning energy consumption, AI emerges as the maestro orchestrating a harmonious balance between efficiency and sustainability.

1.5 Smart Sensors and Data Collection

Revolutionizing Data Dynamics: The Journey of Smart Sensors

The evolution of smart sensors in wastewater treatment signifies a paradigm shift in data dynamics, transcending traditional monitoring methods. Delving into the remarkable journey of smart sensors, we trace their transformation from conventional instruments to advanced, interconnected devices designed for real-time data acquisition [53].

From Conventional to Smart: A Technological Metamorphosis

Smart sensors represent a technological metamorphosis in the landscape of wastewater treatment. The journey begins with the transition from conventional monitoring approaches, marked by periodic sampling and manual measurements, to the era of smart sensors. These sensors, embedded with cutting-edge technologies, usher in an era of continuous, real-time data collection, offering an unprecedented level of accuracy and immediacy [1].

Connected Insights: Real-time Data Acquisition

The integration of smart sensors into wastewater treatment processes introduces a new era of connected insights. These sensors, equipped with connectivity features, enable seamless communication and data transfer [83]. The real-time acquisition of data is not confined to isolated instances but becomes an ongoing process, providing treatment plants with a comprehensive view of key parameters. Smart sensors redefine wastewater monitoring by offering a comprehensive perspective. They go beyond the basics, capturing a spectrum of parameters crucial for effective treatment [55]. From chemical composition and microbial activity to physical characteristics, smart sensors create a holistic understanding of the wastewater matrix [74]. This comprehensive monitoring ensures that treatment plants have nuanced insights, fostering more informed and effective decision-making.

1.5.1 Real-Time Data Transmission and Connectivity

Significance of Instantaneous Data Transmission

The hallmark feature of smart sensors, instantaneous data transmission, is a transformative element in wastewater treatment. We can explore the profound significance of real-time data transmission and its impact on the operational dynamics of treatment facilities.

Empowering Prompt Responses: Real-time Insights

Real-time data transmission empowers treatment facilities with prompt insights into the ever-changing conditions of the wastewater treatment process. Unlike traditional systems where data retrieval is a time-consuming process, smart sensors transmit information instantaneously [44]. This immediacy provides treatment plants with a real-time snapshot of critical parameters, enabling swift responses to fluctuations in influent composition and flow rates. The instantaneous nature of data transmission sets the stage for optimizing wastewater treatment processes. Treatment plants can dynamically adjust their operations based on real-time information, ensuring that each phase of the treatment is precisely tailored to prevailing conditions [18]. This dynamic approach enhances overall treatment efficiency, minimizes operational costs, and contributes to a more sustainable and resilient wastewater treatment infrastructure. Real-time data transmission plays a pivotal role in ensuring regulatory compliance. Treatment facilities are better equipped to monitor parameters that fall within specified standards, promptly identifying deviations [26]. The timely nature of data transmission allows for corrective measures to be implemented swiftly, preventing potential violations and maintaining adherence to environmental regulations.

1.5.2 Applications of Smart Sensors in Water Quality Monitoring

Detecting Diverse Contaminants: Heavy Metals, Nutrients, and Organics

Smart sensors stand as key players in the monitoring of water quality, demonstrating exceptional capabilities in detecting and measuring a spectrum of contaminants. This subsection delves into the diverse applications of smart sensors, emphasizing their proficiency in identifying heavy metals, nutrients, and organic pollutants [29]. The precision in detecting these specific contaminants contributes significantly to the efficacy of water quality monitoring.

Heavy Metal Detection: Precision for Environmental Safety

One of the notable applications of smart sensors lies in their ability to detect heavy metals with precision. From mercury to lead, smart sensors provide accurate measurements, allowing treatment facilities to monitor heavy metal concentrations in real-time [60]. This capability is crucial for ensuring environmental safety, as heavy metals pose significant risks to ecosystems and human health. The data collected by smart sensors enables prompt responses to deviations, mitigating potential environmental hazards [41].

Nutrient Monitoring: Balancing Ecosystem Dynamics

Smart sensors play a vital role in monitoring nutrients, a critical aspect of water quality. The precise measurement of nutrient levels, including nitrogen and phosphorus, aids in balancing ecosystem dynamics. Treatment plants can utilize this information to optimize nutrient removal processes, preventing issues such as eutrophication [14]. The applications of smart sensors in nutrient monitoring contribute to the overall health and sustainability of aquatic environments.

Organic Pollutant Identification: Targeted Interventions

In the realm of organic pollutants, smart sensors excel in their capacity to identify and measure various contaminants. From industrial chemicals to pharmaceutical residues, these sensors offer a targeted approach to monitoring organic pollutants [35]. The ability to focus on specific parameters allows treatment facilities to implement precise interventions, ensuring that the treatment processes are tailored to address the unique challenges posed by organic pollutants.

1.5.3 Integration with Internet of Things (IoT) Platforms

Smart sensors play a pivotal role in the integration with IoT platforms, forming a symbiotic relationship that enhances the capabilities of wastewater treatment systems. We explore how smart sensors seamlessly become integral components of the larger IoT ecosystem, creating a networked environment where data becomes a dynamic and collaborative asset [64]. The essence of smart sensors' integration with IoT platforms lies in collaborative data analysis. It's not merely about the collection of data but the transformative process of turning raw information into actionable insights [20]. Through IoT, smart sensors contribute to a collective pool of data that is analyzed in real-time. The interconnected nature of smart sensors within IoT platforms enhances the scalability and adaptability of wastewater treatment systems [46]. As treatment facilities evolve or expand, the IoT ecosystem can seamlessly accommodate additional sensors and devices. This scalability ensures that the monitoring network can grow in tandem with the changing needs of the treatment plant, optimizing coverage and maintaining comprehensive data collection. The integration of smart sensors with IoT platforms opens avenues for predictive analytics [31]. Predictive analytics empower treatment plants to move from reactive to proactive strategies, anticipating challenges [46] and implementing pre-emptive measures for enhanced operational efficiency. The collaborative data generated by smart sensors within the IoT ecosystem allows for holistic optimization of treatment processes. Data-driven decision-making becomes more informed, considering a multitude of parameters and variables [12]. Treatment facilities can optimize resource allocation, energy consumption, and process efficiency by leveraging the interconnected data provided by smart sensors through the IoT platform.

1.6 Machine Learning in Wastewater Treatment Plant Operation

Revolutionizing Wastewater Treatment with Machine Learning

At its core, machine learning is a powerful computational approach that empowers systems to learn and evolve based on data inputs [78]. ML algorithms, inspired by the human capacity to learn from experience, navigate the complexities of wastewater treatment by adapting to evolving datasets. A key facet introduced here is the dynamic adaptability of ML in wastewater treatment operations. Unlike static systems, ML-equipped treatment plants can adjust and fine-tune their processes in real-time [80]. The algorithms within ML models continuously learn from incoming data, enabling treatment systems to optimize their responses to fluctuating conditions such as variations in influent composition and flow rates.

1.6.1 Data-Driven Decision Making with Machine Learning

Revolutionizing Decision-Making in Wastewater Treatment

Machine learning (ML) instigates a transformative paradigm shift in decision-making processes within wastewater treatment plants, placing a strong emphasis on data-driven approaches [25]. At the core of ML's influence is its ability to process vast datasets swiftly and effectively. The result is a reservoir of actionable insights that empower decision-makers in wastewater treatment. By facilitating decision support through meticulous data analysis, ML enhances the accuracy and efficiency of operational decisions in wastewater treatment [4]. The algorithms' capacity to discern subtle trends and deviations ensures that plant operators can make informed choices, leading to optimized processes and resource utilization.

1.6.2 Predictive Modeling for Treatment Process Optimization

Optimizing Wastewater Treatment Processes Through Predictive Modeling

Predictive modelling using machine learning (ML) establishes itself as a pivotal cornerstone in the quest to optimize treatment processes within wastewater treatment plants. At the heart of ML-driven predictive modelling lies its profound capability to anticipate changes in critical parameters [2]. By deciphering these patterns, the models become adept at predicting influent variations, offering a valuable foresight mechanism for plant operators. The true power of predictive modelling unfolds in its ability to empower plant operators to proactively adjust processes. As the models foresee changes in influent characteristics and pollutant concentrations, operators gain a strategic advantage in adjusting treatment processes in advance [48]. This proactive approach ensures optimal performance, minimizes operational costs and contributes to the overall efficiency of wastewater treatment.

1.6.3 Integration of Supervised Learning in Treatment Processes

Within the intricate domain of wastewater treatment, supervised learning algorithms claim a central role in training models to recognize and respond to specific conditions. Supervised learning's effectiveness hinges on its ability to leverage labelled datasets [43]. In wastewater treatment, this could range from pollutant removal efficiency to sludge dewatering and energy consumption optimization [34]. The utilization of labelled datasets ensures that the algorithm learns to associate specific conditions with desired responses, enhancing the accuracy of its predictions. The application of supervised learning extends across a spectrum of tasks critical to wastewater treatment processes [16]. Moreover, their role in sludge dewatering, a key aspect of treatment plant operations, is explored, emphasizing how supervised learning optimizes this process. Additionally, the section sheds light on how these algorithms can be instrumental in optimizing energy consumption and aligning treatment processes with sustainability goals [66]. Supervised learning's integration ensures continuous learning and adaptation within treatment processes. The algorithms, having been trained on labelled datasets, continuously refine their understanding based on real-time data. This adaptability enables treatment plants to respond to changing conditions with precision, ensuring that the processes are aligned with the desired outcomes [36]. In essence, the seamless integration of supervised learning algorithms aligns wastewater treatment processes with desired outcomes. Whether it's achieving higher pollutant removal rates, optimizing sludge dewatering, or enhancing energy efficiency, supervised learning contributes to targeted and effective operation [27].

1.7 Challenges and Future Perspectives

Ensuring data quality in wastewater treatment processes is a critical aspect of the successful integration of AI applications. This involves implementing pragmatic approaches to enhance data accuracy, reliability, and consistency, laying the groundwork for reliable insights and decision-making [71]. The following strategies are advocated to fortify the foundation of wastewater data:

1. **Regular Calibration of Sensors**: Regular calibration of sensors is fundamental to maintaining data accuracy. Sensors are prone to drift and degradation over time, impacting the reliability of the data they generate. By periodically calibrating sensors, deviations and inaccuracies can be identified and corrected. This ensures that the data collected remains precise and aligns with the actual conditions of the wastewater treatment processes.
2. **Robust Data Validation Protocols**: The implementation of robust data validation protocols is crucial for identifying and rectifying anomalies in the collected data. These protocols involve systematic checks and validations to assess data integrity. Any inconsistencies or outliers are flagged and addressed through predefined procedures. This meticulous validation process contributes to the overall reliability of the dataset and prevents the propagation of erroneous information.
3. **Stringent Quality Assurance Measures**: Stringent quality assurance measures are essential to uphold the reliability of the data. This involves implementing rigorous procedures to verify the accuracy and consistency of the collected information. Quality assurance measures may include regular audits, cross-verification of data with independent sources, and adherence to standardized data collection practices. By enforcing stringent quality controls, the risk of relying on flawed or misleading data is minimized.
4. **Documentation and Metadata Management**: Proper documentation and metadata management play a crucial role in ensuring data reliability. Keeping detailed records of sensor calibrations, validation processes, and any modifications to the data helps in traceability and transparency. Metadata, such as information about the source, timestamp, and conditions during data collection, provides context to the data. This aids in understanding the dataset's history and facilitates informed decision-making.
5. **Continuous Monitoring and Feedback Loop**: Establishing a continuous monitoring system with a feedback loop is essential for ongoing data quality improvement. Real-time monitoring allows for immediate identification of issues, enabling prompt corrective actions. Additionally, a feedback loop involving regular assessments and adjustments to data collection processes ensures adaptive data management, addressing emerging challenges and maintaining the accuracy of the dataset over time.

1.7.1 Emerging Trends and Future Directions: Charting the Course

1.7.1.1 Integration of Cutting-Edge Technologies

In the dynamic landscape of wastewater treatment, the integration of cutting-edge technologies stands as a transformative trend. This section explores recent advancements in materials, processes, and monitoring technologies, redefining conventional treatment approaches. It delves into the potential of innovative technologies to enhance treatment efficiency, reduce environmental impact, and introduce novel methodologies [9]. It analyses the advantages of moving away from the traditional centralized model, highlighting benefits such as enhanced resilience, decreased energy consumption, and heightened adaptability to diverse settings [7]. The discussion provides a comprehensive analysis of how decentralization is reshaping the entire wastewater treatment paradigm. It provides insights into the latest advancements in sensor technologies, including miniaturization, heightened sensitivity, and increased connectivity [76]. The discussion explores the integration of smart sensors into an IoT framework, showcasing the potential for creating a networked ecosystem that facilitates comprehensive data analysis and interpretation.

1.7.1.2 Resilient and Sustainable Practices

The concluding part emphasizes the overarching theme of transitioning towards more adaptive, resilient, and sustainable practices in wastewater treatment. It highlights the commitment of the industry to address environmental concerns and optimize water treatment processes for long-term sustainability [57]. This exploration sets the stage for an industry-wide shift towards sustainable and intelligent water management practices, ensuring the responsible stewardship of this vital resource.

1.8 Ethical and Environmental Considerations

1.8.1 Ethical Implications of AI in Wastewater Treatment

Ensuring fairness and equity stands as a paramount ethical consideration in the integration of AI algorithms within wastewater treatment processes. Fairness assessments take centre stage as systematic evaluations designed to identify and address any unintended discriminatory impacts. The narrative emphasizes the proactive nature of such assessments, advocating for continuous monitoring and adaptation to evolving challenges [40]. This approach ensures that the ethical considerations of fairness and equity are embedded throughout the lifecycle of AI applications in wastewater treatment. Involving diverse perspectives, including those of the affected communities, regulatory bodies, and industry experts, becomes a cornerstone for ensuring that the benefits and burdens of AI applications are distributed equitably [33]. This collaborative approach not only contributes to the development of fair and inclusive AI models but also promotes a shared responsibility for the ethical implications of these technologies in the context of wastewater treatment.

1.8.1.1 Life Cycle Analysis of AI Technologies

In the integration of AI technologies within wastewater treatment, a critical dimension that demands scrutiny is the life cycle analysis. This comprehensive assessment evaluates the environmental impact of AI technologies across various stages, ranging from manufacturing to disposal.

Carbon Footprint Examination: The analysis begins by delving into the carbon footprint, meticulously assessing the amount of carbon dioxide emissions generated throughout the life cycle of AI technologies [17]. This encompasses the manufacturing phase, operational use, and eventual disposal. Quantifying carbon emissions provides a holistic understanding of the ecological footprint associated with the adoption of AI-driven systems.

Energy Consumption Evaluation: Energy consumption emerges as a pivotal consideration. The subsection rigorously evaluates the total energy expended during both the production and operation phases of AI-driven systems [58]. This in-depth examination sheds light on the overall sustainability of these technologies, enabling stakeholders to make informed decisions regarding energy-efficient solutions and practices.

Resource Utilization Scrutiny: A critical aspect of the analysis involves scrutinizing the materials and resources involved in manufacturing AI technologies. This examination assesses the impact of resource utilization, providing insights into the availability and sustainability of essential materials. By navigating the complexities of resource dependencies, the narrative identifies areas for potential optimization and highlights opportunities for using alternative, more sustainable materials.

Advocacy for Sustainability: The overarching narrative strongly advocates for the selection of environmentally sustainable technologies in wastewater treatment. It emphasizes the significance of continually improving the life cycle performance of AI technologies. This involves a commitment to an iterative process of innovation and refinement, aiming to mitigate environmental impact and foster sustainable practices throughout the life cycle.

1.8.1.2 Ecological Footprint of Smart Sensors

The exploration of the ecological footprint of smart sensors within the realm of wastewater treatment unveils a nuanced understanding of the environmental implications associated with these indispensable components. The analysis initiates an examination of the materials used in the manufacturing of smart sensors [58]. It delves into the ecological consequences of acquiring, processing, and utilizing these materials. By scrutinizing the production process, the narrative assesses the environmental impact associated with the extraction of raw materials, manufacturing procedures, and the overall energy footprint involved in sensor production.

Energy Requirements During Operation: Smart sensors' ecological footprint extends beyond their manufacturing phase to their operational life. This part thoroughly evaluates the energy requirements during the sensors' operation within wastewater treatment processes [73]. Understanding the energy dynamics provides insights into the sustainability of sensor usage and identifies opportunities for optimizing energy consumption to minimize environmental impact.

Recycling and Disposal Challenges: Addressing the end-of-life phase, the write-up explores potential challenges related to recycling or disposing of smart sensors. It examines the environmental consequences of different disposal methods and advocates for the development of recycling programs [15]. The focus is on mitigating the environmental impact associated with sensor disposal and fostering responsible waste management practices within the context of smart sensor technologies.

Advocacy for Eco-Friendly Technologies: A central theme throughout the discussion is the advocacy for the adoption of eco-friendly sensor technologies. The narrative underscores the importance of choosing materials and manufacturing processes that align with sustainability goals [3]. By promoting the development and integration of environmentally conscious sensor technologies, the wastewater treatment sector can contribute to minimizing its ecological footprint.

1.9 Conclusion

The conclusion of this comprehensive exploration into the integration of artificial intelligence (AI) and advanced technologies in wastewater treatment encapsulates the key findings and provides insightful recommendations for future research endeavours.

Summary of Key Findings

The journey through this exploration revealed the transformative potential of AI in revolutionizing wastewater treatment systems. The integration of smart sensors and machine learning algorithms has ushered in a new era of efficiency, adaptability, and sustainability. Key findings include:

1. **Efficient Treatment Processes**: AI-driven systems enable real-time monitoring, adaptive control, and predictive modelling, optimizing treatment processes for enhanced efficiency and resource management.
2. **Environmental Considerations**: Ethical and environmental considerations, including fairness, privacy, and life cycle analysis, are integral to responsible AI implementation in wastewater treatment. The ecological footprint of smart sensors and the societal impact of AI technologies underscore the need for a balanced approach.

3. **Challenges and Opportunities**: The exploration delved into challenges such as data quality, model interpretability, and scalability, while concurrently highlighting opportunities to overcome these challenges. The dynamic landscape of AI applications in wastewater treatment opens avenues for continuous improvement and innovation.

Recommendations for Future Research

As the field of AI in wastewater treatment evolves, several avenues for future research emerge:

1. **Interdisciplinary Collaboration**: Encouraging interdisciplinary collaboration between environmental scientists, engineers, AI researchers, and social scientists can enrich the understanding of the holistic impact of AI-driven wastewater treatment.
2. **Further Ethical Inquiry**: Continued ethical inquiry is essential, particularly in addressing biases in AI algorithms, ensuring privacy, and understanding the broader societal impact. Future research should focus on developing frameworks for responsible AI implementation.
3. **Adaptive Technologies**: Research efforts can explore the development of AI technologies that dynamically adapt to changing conditions, further enhancing the resilience of wastewater treatment systems.
4. **Community Engagement**: Investigating effective strategies for community engagement and communication is crucial. Understanding how to foster a positive social license to operate is an ongoing area for exploration.
5. **Technological Innovations**: Exploring emerging technologies and their integration with AI, such as advanced sensors, nanotechnology, or decentralized treatment approaches, holds promise for the continuous improvement of wastewater treatment practices.

In conclusion, this exploration underscores the immense potential of AI in advancing wastewater treatment, offering solutions to critical challenges while posing new questions for researchers to address. The dynamic synergy between technological innovation, ethical considerations, and community engagement will shape the future landscape of intelligent wastewater treatment. As we move forward, a commitment to responsible and sustainable practices will be essential in realizing the full potential of AI for the benefit of both the environment and society.

References

1. Abdelfattah I, El-Shamy AM (2024) Review on the escalating imperative of zero liquid discharge (ZLD) technology for sustainable water management and environmental resilience. J Environ Manage 351:119614. https://doi.org/10.1016/j.jenvman.2023.119614
2. Acosta SM, Amoroso AL, Sant'Anna ÂMO, Junior OC (2022) Predictive modeling in a steelmaking process using optimized relevance vector regression and support vector regression. Ann Oper Res 316(2):905–926.https://doi.org/10.1007/s10479-021-04053-9

3. Agarwal H, Venkat Kumar S, Rajeshkumar S (2017) A review on green synthesis of zinc oxide nanoparticles—an eco-friendly approach. R-EFFIT 3(4):406–413. https://doi.org/10.1016/j.reffit.2017.03.002
4. Alkinani HH, Al-Hameedi ATT, Dunn-Norman S (2020) Data–driven decision–making for lost circulation treatments: a machine learning approach. Energy AI 2:100031. https://doi.org/10.1016/j.egyai.2020.100031
5. Alshehri F, Rahman A (2023) Coupling machine and deep learning with explainable artificial intelligence for improving prediction of groundwater quality and decision-making in arid region, Saudi Arabia. Water 15(12):2298. https://doi.org/10.3390/w15122298
6. Altowayti WAH, Shahir S, Othman N, Eisa TAE, Yafooz WMS, Al-Dhaqm A, Soon CY, Yahya IB, Che Rahim NAN, Abaker M, Ali A (2022) The role of conventional methods and artificial intelligence in the wastewater treatment: a comprehensive review. Processes 10(9):1832. https://doi.org/10.3390/pr10091832
7. Ashraf MA, Yang M, Zhang Y, Denden M, Tlili A, Liu J, Huang R, Burgos D (2021) A systematic review of systematic reviews on blended learning: trends, gaps and future directions. Psychol Res Behav Manag 14:1525–1541. https://doi.org/10.2147/PRBM.S331741
8. Awan U, Shamim S, Khan Z, Zia NU, Shariq SM, Khan MN (2021) Big data analytics capability and decision-making: the role of data-driven insight on circular economy performance. Technol Forecast Soc Chang 168:120766. https://doi.org/10.1016/j.techfore.2021.120766
9. Baum ZJ, Yu X, Ayala PY, Zhao Y, Watkins SP, Zhou Q (2021) Artificial intelligence in chemistry: current trends and future directions. J Chem Inf Model 61(7):3197–3212. https://doi.org/10.1021/acs.jcim.1c00619
10. Bibri SE, Krogstie J, Kaboli A, Alahi A (2024) Smarter eco-cities and their leading-edge artificial intelligence of things solutions for environmental sustainability: a comprehensive systematic review. Environ Sci Ecotechnol 19:100330. https://doi.org/10.1016/j.ese.2023.100330
11. Bosso N, Magelli M, Trinchero R, Zampieri N (2024) Application of machine learning techniques to build digital twins for long train dynamics simulations. Veh Syst Dyn 62(1):21–40. https://doi.org/10.1080/00423114.2023.2174885
12. Bousdekis A, Lepenioti K, Apostolou D, Mentzas G (2021) A review of data-driven decision-making methods for industry 4.0 maintenance applications. Electronics 10(7):828. https://doi.org/10.3390/electronics10070828
13. Carobene A, Padoan A, Cabitza F, Banfi G, Plebani M (2023) Rising adoption of artificial intelligence in scientific publishing: evaluating the role, risks, and ethical implications in paper drafting and review process. Clin Chem Lab Med (CCLM) 62(5). https://doi.org/10.1515/cclm-2023-1136
14. Chang N-B, Imen S, Vannah B (2015) Remote sensing for monitoring surface water quality status and ecosystem state in relation to the nutrient cycle: a 40-year perspective. Crit Rev Environ Sci Technol 45(2):101–166. https://doi.org/10.1080/10643389.2013.829981
15. Chen K-H, Wang H-C, Han J-L, Liu W-Z, Cheng H-Y, Liang B, Wang A-J (2020) The application of footprints for assessing the sustainability of wastewater treatment plants: a review. J Clean Prod 277:124053. https://doi.org/10.1016/j.jclepro.2020.124053
16. Cheng X, Xu R, Wu Y, Tang B, Luo Y, Huang W, Wang F, Fang S, Feng Q, Cheng Y, Cheng S, Luo J (2023) Predicting and evaluating different pretreatment methods on methane production from sludge anaerobic digestion via automated machine learning with ensembled semisupervised learning. ACS ES&T Eng. https://doi.org/10.1021/acsestengg.3c00368
17. Corominas L, Byrne DM, Guest JS, Hospido A, Roux P, Shaw A, Short MD (2020) The application of life cycle assessment (LCA) to wastewater treatment: a best practice guide and critical review. Water Res 184:116058. https://doi.org/10.1016/j.watres.2020.116058
18. Cui D, Wang Z, Liu P, Wang S, Dorrell DG, Li X, Zhan W (2023) Operation optimization approaches of electric vehicle battery swapping and charging station: a literature review. Energy 263:126095. https://doi.org/10.1016/j.energy.2022.126095
19. Cui Z, Zhang J, Wu D, Cai X, Wang H, Zhang W, Chen J (2020) Hybrid many-objective particle swarm optimization algorithm for green coal production problem. Inf Sci 518:256–271. https://doi.org/10.1016/j.ins.2020.01.018

20. Datta SK, Da Costa RPF, Harri J, Bonnet C (2016) Integrating connected vehicles in Internet of Things ecosystems: challenges and solutions. In: IEEE 17th international symposium on a world of wireless, mobile and multimedia networks (WoWMoM). pp 1–6. https://doi.org/10.1109/WoWMoM.2016.7523574
21. Diaz-Elsayed N, Rezaei N, Ndiaye A, Zhang Q (2020) Trends in the environmental and economic sustainability of wastewater-based resource recovery: a review. J Clean Prod 265:121598. https://doi.org/10.1016/j.jclepro.2020.121598
22. Dong G, Chen B, Liu B, Hounjet LJ, Cao Y, Stoyanov SR, Yang M, Zhang B (2022) Advanced oxidation processes in microreactors for water and wastewater treatment: development, challenges, and opportunities. Water Res 211:118047. https://doi.org/10.1016/j.watres.2022.118047
23. Dounis AI, Caraiscos C (2009) Advanced control systems engineering for energy and comfort management in a building environment—a review. Renew Sustain Energy Rev 13(6–7):1246–1261. https://doi.org/10.1016/j.rser.2008.09.015
24. Du X, Hu X, Hu J, Sun Z (2023) An adaptive interactive multi-model navigation method based on UUV. Ocean Eng 267:113217. https://doi.org/10.1016/j.oceaneng.2022.113217
25. Fu C, Xu C, Xue M, Liu W, Yang S (2021) Data-driven decision making based on evidential reasoning approach and machine learning algorithms. Appl Soft Comput 110:107622. https://doi.org/10.1016/j.asoc.2021.107622
26. Fung AWS (2020) Utilizing connectivity and data management system for effective quality management and regulatory compliance in point of care testing. Pract Lab Med 22:e00187. https://doi.org/10.1016/j.plabm.2020.e00187
27. Gabriel I (2020) Artificial intelligence, values, and alignment. Mind Mach 30(3):411–437. https://doi.org/10.1007/s11023-020-09539-2
28. Garrido-Cardenas JA, Esteban-García B, Agüera A, Sánchez-Pérez JA, Manzano-Agugliaro F (2019) Wastewater treatment by advanced oxidation process and their worldwide research trends. Int J Environ Res Public Health 17(1):170. https://doi.org/10.3390/ijerph17010170
29. Garrido-Momparler V, Peris M (2022) Smart sensors in environmental/water quality monitoring using IoT and cloud services. Trends Environ Anal Chem 35:e00173. https://doi.org/10.1016/j.teac.2022.e00173
30. Gaudio MT, Coppola G, Zangari L, Curcio S, Greco S, Chakraborty S (2021) Artificial intelligence-based optimization of industrial membrane processes. Earth Syst Environ 5(2):385–398. https://doi.org/10.1007/s41748-021-00220-x
31. Georgakopoulos D, Jayaraman PP (2016) Internet of things: from internet scale sensing to smart services. Computing 98(10):1041–1058. https://doi.org/10.1007/s00607-016-0510-0
32. Gu Y, Li Y, Li X, Luo P, Wang H, Robinson ZP, Wang X, Wu J, Li F (2017) The feasibility and challenges of energy self-sufficient wastewater treatment plants. Appl Energy 204:1463–1475. https://doi.org/10.1016/j.apenergy.2017.02.069
33. Gunasekaran K, Boopathi S (2023) Artificial intelligence in water treatments and water resource assessments. pp 71–98. https://doi.org/10.4018/978-1-6684-6791-6.ch004
34. Guo H, Jeong K, Lim J, Jo J, Kim YM, Park J, Kim JH, Cho KH (2015) Prediction of effluent concentration in a wastewater treatment plant using machine learning models. J Environ Sci 32:90–101. https://doi.org/10.1016/j.jes.2015.01.007
35. Ha N, Xu K, Ren G, Mitchell A, Ou JZ (2020) Machine learning-enabled smart sensor systems. Adv Intell Syst 2(9). https://doi.org/10.1002/aisy.202000063
36. Heo S, Nam K, Tariq S, Lim JY, Park J, Yoo C (2021) A hybrid machine learning–based multi-objective supervisory control strategy of a full-scale wastewater treatment for cost-effective and sustainable operation under varying influent conditions. J Clean Prod 291:125853. https://doi.org/10.1016/j.jclepro.2021.125853
37. Hlongwane GN, Sekoai PT, Meyyappan M, Moothi K (2019) Simultaneous removal of pollutants from water using nanoparticles: a shift from single pollutant control to multiple pollutant control. Sci Total Environ 656:808–833. https://doi.org/10.1016/j.scitotenv.2018.11.257
38. Iratni A, Chang N-B (2019) Advances in control technologies for wastewater treatment processes: status, challenges, and perspectives. IEEE/CAA J Autom Sinica 6(2):337–363. https://doi.org/10.1109/JAS.2019.1911372

39. Ismail SA, Ang WL, Mohammad AW (2021) Electro-Fenton technology for wastewater treatment: a bibliometric analysis of current research trends, future perspectives and energy consumption analysis. J Water Process Eng 40:101952. https://doi.org/10.1016/j.jwpe.2021.101952
40. Jacobs D, McDaniel T, Varsani A, Halden RU, Forrest S, Lee H (2021) Wastewater monitoring raises privacy and ethical considerations. IEEE Trans Technol Soc 2(3):116–121. https://doi.org/10.1109/TTS.2021.3073886
41. Jain R, Thakur A, Kaur P, Kim K-H, Devi P (2020) Advances in imaging-assisted sensing techniques for heavy metals in water: trends, challenges, and opportunities. TrAC, Trends Anal Chem 123:115758. https://doi.org/10.1016/j.trac.2019.115758
42. Kamyab H, Khademi T, Chelliapan S, SaberiKamarposhti M, Rezania S, Yusuf M, Farajnezhad M, Abbas M, Hun Jeon B, Ahn Y (2023) The latest innovative avenues for the utilization of artificial intelligence and big data analytics in water resource management. Results Eng 20:101566. https://doi.org/10.1016/j.rineng.2023.101566
43. Kang B-S, Park S-C (2000) Integrated machine learning approaches for complementing statistical process control procedures. Decis Support Syst 29(1):59–72. https://doi.org/10.1016/S0167-9236(00)00063-4
44. Karn AL, Pandya S, Mehbodniya A, Arslan F, Sharma DK, Phasinam K, Aftab MN, Rajan R, Bommisetti RK, Sengan S (2023) An integrated approach for sustainable development of wastewater treatment and management system using IoT in smart cities. Soft Comput 27(8):5159–5175. https://doi.org/10.1007/s00500-021-06244-9
45. Khurana D, Koli A, Khatter K, Singh S (2023) Natural language processing: state of the art, current trends and challenges. Multimedia Tools Appl 82(3):3713–3744. https://doi.org/10.1007/s11042-022-13428-4
46. Kumar S, Tiwari P, Zymbler M (2019) Internet of Things is a revolutionary approach for future technology enhancement: a review. J Big Data 6(1):111. https://doi.org/10.1186/s40537-019-0268-2
47. Kumari S, Chowdhry J, Chandra Garg M (2024) AI-enhanced adsorption modeling: challenges, applications, and bibliographic analysis. J Environ Manage 351:119968. https://doi.org/10.1016/j.jenvman.2023.119968
48. Lin H, Wang J, Long AC, Clifford MJ, Harrison P (2007) Predictive modelling for optimization of textile composite forming. Compos Sci Technol 67(15–16):3242–3252. https://doi.org/10.1016/j.compscitech.2007.03.040
49. Liu J, Ma L, Wang Q (2023) Energy management method of integrated energy system based on collaborative optimization of distributed flexible resources. Energy 264:125981. https://doi.org/10.1016/j.energy.2022.125981
50. Longo L, Goebel R, Lecue F, Kieseberg P, Holzinger A (2020) Explainable artificial intelligence: concepts, applications, research challenges and visions. pp 1–16. https://doi.org/10.1007/978-3-030-57321-8_1
51. Lotfollahi M, Klimovskaia Susmelj A, De Donno C, Hetzel L, Ji Y, Ibarra IL, Srivatsan SR, Naghipourfar M, Daza RM, Martin B, Shendure J, McFaline-Figueroa JL, Boyeau P, Wolf FA, Yakubova N, Günnemann S, Trapnell C, Lopez-Paz D, Theis FJ (2023) Predicting cellular responses to complex perturbations in high-throughput screens. Mol Syst Biol 19(6). https://doi.org/10.15252/msb.202211517
52. Maiurova A, Kurniawan TA, Kustikova M, Bykovskaia E, Othman MHD, Singh D, Goh HH (2022) Promoting digital transformation in waste collection service and waste recycling in Moscow (Russia): applying a circular economy paradigm to mitigate climate change impacts on the environment. J Clean Prod 354:131604. https://doi.org/10.1016/j.jclepro.2022.131604
53. Maleki R, Jahromi AM, Ghasemy E, Khedri M (2022) Smart sensing technologies for wastewater treatment plants. In: Artificial intelligence and data science in environmental sensing. Elsevier, pp 1–17. https://doi.org/10.1016/B978-0-323-90508-4.00003-4
54. Malviya A, Jaspal D (2021) Artificial intelligence as an upcoming technology in wastewater treatment: a comprehensive review. Environ Technol Rev 10(1):177–187. https://doi.org/10.1080/21622515.2021.1913242

55. Martínez R, Vela N, el Aatik A, Murray E, Roche P, Navarro JM (2020) On the use of an IoT integrated system for water quality monitoring and management in wastewater treatment plants. Water 12(4):1096. https://doi.org/10.3390/w12041096
56. Mehmood K, Verleye K, De Keyser A, Larivière B (2023) Piloting personalization research through data-rich environments: a literature review and future research agenda. J Serv Manag 34(3):520–552. https://doi.org/10.1108/JOSM-10-2021-0405
57. Murtagh N, Scott L, Fan J (2020) Sustainable and resilient construction: current status and future challenges. J Clean Prod 268:122264. https://doi.org/10.1016/j.jclepro.2020.122264
58. Nakatsuka N, Kishita Y, Kurafuchi T, Akamatsu F (2020) Integrating wastewater treatment and incineration plants for energy-efficient urban biomass utilization: a life cycle analysis. J Clean Prod 243:118448. https://doi.org/10.1016/j.jclepro.2019.118448
59. Niu G, Yang B-S, Pecht M (2010) Development of an optimized condition-based maintenance system by data fusion and reliability-centered maintenance. Reliab Eng Syst Saf 95(7):786–796. https://doi.org/10.1016/j.ress.2010.02.016
60. Oukhatar A, Bakhouya M, El Ouadghiri D (2021) Electromagnetic-based wireless nano-sensors network: architectures and applications. J Commun 8–19. https://doi.org/10.12720/jcm.16.1.8-19
61. Pan Y-R, Wang X, Ren ZJ, Hu C, Liu J, Butler D (2019) Characterization of implementation limits and identification of optimization strategies for sustainable water resource recovery through life cycle impact analysis. Environ Int 133:105266. https://doi.org/10.1016/j.envint.2019.105266
62. Patel V, Chesmore A, Legner CM, Pandey S (2022) Trends in workplace wearable technologies and connected-worker solutions for next-generation occupational safety, health, and productivity. Adv Intell Syst 4(1). https://doi.org/10.1002/aisy.202100099
63. Patra RC, Saritha G, Raghuwanshi GA, Parthiban P, Ashok V (2023) Recent advances, challenges, and applications of deep learning in healthcare systems for medical diagnosis and treatment. In: 3rd international conference on technological advancements in computational sciences (ICTACS). pp 716–724. https://doi.org/10.1109/ICTACS59847.2023.10390124
64. Pradeep P, Krishnamoorthy S, Vasilakos AV (2021) A holistic approach to a context-aware IoT ecosystem with adaptive ubiquitous middleware. Pervasive Mob Comput 72:101342. https://doi.org/10.1016/j.pmcj.2021.101342
65. Puyol D, Batstone DJ, Hülsen T, Astals S, Peces M, Krömer JO (2017) Resource recovery from wastewater by biological technologies: opportunities, challenges, and prospects. Front Microbiol 7.https://doi.org/10.3389/fmicb.2016.02106
66. Roh Y, Heo G, Whang SE (2021) A survey on data collection for machine learning: a big data—AI integration perspective. IEEE Trans Knowl Data Eng 33(4):1328–1347. https://doi.org/10.1109/TKDE.2019.2946162
67. Rojek I, Jasiulewicz-Kaczmarek M, Piechowski M, Mikołajewski D (2023) An artificial intelligence approach for improving maintenance to supervise machine failures and support their repair. Appl Sci 13(8):4971. https://doi.org/10.3390/app13084971
68. Rout PR, Shahid MK, Dash RR, Bhunia P, Liu D, Varjani S, Zhang TC, Surampalli RY (2021) Nutrient removal from domestic wastewater: a comprehensive review on conventional and advanced technologies. J Environ Manage 296:113246. https://doi.org/10.1016/j.jenvman.2021.113246
69. Safeer S, Pandey RP, Rehman B, Safdar T, Ahmad I, Hasan SW, Ullah A (2022) A review of artificial intelligence in water purification and wastewater treatment: recent advancements. J Water Process Eng 49:102974. https://doi.org/10.1016/j.jwpe.2022.102974
70. Sandhya H, Varghese B (2023) Unleashing AI's potential for optimal student learning in education. pp 99–114. https://doi.org/10.4018/979-8-3693-0205-7.ch005
71. Saravanan A, Deivayanai VC, Kumar PS, Rangasamy G, Hemavathy RV, Harshana T, Gayathri N, Alagumalai K (2022) A detailed review on advanced oxidation process in treatment of wastewater: mechanism, challenges and future outlook. Chemosphere 308:136524. https://doi.org/10.1016/j.chemosphere.2022.136524

72. Shah AI, Din Dar MU, Bhat RA, Singh JP, Singh K, Bhat SA (2020) Prospectives and challenges of wastewater treatment technologies to combat contaminants of emerging concerns. Ecol Eng 152:105882. https://doi.org/10.1016/j.ecoleng.2020.105882
73. Sharma R, Verma N, Lugani Y, Kumar S, Asadnia M (2021) Conventional and advanced techniques of wastewater monitoring and treatment. In: Green sustainable process for chemical and environmental engineering and science. Elsevier, pp 1–48. https://doi.org/10.1016/B978-0-12-821883-9.00009-6
74. Singh BJ, Chakraborty A, Sehgal R (2023) A systematic review of industrial wastewater management: evaluating challenges and enablers. J Environ Manage 348:119230. https://doi.org/10.1016/j.jenvman.2023.119230
75. Tejaswini V, Sathya Babu K, Sahoo B (2024) Depression detection from social media text analysis using natural language processing techniques and hybrid deep learning model. ACM Trans Asian Low-Resour Lang Inf Process 23(1):1–20. https://doi.org/10.1145/3569580
76. Ullo SL, Sinha GR (2020) Advances in smart environment monitoring systems using IoT and sensors. Sensors 20(11):3113. https://doi.org/10.3390/s20113113
77. Varun S, George NM, Chandran AM, Varghese LA, Mural PKS (2023) Multifaceted PVDF nanofibers in energy, water and sensors: a contemporary review (2018–2022) and future perspective. J Fluorine Chem 265:110064. https://doi.org/10.1016/j.jfluchem.2022.110064
78. Wang D, Thunéll S, Lindberg U, Jiang L, Trygg J, Tysklind M, Souihi N (2021) A machine learning framework to improve effluent quality control in wastewater treatment plants. Sci Total Environ 784:147138. https://doi.org/10.1016/j.scitotenv.2021.147138
79. Wang X, Han Y, Leung VCM, Niyato D, Yan X, Chen X (2020) Fundamentals of artificial intelligence. In: Edge AI. Springer, Singapore, pp 33–47. https://doi.org/10.1007/978-981-15-6186-3_3
80. Wu X, Zheng Z, Wang L, Li X, Yang X, He J (2023) Coupling process-based modeling with machine learning for long-term simulation of wastewater treatment plant operations. J Environ Manage 341:118116. https://doi.org/10.1016/j.jenvman.2023.118116
81. Xu P, Ji X, Li M, Lu W (2023) Small data machine learning in materials science. NPJ Comput Mater 9(1):42. https://doi.org/10.1038/s41524-023-01000-z
82. Zhang D-G, Ni C-H, Zhang J, Zhang T, Zhang Z-H (2023) New method of vehicle cooperative communication based on fuzzy logic and signaling game strategy. Futur Gener Comput Syst 142:131–149. https://doi.org/10.1016/j.future.2022.12.039
83. Zhang W, Tooker NB, Mueller AV (2020) Enabling wastewater treatment process automation: leveraging innovations in real-time sensing, data analysis, and online controls. Environ Sci: Water Res Technol 6(11):2973–2992. https://doi.org/10.1039/D0EW00394H

Chapter 2
Introduction to Artificial Intelligence in Optimization of Wastewater Treatment Processes

Shubo Zhang, Jinfeng Wang, and Hongqiang Ren

Abstract Wastewater treatment is an important and challenging task for environmental protection and resource recovery. Artificial intelligence (AI) techniques can provide effective solutions for optimizing wastewater treatment processes. In this paper, we review the recent research progress on AI methods for improving wastewater treatment efficiency, reducing wastewater treatment costs, and minimizing wastewater treatment emissions. We compare and evaluate different AI methods and models, such as neural networks, fuzzy logic, genetic algorithms, and their applications and performance in various wastewater treatment process optimizations. Furthermore, we review the multi-objective optimization problems based on artificial neural networks and genetic algorithms, including methods, principles, pareto sets selection and evaluation of solving multi-objective optimization problems. In this study, we present the prediction and optimization results of the model, and analyze the model performance and advantages and disadvantages. We conclude the limitations and improvement directions of this paper, and the future development trends and challenges of AI in wastewater treatment field.

Keywords Artificial intelligence (AI) · Wastewater treatment optimization · Neural networks · Genetic algorithms · Multi-objective optimization · Environmental protection

---

S. Zhang · J. Wang (✉) · H. Ren
State Key Laboratory of Pollution Control and Resource Reuse, School of the Environment, Nanjing University, Nanjing 210023, Jiangsu, China
e-mail: jfwang@nju.edu.cn

M. C. Garg (ed.), *The AI Cleanse: Transforming Wastewater Treatment Through Artificial Intelligence*, Springer Water, https://doi.org/10.1007/978-3-031-67237-8_2

2.1 Introduction

2.1.1 *The Importance and Challenges of Wastewater Treatment*

Wastewater treatment plays a pivotal role in safeguarding the environment, as untreated wastewater can discharge an array of pollutants and contaminants into natural water bodies [87]. Over time, the composition of these pollutants has become increasingly diverse, encompassing hazardous chemicals, pathogenic microorganisms, pharmaceuticals, heavy metals, as well as various organic compounds [54, 61]. Consequently, the treatment process has grown progressively intricate. The release of such contaminants can inflict severe harm on aquatic ecosystems, leading to water quality degradation, harm to aquatic life, and the disruption of delicate ecosystems [30]. Moreover, inadequate wastewater treatment directly jeopardizes public health and safety. Pathogens and disease-causing microorganisms present in wastewater can infiltrate drinking water sources, resulting in waterborne diseases such as cholera, typhoid, and gastroenteritis [6]. Therefore, proper wastewater treatment is indispensable in averting such outbreaks and ensuring the integrity of drinking water supplies.

Wastewater is not merely a waste product; it contains valuable resources. Through advanced treatment processes, it becomes possible to recover vital nutrients like phosphorus and nitrogen, which can be repurposed as fertilizers [70]. Additionally, energy can be harnessed from wastewater through anaerobic digestion and other techniques, contributing to sustainable energy generation [28]. Thus, effective wastewater treatment aligns with the principles of the circular economy by recuperating resources from what would otherwise be considered waste. The challenges in wastewater treatment are manifold and continually evolving. Addressing this diversity necessitates adaptable and targeted approaches. For example, methods that combine biochar and microorganisms to remove various contaminants [53]. Furthermore, the treatment processes themselves are highly intricate, comprising physical, chemical, and biological steps [1]. Primary treatment, which involves the removal of solids, is followed by secondary treatment, which is a biological process [65]. Tertiary treatment includes advanced filtration and disinfection [26]. Coordinating these processes and maintaining consistent treatment efficiency represents a formidable challenge.

Over the past year, environmental concerns about water pollution have gradually been raised by all sectors of the community [22]. As global population growth and urbanization trends lead to increasing wastewater generation, the pressure on treatment facilities has increased [72]. However, environmental regulations and discharge standards have become increasingly stringent, and this demand has driven the development of wastewater treatment facilities to accommodate larger loads to maintain treatment standards. Therefore, continuously improved treatment processes are necessary to mitigate the environmental impacts of wastewater. Additionally, wastewater treatment is energy-intensive, primarily due to aeration, pumping, and the need to treat complex contaminants [10]. Escalating energy costs and the imperative of sustainable practices optimizing energy use and overall treatment costs a pressing concern.

2.1.2 Advantages and Applications of Artificial Intelligence Technology in Wastewater Treatment

Artificial Intelligence (AI) technology is becoming increasingly integrated with the wastewater treatment field, thanks to its good ability to solve nonlinear complexity problems, and it has demonstrated remarkable advantages and transformative potential [74]. As one of the top agendas for both developing and developed countries, in this section, we delve into the pivotal role of AI in optimizing wastewater treatment processes, with a specific focus on practical applications and the benefits it offers.

AI technology brings several compelling advantages to the realm of wastewater treatment, it has been widely concerned by researchers. Chief among these is the ability to solve unknown problems based on existing big data [67]. The need for experimental measurements is eliminated by machine learning and neural networks, which process and analyze large datasets and systematically learn the relationships between variables in historical data [68]. In the context of wastewater treatment, this function is also becoming more widely used because it supports real-time data analysis [58].

One of the primary applications of AI in wastewater treatment is process algorithm optimization [89]. AI algorithms can predict trends by recognizing and learning from complex patterns in wastewater treatment data, adjusting to changing conditions based on feedback, and using this capability, AI facilitates precise process optimization [40]. For instance, it can continuously adjust variables such as chemical dosing rates, aeration levels, and flow rates to maximize treatment efficiency while minimizing resource consumption. The result is more efficient and cost-effective wastewater treatment [4]. Another notable application is predictive maintenance. Machine learning algorithms can anticipate equipment failures and maintenance needs by analyzing historical equipment data, to improve efficiency and sustainability [49]. This proactive approach helps prevent unexpected breakdowns, reduce downtime, and ensure the uninterrupted operation of wastewater treatment plants.

AI also plays a key role in reducing human error [57]. By automating decision-making and control processes, AI systems minimize the risk of costly errors and environmental accidents [12]. They make precise adjustments based on data and predefined rules, improving the overall reliability of wastewater treatment operations. However, different stages of wastewater treatment require different algorithmic models because the process parameters are different for each level of treatment and further research is needed [33].

2.1.3 Brief on Artificial Intelligence Optimization of Wastewater Treatment Processes

AI is a transformative force in the quest to improve efficiency. AI systems could to continuously monitor and analyze a variety of data sources [2]. The ability to optimize the operational parameters of a wastewater treatment plant in real time, to monitor effluent quality, to design advanced treatments, to control gas emissions, etc., so that the treatment process always operates at peak performance [71, 84]. To maximize pollutant removal while minimizing energy consumption and resource use. The result is a significant increase in operational efficiency, which translates into higher processing throughput and enhanced compliance with strict regulatory standards.

Simultaneously, AI significantly contributes to the overarching goal of reducing operational costs in wastewater treatment [62]. Through predictive maintenance and the judicious allocation of resources, AI systems enable substantial cost savings. By predicting equipment failures in advance, minimizing costly emergency repairs and associated downtime. Additionally, AI-driven energy management strategies lead to reduced energy consumption, resulting in considerable long-term cost reductions. AI optimizes resource recovery processes, such as anaerobic digestion, to yield valuable byproducts like biogas and nutrients [18, 76]. These reclaimed resources offer the potential to offset treatment costs and align with sustainability objectives. Finally, AI plays a central role in the endeavor to minimize emissions from wastewater treatment processes. This encompasses greenhouse gases and various pollutants of environmental concern [5, 60]. In summary, AI has become the most valuable research direction in the field of wastewater treatment process optimization.

In this work, the optimization of wastewater treatment process by AI is firstly reviewed in two separate aspects, namely, enhancing wastewater treatment efficiency and controlling cost. Then, based on the results of the review, it is found that the optimization of wastewater treatment processes often contains many conflicting optimization objectives, such as effluent quality with respect to cost and energy consumption. Multi-objective optimization (MOO) has been the main means of saving energy and improving efficiency of wastewater treatment processes [87]. Therefore, we overview the main principles of MOO algorithms, as well as common types of algorithms. Intelligent optimization algorithms combined with AI, such as biologically inspired algorithms, have better performance in solving non-convex optimization problems. Finally, we summarize the challenges faced by AI and MOO algorithms in wastewater treatment. This work presents a systematic review of the application of AI to wastewater treatment process optimization to provide comprehensive theoretical support for future research in this area.

2.2 Research Progress of Artificial Intelligence to Optimize Wastewater Treatment Process

To summarize, AI as a powerful tool can solve many complex problems based on good dataset. Wastewater treatment process is a typical black box problem with complex influencing factors and dynamic changes, which is very suitable for using AI to solve the problems it faces, especially in terms of improving the efficiency of wastewater treatment, controlling costs, and reducing its environmental impact.

2.2.1 Artificial Intelligence to Improve Wastewater Treatment Efficiency

The main purpose of wastewater treatment is to remove the pollutants contained in wastewater generated from domestic and industrial sources through chemical-physical or biological methods, so that the effluent water quality meets the standards for discharge or reuse. However, the effectiveness of pollutant removal depends on several factors including the quality of the influent water, environmental factors, process parameters and the type of process [64]. Therefore, optimizing wastewater treatment efficiency is a very broad optimization goal. For wastewater treatment process, understanding the future variations of water quality indicators and timely detection of the operational status of wastewater treatment equipment would be beneficial for adjusting and optimizing the workflow and parameters in a timely manner, thus further improving the efficiency of wastewater treatment [48].

Application of AI to testing and predicting effluent quality of wastewater treatment processes is of widespread interest to scientists [63]. Table 2.1 shows some of the studies for water quality prediction. This integrated model effectively predicts wastewater quality with higher accuracy and better universality. COD can measure the degree of water pollution and is one of the important indicators for evaluating the efficiency of wastewater treatment. Kundu et al. established an artificial neural network (ANN) model using the Levenberg–Marquardt training algorithm to predict the removal efficiency of chemical oxygen demand (COD) and NH^{4+}-N under three conditions in an aerobic-anoxic sequence. The model achieved a minimum error rate of 3.33% and exhibited strong learning capabilities for nonlinear relationships [39]. Alavi et al. designed a kernel-based Extreme Learning Machine (KELM) combined with the intelligent optimization algorithm called Salp Swarm Algorithm, referred to as KELM-SSA, to predict the inlet COD [3]. This method exhibits excellent predictive reliability with a minimum root mean square error of 0.058 and an average absolute error of 0.044. Liu et al. introduced time lag coefficients in the extreme learning machine (ELM) and implemented dynamic feature extraction to achieve better prediction accuracy in the treatment of papermaking wastewater, considering the complexity of the wastewater treatment process and the nonlinearity of the data.

This approach is named Dynamic Kernel Extreme Learning Machine (DKELM) [44].

In addition to predicting COD concentrations, machine learning can also be applied to predict other indicators such as total nitrogen (TN), total phosphorus (TP), and biochemical oxygen demand (BOD). Safder et al. proposed a multi-attention-based gated recurrent unit (MAGRU) algorithm to predict the TN concentration in effluent several hours in advance. The method achieves an accuracy of 96.3% in predicting total nitrogen three hours in advance [63]. Fu et al. introduced wavelet denoising technology in the adaptive neuro-fuzzy inference system (ANFIS) to improve the prediction accuracy by addressing the problem of over-prediction caused by uneven distribution of randomly sampled training data. The results show that this method performs well in predicting chloride, sulfate, conductivity, and total dissolved solids (TDS) [19]. Yu et al. compared three methods, including extreme learning machine (ELM), least squares support vector machine (LS-SVM), and back-propagation neural network (BPNN), in predicting the outlet total phosphorus (TP) concentration [82]. The results indicate that ELM has the best predictive performance, outperforming BPNN and LS-SVM. Abdulaziz et al. demonstrated that the random forest (RF) algorithm has excellent performance in predicting effluent BOD_5 concentration [59].

The above studies only predicted the degradation effects of the target pollutants, but AI can do more than that. Zaghloul and Achari et al. developed an integrated model using ANNs, ANFIs, and SVM regression for simulating a full-scale biological nutrient removal wastewater treatment process, and predicted 15 parameters [84]. The method achieves an accuracy of 96.3% in predicting total nitrogen three hours in advance.

Table 2.1 Summary of predictive indicators and it's algorithms performance

Predictive indicators	Methods	Performance	Reference
COD	ANN	RMSE $\leq$ 0.033, R^2>0.900	[39]
COD	KELM-SSA	RMSE = 0.058, MAE = 0.044	[39]
COD	DKELM	R^2 = 0.9057, RMSE = 2.577	[44]
TN	MAGRU	MAE = 0.28, Accuracy = 96.3%	[63]
TDS, conductivity	ANFIS	R^2 > 0.975	[19]
TP	ELM	MSE = 4.0259e−5, R^2 = 0.9859, MAPE = 1.054	[82]
BOD_5	RF	MAE = 6.56, R^2 = 0.9200	[59]

2.2.2 Artificial Intelligence to Reduce Wastewater Treatment Costs

All around the world, there are strict standards for the placement and reuse of treated wastewater, which makes it necessary to control the quality of the effluent. Process optimization is the main way to reduce treatment costs while ensuring the quality of the effluent. The main costs that can be optimized for wastewater treatment processes include energy consumption, chemical consumption, and maintenance cost. The energy consumption of the wastewater treatment process is the main reason for its impact on the environment and operating costs, most notably in relation to both aeration and pumping devices. Also, chemical consumption is another component of operational cost in wastewater treatment processes, especially for processes that involve coagulation/flocculation or disinfection. Maintenance cost is also a significant component of operational cost in wastewater treatment processes, which is related to the frequency and intensity of maintenance activities such as cleaning or repairing [56]. AI can provide optimal cost management solutions for wastewater treatment processes.

The high energy cost of wastewater treatment plants has a huge optimization space. Torregrossa et al. collected a database of 317 Northwest European wastewater treatment plants, and established a prediction model of energy cost of wastewater treatment plants by machine learning methods such as RF and ANN, which lays a good foundation for optimizing the cost of wastewater treatment process [73]. Guo et al. by combining convolutional neural networks (CNN) and recurrent neural networks (RNN), we propose a PC-CR platform for automatic ventilation control that is more efficient and stable than traditional models [27]. Wang et al. proposed a machine learning framework based on varying sliding layer (VSL) is used for intelligent ventilation control. Compared with traditional methods such as preset dissolved oxygen (DO), ventilation requirements are reduced by 16.12%. Their study further demonstrates the potential of AI in reducing energy consumption and greenhouse gas emissions in wastewater treatment [78].

In terms of saving chemical dosing, Lu et al. developed a prediction model for the efficiency of chitosan-based flocculants (CBFs) removal of heavy metals using RF and identified the influence of each influencing factor on the action of CBFs, which provided guidance for saving the use of CBFs [46]. Li et al. obtained the optimal conditions for the use of polymerized aluminum chloride as a flocculant by GA quasi to maximize the flocculation effect and reduce the use of polymeric aluminum chloride in deinking wastewater [41].

2.2.3 Common Algorithms for Optimizing Wastewater Treatment Processes

There is currently no universal artificial intelligence model in the field of wastewater treatment, so it is necessary to choose appropriate algorithms based on specific circumstances. Eight commonly used artificial intelligence algorithms in water quality assessment systems are ANN, ANFIS, Support Vector Machines (SVM), Particle Swarm Optimization (PSO), GA, General Regression Neural Network (GRNN), Multilayer Perceptron (MLP), and Multiple Linear Regression (MLR) [48]. These algorithms can be used in various aspects such as water treatment monitoring, drinking water treatment, wastewater treatment, water quality prediction, and fault detection. There is also a growing trend of exploring combined methods, and studies have reported that hybrid methods have stability and accuracy advantages over single methods. For example, in Nourani et al.'s study [55], researchers used ANN-based algorithms, specifically Feed Forward Neural Network (FFNN) and ANFIS, to predict BOD, COD, and TN in the Nicosia wastewater treatment plant. The results showed that ANFIS had better prediction accuracy compared to a single model. In the prediction of BOD, combining several artificial intelligence algorithms improved the accuracy by 14–26%.

Currently, the most used machine learning models in wastewater treatment are ANN and SVM (ref 2). Therefore, this section mainly focuses on these two models. ANN typically consists of three layers: the input layer, hidden layer(s), and output layer. The input layer receives relevant data for predicting indicators, and the information is propagated layer by layer through node connections until the output layer produces the prediction result. It is important to note that data can only propagate vertically between layers and cannot propagate horizontally or in reverse between layers. The hidden layer usually contains one layer, but there are cases where multiple layers are used [45].

In addition to predicting influent and effluent concentrations of BOD, COD, and TS in wastewater treatment plants [45], ANN and SVM could provide performance prediction and operational control in the aerobic biological process of wastewater treatment plants. They can also be applied to fault detection in wastewater treatment systems. Fault detection methods can generally be categorized into three types: statistical methods, learning models, and time series analysis (ref 4). Currently, statistical methods such as Principal Component Analysis (PCA) and Kernel PCA are most used for fault detection. Learning models treat fault detection as a binary classification problem and utilize methods such as SVM, Random Forest, and ANNs. However, statistical methods require a large amount of historical data (quantitative data, qualitative data, and process data) and are generally only suitable for simple or linear fault problems. They have significant limitations when dealing with complex, uncertain, and nonlinear data. ANN and SVM, as typical supervised learning models, can transform the fault problems of wastewater treatment equipment into classification problems based on historical data, thereby achieving reliable detection results [32].

Harrou et al. designed a wastewater equipment anomaly alarm system called DBN-OCSVM based on a one-class support vector machine. This system combines the powerful handling and learning abilities of deep belief networks (DBN) for nonlinear data and the excellent classification and prediction abilities of OCSVM [29]. It can raise early alerts for abnormal conditions. The system has been successfully implemented in wastewater treatment plants in the Golden area of the United States and has achieved satisfactory results.

In general, artificial intelligence can make full use of data for nonlinear regression, classification, and prediction, making it one of the important approaches in solving complex issues in wastewater treatment processes. It has broad prospects for application in optimizing the performance of wastewater treatment plants, and there is an increasing number of combined algorithms being applied in wastewater treatment [45]. However, there are also challenges that need to be addressed in using machine learning to optimize the performance of wastewater treatment plants. Firstly, many machine learning algorithms exhibit black-box or gray-box characteristics, making it difficult to intuitively explain the relationships between data. This lack of interpretability may impact the reliability of system operation to some extent. Secondly, certain data parameters that are closely related to wastewater treatment efficiency, such as environmental conditions and meteorological data, cannot be directly collected through sensors, making their quantification challenging [82].

2.3 Multi-objective Optimization Problems in Wastewater Treatment Processes

It is not difficult to see from the above that the optimization of wastewater treatment process is often a MOO problem, and the intelligent optimization algorithm has good robustness, adaptability, and good solution set convergence and uniformity. It is the best solution to solve the MOO problem in the wastewater treatment process [87]. Compared with single-objective optimization problems, many real-world problems have multi-objective characteristics [17, 81]. Solutions to multi-objective optimization problems are often not mathematically interpretable. Because of the complexity of the need to satisfy multi-objective requirements, MOO algorithms are born. Decision makers can solve these problems through MOO algorithms.

2.3.1 The Essence of the Multi-objective Problem

Many of the above optimization objectives are against each other, such as the concentration of DO in the aerator and the minimum cost, making simple methods unable to satisfy a multi-objective optimal solution. The method of finding the optimal solution when multiple objectives conflict with each other is called the MOO algorithm, and the set of all non-dominated solutions obtained by using this algorithm is called the Pareto set, which contains a variety of possibilities that can satisfy the MOO [17]. The algorithm gives a range of Pareto solution sets by evaluating the degree to which the solutions satisfy the objective function, which for general MOO is like the following equation:

$$Min_{\mathrm{u,tf}} J\left(x, u, t_f\right) = [J_1, J_2, \ldots, J_n]^{\mathrm{T}} \tag{2.1}$$

where J stands for the objective function, the left side of the equation indicates the objective function that requires minimal words, and the right side indicates multiple objectives, t_f is the time horizon, x and u represent state variables and decision variables, respectively [31].

There are several definitions in a MOO problem: (1) A solution vector is said to feasible solution if it lies within the boundaries of the decision variables and it satisfies both equality and inequality constraints; (2) All feasible solutions form a feasible domain and all infeasible solutions are called infeasible domains; (3) For two variables $\boldsymbol{a} = [a_1, a_2, \ldots, a_{\mathrm{n}}{}^{\mathrm{T}}]$ and $\boldsymbol{b} = [b_1, b_2, \ldots, b_{\mathrm{n}}{}^{\mathrm{T}}]$ of the decision variable space, if $\forall_i \in \{1, 2, \ldots, \mathrm{m}\}, f_j(\boldsymbol{a}) \leq f_j(\boldsymbol{b})$, and $\exists_j \in \{1, 2, \ldots, \mathrm{m}\}, f_j(\boldsymbol{a}) < f_j(\boldsymbol{b})$, is called b being dominated by a, expressed as $\mathrm{b} \prec \mathrm{a}$; (4) For two variables $\boldsymbol{g} = \{g_1, g_2, \ldots, g_{\mathrm{n}}{}^{\mathrm{T}}\}$ and $\boldsymbol{h} = \{h_1, h_2, \ldots, h_{\mathrm{n}}{}^{\mathrm{T}}\}$ of the objective function space, if $\forall_i \in \{1, 2, \ldots, \mathrm{m}\}, \boldsymbol{g}_i \leq \boldsymbol{h}_i$, and $\exists_j \in \{1, 2, \ldots, \mathrm{m}\}, \boldsymbol{g}_i < \boldsymbol{h}_i$ is called h being dominated by g, expressed as $\mathrm{h} \prec \mathrm{g}$; (5) A solution is a global Pareto optimal solution if there is no dominated solution in the space of decision variables. All global Pareto optimal solutions constitute a global Pareto optimal set; (6) The Pareto optimal set presented in the objective function space is called Pareto optimal front [43].

To better understand the MOO problem, Fig. 2.1 shows the relationship between the objective space and the decision space, and how to understand the Pareto optimal solution. The left panel is a one-dimensional decision space and the right panel is the objective function space. The decision space will form a mapping relationship on the objective function space. The shaded area is the infeasible domain, which consists of infeasible solutions. Points A and B are non-dominated solutions on the Pareto front and are the two globally optimal Pareto solutions, while E is an infeasible solution, and D, F, and C are all dominated solutions. Briefly explain the principle of selection of Pareto solution set i.e., to obtain solutions as close as possible to the Pareto frontiers and the solution set is uniformly distributed. After determining the Pareto solution set, the final solution is determined according to the decision maker's needs to achieve a purposeful MOO problem.

2.3.2 Multi-objective Intelligent Optimization Algorithm

Analytic and numerical methods are two common approaches to solving multi-objective problems. In practical problems, numerical methods are more compatible with applications in practice by iterating to obtain an approximate solution,

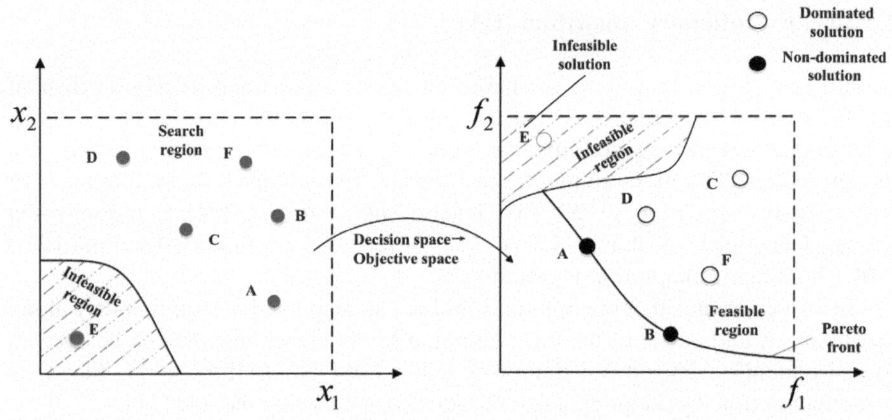

Fig. 2.1 Optimization analysis diagram for bi-objective problems [15]. Copyright 2017, Elsevier

since practical problems often do not have an explicit expression. However, analytical methods can obtain more accurate results by derivation through rigorous mathematical formulas, but do not fit most practical problems.

Intelligent numerical methods based on heuristic search effectively avoid the problems of local optimal solutions and large computation volume that are more likely to occur in traditional numerical methods by mimicking the behavior of nature, biological responses, and communication mechanisms. The classification of common heuristic optimization algorithms is shown in Fig. 2.2. Heuristic optimization algorithms are subdivided into physically inspired algorithms, biologically inspired algorithms, socially inspired algorithms, and geographically inspired algorithms based on their principles. These methods are also known as intelligent optimization algorithms.

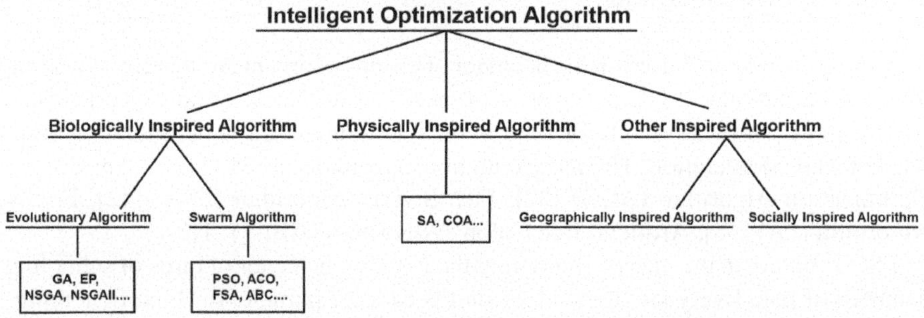

Fig. 2.2 Common intelligent optimization algorithms

2.3.2.1 Evolutionary Algorithm (EA)

Evolutionary algorithms (EA) are based on the environmental selection effect of the environment on organisms which is also known as the survival of the fittest process in a stochastic search manner, good self-organization and adaptability are its advantages. The more common ones include Evolutionary Programming (EP) [69], Evolutionary Strategy (ES) [35], Genetic Algorithm (GA) [34], Non-dominated Sorting Genetic Algorithm (NSGA) [11], Differential Evolution Algorithm (DE) [50], Membrane Computing (P system).

GA can guarantee that the optimal solution can be passed on continuously along the iterations and is one of the most common MOO algorithms. NSGA has been a focus of attention since it was first proposed in 1994. NSGAII in 2000 reduced the computation time by adopting a fast non-dominated sorting method [11].

P-systems are based on aspects inspired by the structure, function, and interactions of biological cells, and are mainly classified into neural-like p-systems, tissue-like p-systems and cell-like p-systems. By placing the study subject in a virtual p-system of the three modes described above, each membrane or each cell within the cell represents a different layer. By simulating biological activities within the cell such as lysis and division, the final evolutionary results are output through the outermost skin membrane [9].

However, recently many researchers have focused more on hybrid algorithms of EA with other algorithms that tend to have better performance. Gui et al. found that the normal vibration distribution search-based differential evolution algorithm (NVSA) can perform the linear registration task better [25]. Jhansi et al. proposed the Whale Optimization Algorithm with Genetic Algorithm (WOA-GA) over single algorithm both have better performance in identifying diseased fish.

2.3.2.2 Swarm Algorithm

Swarm algorithms are based on the behavior of groups of organisms thereby inventing biological heuristics. To meet the survival needs of the flock, groups of organisms such as flocks of birds search for food in the environment through mutual communication and co-operation. The more common algorithms are PSO [20], Ant Colony Optimization Algorithm (ACO) [52], Fish Swarm Algorithm (FSA) [42], Firefly Algorithm (FA) [38], Artificial Bee Colony Algorithm (ABC) [21].

PSO is based on the idea of simulating the flocking behavior of birds or schooling behavior of fish. Everyone in the algorithm is called a particle. Each particle represents a potential solution to the problem and moves within the solution space. The particles adjust their positions and velocities based on their individual best-known solutions and the global best-known solution among the entire group. This collaboration and information sharing among particles allow them to collectively explore and discover better solutions in the search space. Multi-objective particle swarm optimization (MOPSO) algorithms, proposed by combining adaptive strategies with external populations, are one of the most widespread means of solving multi-objective problems today [14].

ACO is inspired by the foraging behavior of ants. In ACO, ants release pheromones and rely on pheromone concentration to choose paths. Paths with higher pheromone concentrations are considered better. When ants choose paths, they tend to prefer those with higher pheromone concentrations. This behavior enables ACO to search for good solutions in the problem space. The accumulated pheromones represent the quality of solutions.

FSA simulates the collective behavior of fish in a school. Each fish represents a potential solution, and fish adjust their positions and velocities based on the objective function values and the positions of other fish. Fish in the swarm may attract or repel each other as they search for better solutions. FSA iterates in the solution space, attempting to find the optimal solution to the problem.

ABC contains three types of virtual bees, where the employed bee represents a solution, searches the objective function space again, and evaluates its own quality by comparing the objective function values. Observer bees do not have their own nests, it is mainly used to evaluate the nests of hired bees, the better nests are more likely to be selected. When the result does not improve after iteration, the hired bee will be transformed into a scout bee, and the above steps will be repeated from the unexplored space, preventing the generation of locally optimal solutions.

2.3.2.3 Others Inspired Algorithms

Physically inspired algorithms are optimization algorithms developed based on the behaviors of matter based on the laws of physics. Common algorithms include Simulated Annealing (SA) [36], chaotic optimization algorithms (COA) etc.

Simulated annealing algorithms are inspired based on the process of gradual cooling of the material after heating. In SA all the solutions are related to the value of energy during the annealing process. The worse solution is accepted given a certain probability, thus moving away from the local optimal solution to reach the global optimal solution. In addition to the inspired algorithms mentioned above, geographically inspired algorithms and socially inspired algorithms are also included.

2.3.3 Evaluation Metrics and Pareto Sets for Multi-objective Optimization

2.3.3.1 Performance Evaluation Metrics for Multi-objective Optimization Algorithms

Based on the above, the solution of MOO is a near-optimal solution close to the Pareto Front. Typically, there are three main aspects of evaluating the approximate solution, which are Convergence, Distribution and Balance. A good optimization algorithm can find an optimal solution close to the Pareto Front in a finite number of iterations, which is the Convergence of the algorithm [88]. Balance indicates whether the generated Pareto Front solutions are evenly distributed in the objective function space, which should theoretically avoid overconcentration on a given objective. The final solution of the MOO problem needs to cover the entire Pareto front as much as possible; this is the Distribution nature of the algorithm [79].

In addition, there are many other ways of evaluating MOO algorithms for different needs. (1) Error rate is used to detect the ratio of the unincluded approximate Pareto optimal solution to the overall [13]. A higher error rate indicates that the more Pareto approximations are outside the true solution, the worse the performance of the algorithm, and when the error rate is 0, the true solution contains all Pareto approximations. (2) Hypervolume Evaluation Index is a common performance metric in the evaluation of MOO algorithms to measure how well the solution set generated by the algorithm covers the objective function space. It is a quantitative metric commonly used in MOO problems to measure the diversity of solution sets and the volume of non-inferior solutions (Pareto Front) contained in the set [75]. The formula for this indicator is given below:

$$\begin{aligned} &\mathit{Hypervolume\ Evaluation\ Index} \\ &= \iint \ldots \int (\mathit{Pareto\ Front} - \mathit{Solution\ Set})\mathrm{d}f_1\mathrm{d}f_2\ldots\mathrm{d}f_k \end{aligned} \tag{2.2}$$

The Hypervolume evaluation index usually has a value between 0 and 1, where 1 indicates that the solution set completely covers the Pareto Front and 0 indicates that the solution set does not cover the Pareto Front. Thus, a higher Hypervolume value indicates that the solution set generated by the algorithm better covers the non-inferior solutions in the multi-objective space, suggesting that the algorithm performs better. (3) Generational Distance is a commonly used metric for evaluating the distance between the set of solutions generated by a MOO algorithm and the Pareto Front. The Generational Distance is used to measure the performance of an algorithm to determine how close the set of solutions generated by the algorithm is to the Generational Distance. Generational Distance is often expressed using the Euclidean distance formula:

$$\mathit{Generational\ Distance} = \sqrt{\frac{1}{N}\sum_{i=1}^{N} d(i)^2} \tag{2.3}$$

where d_i denotes the distance of the ith solution to the Pareto Front and N denotes the number of solutions in the solution set. When the Generational Distance result is close to zero, it means that the solution set is closer to the optimal solution and the algorithm performs better [24]. Like Generational Distance is (4) Inverted Generational Distance, which is calculated in a simpler way by eliminating the squaring and squaring operations in the above equation, and the principle is similar to Generational

Distance. In addition, (5) Maximum Spread is also used to measure the degree of dispersion of the different solutions again the target space, responding to the indicator of the homogeneity of the solution set [23]. (6) Convergence Metric Index is used to assess the convergence of the algorithm and the speed of convergence. When comparing good and bad solution sets of multiple algorithms, the coverage of multiple solution sets can also be assessed by (7) Coverage of two set [90].

2.3.3.2 Pareto Sets Trade-Offs Methods

How the decision maker weighs the solution sets of the MOO algorithms and finally obtains an optimization solution that can be applied is the most critical part of the MOO problem. Many traditional methods rely on simplifying a MOO problem into a single-objective optimization problem. The single-objective problem is solved based on different methods and the results are compared to select the final decision. Examples include a priori methods such as the Weighting method and interaction methods such as the Boundary Intersection method [15]. We count the common Pareto Sets trade-offs methods within the Fig. 2.3.

Priori Methods and Interactive Methods. A priori methods refer to the simplification of a multi-objective problem for solution by subjectively evaluating the objective function through prior knowledge, experience, and relevant research results. The most common a priori method is the Weighted Sum Method. Depending on the needs of the decision maker, Constraint Method, Dictionary Ordering Method, Constrained Scalarizing Method and Objective Programming Method are also common.

Although a priori methods are simple and easy to understand and compute, there are great limitations. For example, in Weighted Sum Method, the setting of the weights of each objective function and the determination of the order of the objective function in the Dictionary Ordering Method can only rely on a priori experience to solve the problem, which has a better performance in solving the convex optimization

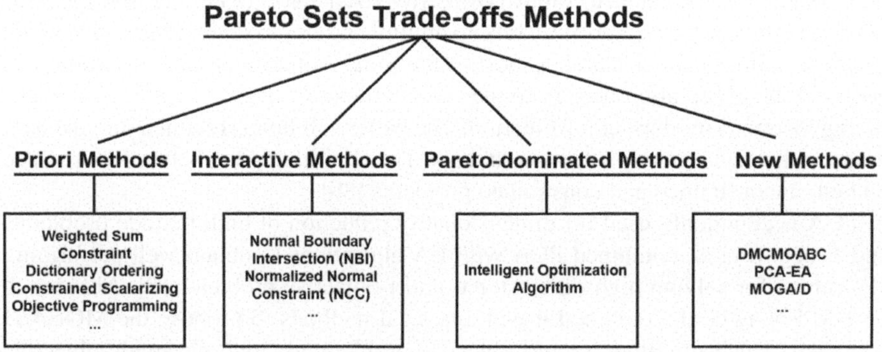

Fig. 2.3 Common Pareto sets trade-offs methods

problem, and its ability to solve many non-linear and complex problems in the field of wastewater treatment is relatively limited [51, 77].

To avoid the limitations of a priori methods, interactive methods that fully incorporate a priori experience, uncertainty, etc. have better capabilities in solving high-dimensional non-convex optimization problems. The core of such methods is to break down the multi-objective problem into a series of decision-making steps, and continuously adjust the weights through the feedback of the solution to find an approximate Pareto solution.

The two more common interactive methods are Normal Boundary Intersection (NBI) and Normalized Normal Constraint (NNC). NBI was first proposed in 1998, and the method gradually determines the Pareto solution by continuously searching for the intersection of the gradient direction with the Pareto front [16]. The advantage of NBI is the formation of uniform Pareto fronts, again on solving multi-objective non-convex optimization problems due to a priori methods. In wastewater treatment, some researchers have used NBI to analyze the prediction of the formation of bacterial and algal symbiotic system, through NBI to analyze and optimize various parameters of the formation process of bacterial and algal symbiosis to simulate the formation of the system [66]. NNC has similar advantages to NBI.

Pareto-dominated Methods and New Methods. Since the NSGA was first proposed for application to MOO problems, modern intelligent optimization approaches to Pareto solution sets have continued to emerge over time. Intelligent optimization methods based on the Pareto dominance approach are becoming more and more common when confronted with complex engineering problems, especially on the optimization of wastewater treatment processes (Fig. 2.2). They generally have an excellent ability to solve non-convex optimization problems. However, intelligent optimization methods still have some shortcomings, such as lower efficiency and the loss of non-inferior solutions in some cases, which can affect the performance of the algorithm in terms of time and accuracy [47]. To avoid making the algorithm faster and more accurate, many new methods have been created.

Xiang et al. improved the ABC algorithm and proposed a dynamic multi-colony multi-objective artificial bee colony (DMCMOABC) algorithm to improve the optimization efficiency. The algorithm sets up multiple groups to share information while conducting independent searches most of the time. At the same time the employed bees and the observation bees keep the same number of bees to explore the neighboring locations by sharing information. The better solution is retained into the next iteration through a greedy selection strategy. The algorithm gives better performance on both unconstrained and constrained problems [80].

PCA is commonly used for dimensionality reduction of high latitude problems, and Brockoff et al. combined PCA with EA algorithms to obtain well-performing algorithms for solving high-dimensional multi-objective problems [7]. Also based on EA, Zhang et al. combined it with classical methods to propose the MOEA/D approach, which simplifies a multi-objective problem into multiple single-objective problems by classical methods and solves them by EA [85]. MOO algorithms based on the theory of fuzzy logic have likewise received extensive attention, and these methods can minimize the disadvantage of conflicting objective functions [37].

2.3.4 Application of Multi-objective Optimization Algorithm in Wastewater Treatment Process

Wastewater treatment process is a complex multi-objective problem, including treatment cost, treatment efficiency, effluent quality, residual sludge discharge, and energy conservation, etc. These objectives are often in conflict, such as cost and effluent quality.

Caligan et al. constructed a sludge energy network model based on a bi-objective optimization algorithm to optimize the environmental impacts with disposal costs. The model was applied to two Philippine wastewater treatment plants, and both CO_2 emissions and costs were successfully optimized for minimization [8].

Zhang et al. used simulation data from the ASM model to optimize the effluent water quality (COD, NH^{4+}-N, NO^{3-}-N) and cost using Support Vector Machine SVM combined with GA to achieve the improvement of effluent water quality under the constraint of acceptable cost [86].

In addition to optimization in wastewater treatment process, there are many other aspects related to wastewater, for example, Yuan et al. found the optimal relationship between wastewater discharge permit allocation and economic benefits by weighing wastewater discharge permits against economic benefits through a compromise programming approach, and then solving the problem through an improved PSO algorithm [83].

2.4 Conclusions and Future Perspectives

2.4.1 Advantages and Innovations of Artificial Intelligence in Wastewater Treatment Process Optimization

The integration of AI into wastewater treatment processes has ushered in a paradigm shift with a multitude of advantages and groundbreaking innovations. In an era where water scarcity and environmental concerns are paramount, the optimization of wastewater treatment has evolved from a necessity to a strategic advantage. AI's role in this domain has been pivotal. One of its standout advantages is the substantial enhancement of efficiency. AI algorithms possess the ability to analyze copious amounts of data swiftly and accurately. This translates into optimized wastewater treatment processes by identifying and addressing bottlenecks, fine-tuning chemical dosages, and optimizing resource utilization. Consequently, this leads to tangible benefits, including reduced energy consumption, lowered operational costs, and a significantly diminished environmental footprint.

Furthermore, the advent of AI has introduced real-time monitoring and control systems that are revolutionizing wastewater treatment. These systems leverage AI's capacity to process data in real-time, facilitating continuous adjustments to treatment processes. The capability to swiftly adapt to changing conditions and promptly respond to compliance requirements represents a quantum leap in wastewater treatment. AI-driven solutions empower wastewater treatment plants to meet stringent environmental standards while maximizing resource efficiency.

AI has demonstrated remarkable prowess in predictive modeling. Through advanced machine learning techniques, particularly deep learning and neural networks, AI can predict and optimize complex wastewater treatment processes with unprecedented accuracy. The ability to predict system behavior, anticipate issues, and proactively optimize processes has not only streamlined operations but also minimized downtime and maintenance costs. The adaptability and scalability of AI are invaluable in the context of wastewater treatment. AI models can be tailored to suit the specific needs and complexities of different treatment plants, allowing for a wide range of applications across various scales. This adaptability ensures that both smaller and larger treatment facilities can benefit from AI-driven optimization.

In terms of innovation, AI has spearheaded the development of intelligent monitoring systems and advanced sensors. These innovations deliver high-quality, real-time data, enabling AI algorithms to make more informed decisions and fine-tune treatment processes. Furthermore, the concept of explainable AI is gaining prominence, ensuring that the decision-making process of AI models is transparent and comprehensible, instilling trust among operators and regulators.

In conclusion, the incorporation of AI into wastewater treatment processes is synonymous with efficiency enhancement, real-time control, predictive modeling, adaptability, and innovation. It holds the promise of not only mitigating the environmental impact of wastewater treatment but also optimizing resource utilization and reducing operational costs. As AI continues to evolve and mature in this domain, it is poised to play a pivotal role in addressing the critical challenges of water resource management and environmental sustainability.

2.4.2 Challenges and Development Trends of Artificial Intelligence in Wastewater Treatment Process Optimization

AI is expected to improve wastewater treatment efficiency, control treatment costs and reduce environmental impacts by optimizing wastewater treatment processes. However, AI still faces several spacing challenges when optimizing wastewater treatment processes.

Taking machine learning as an example, the quantity and quality of data is an important factor that affects the final accuracy of machine learning, and data sets with unequal lengths, imbalances, and outliers can reduce the performance of the algorithms. This requires good data collection and preprocessing capabilities. In this regard, real-time inspection is combined with AI to collect a large amount of data simultaneously from all aspects of the wastewater treatment process and process and compute it in real time through scientific means such as IoT. This also necessitates more research efforts in instrument development such as sensor development.

Second, the wastewater treatment process is more akin to a black box problem, a complex problem with a lot of uncertainties and influencing factors. Although the accuracy of AI in predicting water quality and other directions can give a statistically satisfactory figure, the underlying mechanism is opaque. This also requires more exploration of the underlying mechanisms of material transformation in the wastewater treatment process, which can then be combined with AI to obtain more convincing results.

Third, based on the huge amount of data, many AIs require computers with high arithmetic power to optimize the wastewater treatment process, especially deep learning algorithms. The lack of computing power in conventional wastewater treatment plants may limit the combination of AI and wastewater treatment.

Finally, cost is one of the main objectives of wastewater treatment optimization, but many studies have focused only on the costs incurred in the wastewater treatment process, which is not comprehensive enough. The cost of installing advanced sensors, computers with high arithmetic efficiency and the labor cost of mastering AI have not been fully considered in many studies, and it remains to be further researched whether the cost savings generated by the optimization can actually generate economic benefits.

However, with the continuous development of science and technology, all these problems may be solved. The continuous development of intelligent monitoring systems and sensor technology has made it increasingly possible to provide high-quality data in real time. In addition, the continuous innovation of AI algorithms is constantly adapting to the optimization process in various industries, and achieving real-time control and even implementing optimization is gradually becoming possible.

2.4.3 Main Findings and Contributions of This Review

In this work, the optimization of wastewater treatment process by AI is firstly reviewed in two separate aspects, namely, enhancing wastewater treatment efficiency and controlling cost. Then, based on the results of the review, it is found that the optimization of wastewater treatment processes often contains many conflicting optimization objectives, such as effluent quality with respect to cost and energy consumption. Therefore, we overview the main principles of MOO algorithms, as well as common types of algorithms. Intelligent optimization algorithms combined with AI, such as biologically inspired algorithms, have better performance in solving non-convex optimization problems. Finally, we summarize the challenges faced by AI and MOO algorithms in wastewater treatment. This work presents a systematic review of the application of AI to wastewater treatment process optimization to provide comprehensive theoretical support for future research in this area.

References

1. Ahmed SF, Mofijur M, Nuzhat S, Chowdhury AT, Rafa N, Uddin MA, Inayat A, Mahlia TMI, Ong HC, Chia WY, Show PL (2021) Recent developments in physical, biological, chemical, and hybrid treatment techniques for removing emerging contaminants from wastewater. J Hazard Mater 416.https://doi.org/10.1016/j.jhazmat.2021.125912
2. Ahmed Z, Mohamed K, Zeeshan S, Dong XQ (2020) Artificial intelligence with multi-functional machine learning platform development for better healthcare and precision medicine. Database-Oxford. https://doi.org/10.1093/database/baaa010
3. Alavi J, Ewees AA, Ansari S, Shahid S, Yaseen ZM (2022) A new insight for real-time wastewater quality prediction using hybridized kernel-based extreme learning machines with advanced optimization algorithms. Environ Sci Pollut R 29(14):20496–20516. https://doi.org/10.1007/s11356-021-17190-2
4. Asadi A, Verma A, Yang K, Mejabi B (2017) Wastewater treatment aeration process optimization: a data mining approach. J Environ Manage 203:630–639. https://doi.org/10.1016/j.jenvman.2016.07.047
5. Bellamoli F, Di Iorio M, Vian M, Melgani F (2023) Machine learning methods for anomaly classification in wastewater treatment plants. J Environ Manage 344.https://doi.org/10.1016/j.jenvman.2023.118594
6. Bohara RA, Throat ND, Mulla NA, Pawar SH (2017) Surface-modified cobalt ferrite nanoparticles for rapid capture, detection, and removal of pathogens: a potential material for water purification. Appl Biochem Biotech 182(2):598–608. https://doi.org/10.1007/s12010-016-2347-6
7. Brockhoff D, Zitzler E (2006) Are all objectives necessary? On dimensionality reduction in evolutionary multiobjective optimization
8. Caligan CJA, Garcia MMS, Mitra JL, San Juan JLG (2022) Multi-objective optimization for a wastewater treatment plant and sludge-to-energy network. J Clean Prod 368.https://doi.org/10.1016/j.jclepro.2022.133047
9. Campos M, San Millán A, Sempere JM, Lanza VF, Coque TM, Llorens C, Baquero F (2020) Simulating the influence of conjugative-plasmid kinetic values on the multilevel dynamics of antimicrobial resistance in a membrane computing model. Antimicrob Agents Ch 64(8). https://doi.org/10.1128/AAC.00593-20
10. Cardenes I, Hall JW, Eyre N, Majid A, Jarvis S (2020) Quantifying the energy consumption and greenhouse gas emissions of changing wastewater quality standards. Water Sci Technol 81(6):1283–1295. https://doi.org/10.2166/wst.2020.228
11. Chen WL, Lu XW, Yao CH (2015) Optimal strategies evaluated by multi-objective optimization method for improving the performance of a novel cycle operating activated sludge process. Chem Eng J 260:492–502. https://doi.org/10.1016/j.cej.2014.08.087
12. Chenar SS, Deng ZQ (2018) Development of artificial intelligence approach to forecasting oyster norovirus outbreaks along Gulf of Mexico coast. Environ Int 111:212–223. https://doi.org/10.1016/j.envint.2017.11.032
13. Coello CAC (2007) Evolutionary algorithms for solving multi-objective problems. Springer
14. Coello CAC, Pulido GT, Lechuga MS (2004) Handling multiple objectives with particle swarm optimization. IEEE Trans Evol Comput 8(3):256–279. https://doi.org/10.1109/tevc.2004.826067
15. Cui YF, Geng ZQ, Zhu QX, Han YM (2017) Review: multi-objective optimization methods and application in energy saving. Energy 125:681–704. https://doi.org/10.1016/j.energy.2017.02.174
16. Das I, Dennis JE (1998) Normal-boundary intersection: a new method for generating the Pareto surface in nonlinear multicriteria optimization problems. SIAM J Optimiz 8(3):631–657. https://doi.org/10.1137/S1052623496307510
17. Deepak M, Rustum R (2023) Review of latest advances in nature-inspired algorithms for optimization of activated sludge processes. Processes 11(1). https://doi.org/10.3390/pr11010077

18. Fang BB, Yu JC, Chen ZH, Osman AI, Farghali M, Ihara I, Hamza EH, Rooney DW, Yap PS (2023) Artificial intelligence for waste management in smart cities: a review. Environ Chem Lett 21(4):1959–1989. https://doi.org/10.1007/s10311-023-01604-3
19. Fu Z, Cheng J, Yang M, Batista J, Jiang Y (2020) Wastewater discharge quality prediction using stratified sampling and wavelet de-noising ANFIS model. Comput Electr Eng 85.https://doi.org/10.1016/j.compeleceng.2020.106701
20. Gad AG (2023) Particle swarm optimization algorithm and its applications: a systematic review. Arch Comput Method Eng 30(5):3471–3471. https://doi.org/10.1007/s11831-022-09762-3
21. Gao WF, Liu SY, Huang LL (2012) A global best artificial bee colony algorithm for global optimization. J Comput Appl Math 236(11):2741–2753. https://doi.org/10.1016/j.cam.2012.01.013
22. Gkika DA, Mitropoulos AC, Lambropoulou DA, Kalavrouziotis IK, Kyzas GZ (2022) Cosmetic wastewater treatment technologies: a review. Environ Sci Pollut Res 29(50):75223–75247. https://doi.org/10.1007/s11356-022-23045-1
23. Goodarzian F, Hosseini-Nasab H, Muñuzuri J, Fakhrzad MB (2020) A multi-objective pharmaceutical supply chain network based on a robust fuzzy model: a comparison of meta-heuristics. Appl Soft Comput 92.https://doi.org/10.1016/j.asoc.2020.106331
24. Goodarzian F, Taleizadeh AA, Ghasemi P, Abraham A (2021) An integrated sustainable medical supply chain network during COVID-19. Eng Appl Artif Intell 100:104188. https://doi.org/10.1016/j.engappai.2021.104188
25. Gui P, He FZ, Ling BWK, Zhang DY, Ge ZY (2023) Normal vibration distribution search-based differential evolution algorithm for multimodal biomedical image registration. Neural Comput Appl 35(22):16223–16245. https://doi.org/10.1007/s00521-023-08649-z
26. Guo W, Li J, Luo MY, Mao Y, Yu XY, Elskens M, Baeyens W, Gao Y (2022) Estrogenic activity and ecological risk of steroids, bisphenol A and phthalates after secondary and tertiary sewage treatment processes. Water Res 214.https://doi.org/10.1016/j.watres.2022.118189
27. Guo ZW, Du BX, Wang JH, Shen Y, Li Q, Feng D, Gao X, Wang H (2020) Data-driven prediction and control of wastewater treatment process through the combination of convolutional neural network and recurrent neural network. RSC Adv 10(23):13410–13419. https://doi.org/10.1039/d0ra00736f
28. Hao XD, Wu DQ, Li J, Liu RB, Van Loosdrecht M (2022) Making waves: a sea change in treating wastewater—why thermodynamics supports resource recovery and recycling. Water Res 218.https://doi.org/10.1016/j.watres.2022.118516
29. Harrou F, Dairi A, Sun Y, Senouci M (2018) Statistical monitoring of a wastewater treatment plant: a case study. J Environ Manage 223:807–814. https://doi.org/10.1016/j.jenvman.2018.06.087
30. Hojjati-Najafabadi A, Mansoorianfar M, Liang TX, Shahin K, Karimi-Maleh H (2022) A review on magnetic sensors for monitoring of hazardous pollutants in water resources. Sci Total Environ 824.https://doi.org/10.1016/j.scitotenv.2022.153844
31. Hreiz R, Latifi MA, Roche N (2015) Optimal design and operation of activated sludge processes: state-of-the-art. Chem Eng J 281:900–920. https://doi.org/10.1016/j.cej.2015.06.125
32. Izquierdo J, López PA, Martínez FJ, Pérez R (2007) Fault detection in water supply systems using hybrid (theory and data-driven) modelling. Math Comput Model 46(3–4):341–350. https://doi.org/10.1016/j.mcm.2006.11.013
33. Kamali M, Appels L, Yu XB, Aminabhavi TM, Dewil R (2021) Artificial intelligence as a sustainable tool in wastewater treatment using membrane bioreactors. Chem Eng J 417.https://doi.org/10.1016/j.cej.2020.128070
34. Katoch S, Chauhan SS, Kumar V (2021) A review on genetic algorithm: past, present, and future. Multimed Tools Appl 80(5):8091–8126. https://doi.org/10.1007/s11042-020-10139-6
35. Kazan K, Lyons R (2016) The link between flowering time and stress tolerance. J Exp Bot 67(1):47–60. https://doi.org/10.1093/jxb/erv441
36. Kirkpatrick S, Gelatt CD Jr, Vecchi MP (1983) Optimization by simulated annealing. Science 220(4598):671–680. https://doi.org/10.1126/science.220.4598.671

37. Koduru P, Das S, Welch SM (2007) Multi-objective hybrid PSO using μ-fuzzy dominance. Association for Computing Machinery, London, England, pp 853–860
38. Kumar V, Kumar D (2021) A systematic review on firefly algorithm: past, present, and future. Arch Comput Method Eng 28(4):3269–3291. https://doi.org/10.1007/s11831-020-09498-y
39. Kundu P, Debsarkar A, Mukherjee S (2013) Artificial neural network modeling for biological removal of organic carbon and nitrogen from slaughterhouse wastewater in a sequencing batch reactor. 2013:13–13
40. Li F, Su Z, Wang GM (2022) An effective dynamic immune optimization control for the wastewater treatment process. Environ Sci Pollut Res 29(53):79734–79734. https://doi.org/10.1007/s11356-022-18911-x
41. Li M, Hu K, Wang J (2021) Study on optimal conditions of flocculation in deinking wastewater treatment. 68(1):1–14
42. Li TH, Xie SS, Liu SP, Xiao L, Jia WZ, He DW (2018) A fault detection optimization method based on chaos adaptive artificial fish swarm algorithm on distributed control system. Proc Inst Mech Eng I-J Syst 232(9):1182–1193. https://doi.org/10.1177/0959651818777678
43. Li XB (2012) Study of multi-objective optimization and multi-attribute decision making for economic and environmental power dispatch. Electr Pow Syst Res 82(1):98. https://doi.org/10.1016/j.epsr.2011.09.021
44. Liu HB, Zhang YC, Zhang H (2020) Prediction of effluent quality in papermaking wastewater treatment processes using dynamic kernel-based extreme learning machine. Process Biochem 97:72–79. https://doi.org/10.1016/j.procbio.2020.06.020
45. Liu Q, Ibeas A, Vilanova R (2015) Neural network identification of wastewater treatment plants. 23rd mediterranean conference on control and aAutomation (MED). pp 840–846
46. Lu C, Xu ZX, Dong B, Zhang YH, Wang M, Zeng YF, Zhang C (2022) Machine learning for the prediction of heavy metal removal by chitosan-based flocculants. Carbohyd Polym 285.https://doi.org/10.1016/j.carbpol.2022.119240
47. Lu L, Yao B (2013) A performance oriented multi-loop constrained adaptive robust tracking control of linear motor drive systems: theory and experiments. In: Proceedings of the ASME 5th annual dynamic systems and control division conference and JSME 11th motion and vibration conference, DSCC 2012, vol 1, 85
48. Malviya A, Jaspal D (2021) Artificial intelligence as an upcoming technology in wastewater treatment: a comprehensive review. 10(1):177–187
49. Mamandipoor B, Majd M, Sheikhalishahi S, Modena C, Osmani V (2020) Monitoring and detecting faults in wastewater treatment plants using deep learning. Environ Monit Assess 192(2). https://doi.org/10.1007/s10661-020-8064-1
50. Mandavi S, Rahnamayan S, Deb K (2018) Opposition based learning: a literature review. Swarm Evol Comput 39:1–23. https://doi.org/10.1016/j.swevo.2017.09.010
51. Marler RT, Arora JS (2010) The weighted sum method for multi-objective optimization: new insights. Struct Multidisc Optim 41(6):853–862. https://doi.org/10.1007/s00158-009-0460-7
52. Mohan BC, Baskaran R (2012) A survey: ant colony optimization based recent research and implementation on several engineering domain. Expert Syst Appl 39(4):4618–4627. https://doi.org/10.1016/j.eswa.2011.09.076
53. Mukherjee S, Sarkar B, Aralappanavar VK, Mukhopadhyay R, Basak BB, Srivastava P, Marchut-Mikolajczyk O, Bhatnagar A, Semple KT, Bolan N (2022) Biochar-microorganism interactions for organic pollutant remediation: challenges and perspectives. Environ Pollut 308.https://doi.org/10.1016/j.envpol.2022.119609
54. Nguyen PY, Carvalho G, Reis MAM, Oehmen A (2021) A review of the biotransformations of priority pharmaceuticals in biological wastewater treatment processes. Water Res 188.https://doi.org/10.1016/j.watres.2020.116446
55. Nourani V, Elkiran G, Abba SI (2018) Wastewater treatment plant performance analysis using artificial intelligence—an ensemble approach. Water Sci Technol 78(10):2064–2076. https://doi.org/10.2166/wst.2018.477
56. Pérez V, Moltó JL, Lebrero R, Muñoz R (2021) Ectoine production from biogas in waste treatment facilities: a techno-economic and sensitivity analysis. ACS Sustain Chem Eng 9(51):17371–17380. https://doi.org/10.1021/acssuschemeng.1c06772

57. Petsiou DP, Martinos A, Spinos D (2023) Applications of artificial intelligence in temporal bone imaging: advances and future challenges. Cureus 15(9):e44591. https://doi.org/10.7759/cureus.44591
58. Pham QB, Gaya MS, Abba SI, Abdulkadir RA, Esmaili P, Linh NTT, Sharma C, Malik A, Khoi DN, Dung TD, Linh DQ (2020) Modeling of Bunus regional sewage treatment plant using machine learning approaches. Desalin Water Treat 203:80–90. https://doi.org/10.5004/dwt.2020.26160
59. Qambar AS, Khalidy MMMA (2023) Development of local and global wastewater biochemical oxygen demand real-time prediction models using supervised machine learning algorithms. Eng Appl Artif Intel 118.https://doi.org/10.1016/j.engappai.2022.105709
60. Ranieri E, D'Onghia G, Lopopolo L, Gikas P, Ranieri F, Gika E, Spagnolo V, Ranieri AC (2023) Evaluation of greenhouse gas emissions from aerobic and anaerobic wastewater treatment plants in Southeast of Italy. J Environ Manage 337.https://doi.org/10.1016/j.jenvman.2023.117767
61. Rathi BS, Kumar PS, Vo DVN (2021) Critical review on hazardous pollutants in water environment: occurrence, monitoring, fate, removal technologies and risk assessment. Sci Total Environ 797.https://doi.org/10.1016/j.scitotenv.2021.149134
62. Rojek I, Mikolajewski D, Kopowski J, Kotlarz P, Piechowiak M, Dostatni E (2021) Reducing waste in 3D printing using a neural network based on an own elbow exoskeleton. Materials 14(17). https://doi.org/10.3390/ma14175074
63. Safder U, Kim J, Pak G, Rhee G, You K (2022) Investigating machine learning applications for effective real-time water quality parameter monitoring in full-scale wastewater treatment plants. Water 14(19). https://doi.org/10.3390/w14193147
64. Safeer S, Pandey RP, Rehman B, Safdar T, Ahmad I, Hasan SW, Ullah A (2022) A review of artificial intelligence in water purification and wastewater treatment: recent advancements. 49:102974
65. Salaudeen T, Okoh O, Agunbiade F, Okoh A (2018) Fate and impact of phthalates in activated sludge treated municipal wastewater on the water bodies in the Eastern Cape, South Africa. Chemosphere 203:336–344. https://doi.org/10.1016/j.chemosphere.2018.03.176
66. Sánchez-Zurano A, Guzmán JL, Acién FG, Fernández-Sevilla JM (2021) An interactive tool for simulation of biological models into the wastewater treatment with microalgae. Front Env Sci 9.https://doi.org/10.3389/fenvs.2021.721324
67. Sandeep B, Huang X, Xiao ZW (2023) Artificial intelligence in heart failure improving the efficiency or dependency on it? Letter regarding the article artificial intelligence and heart failure: a state-of-the-art review. Eur J Heart Fail. https://doi.org/10.1002/ejhf.3017
68. Shahbeik H, Rafiee S, Shafizadeh A, Jeddi D, Jafary T, Lam SS, Pan JT, Tabatabaei M, Aghbashlo M (2022) Characterizing sludge pyrolysis by machine learning: towards sustainable bioenergy production from wastes. Renew Energ 199:1078–1092. https://doi.org/10.1016/j.renene.2022.09.022
69. Slowik A, Kwasnicka H (2020) Evolutionary algorithms and their applications to engineering problems. Neural Comput Appl 32(16):12363–12379. https://doi.org/10.1007/s00521-020-04832-8
70. Sniatala B, Kurniawan TA, Sobotka D, Makinia J, Othman D (2023) Macro-nutrients recovery from liquid waste as a sustainable resource for production of recovered mineral fertilizer: uncovering alternative options to sustain global food security cost-effectively. Sci Total Environ 856.https://doi.org/10.1016/j.scitotenv.2022.159283
71. Taoufik N, Boumya W, Achak M, Chennouk H, Dewil R, Barka N (2022) The state of art on the prediction of efficiency and modeling of the processes of pollutants removal based on machine learning. Sci Total Environ 807.https://doi.org/10.1016/j.scitotenv.2021.150554
72. Tian JH, Yan C, Alcega SG, Hassard F, Tyrrel S, Coulon F, Nasir ZA (2022) Detection and characterization of bioaerosol emissions from wastewater treatment plants: challenges and opportunities. Front Microbiol 13. https://doi.org/10.3389/fmicb.2022.958514
73. Torregrossa D, Leopold U, Hernández-Sancho F, Hansen J (2018) Machine learning for energy cost modelling in wastewater treatment plants. J Environ Manage 223:1061–1067. https://doi.org/10.1016/j.jenvman.2018.06.092

74. Turek M, Heiden W, Riesen A, Chhabda TA, Schubert J, Zander W, Krüger P, Keusgen M, Schöning MJ (2009) Artificial intelligence/fuzzy logic method for analysis of combined signals from heavy metal chemical sensors. Electrochim Acta 54(25):6082–6088. https://doi.org/10.1016/j.electacta.2009.03.035
75. Verhellen J (2022) Graph-based molecular Pareto optimisation. Chem Sci 13(25):7526–7535. https://doi.org/10.1039/d2sc00821a
76. Wainaina S, Awasthi MK, Sarsaiya S, Chen HY, Singh E, Kumar A, Ravindran B, Awasthi SK, Liu T, Duan YM, Kumar S, Zhang ZQ, Taherzadeh MJ (2020) Resource recovery and circular economy from organic solid waste using aerobic and anaerobic digestion technologies. Bioresource Technol 301.https://doi.org/10.1016/j.biortech.2020.122778
77. Wang R, Zhou ZB, Ishibuchi H, Liao TJ, Zhang T (2018) Localized weighted sum method for many-objective optimization. IEEE Trans Evol Comput 22(1):3–18. https://doi.org/10.1109/Tevc.2016.2611642
78. Wang YQ, Wang HC, Song YP, Zhou SQ, Li QN, Liang B, Liu WZ, Zhao YW, Wang AJ (2023) Machine learning framework for intelligent aeration control in wastewater treatment plants: automatic feature engineering based on variation sliding layer. Water Res 246:120676. https://doi.org/10.1016/j.watres.2023.120676
79. Wu H, Wang K, Xu L (2022) How can age-based vaccine allocation strategies be optimized? A multi-objective optimization framework. Front Public Health 10:934891. https://doi.org/10.3389/fpubh.2022.934891
80. Xiang Y, Zhou YR (2015) A dynamic multi-colony artificial bee colony algorithm for multi-objective optimization. Appl Soft Comput 35:766–785. https://doi.org/10.1016/j.asoc.2015.06.033
81. Xiu JP, He Q, Yang ZQ, Liu C (2016) Research on a multi-objective constrained optimization evolutionary algorithm. In: 4th international conference on cloud computing and intelligence systems (CCIS). pp 282–286
82. Yu TT, Bai Y (2018) A comparative study of extreme learning machine, least squares support vector machine, back propagation neural network for outlet total phosphorus prediction. Prognostics and system health management conference (PHM-Chongqing). pp 717–722.https://doi.org/10.1109/PHM-Chongqing.2018.00128
83. Yuan MK, Chen XD, Liu GZ, Ren HR (2022) Coordinated allocation of water resources and wastewater emission permits based on multi-objective optimization model: from the perspective of conflict between equity and economic benefits. J Clean Prod 372.https://doi.org/10.1016/j.jclepro.2022.133733
84. Zaghloul MS, Achari G (2022) Application of machine learning techniques to model a full-scale wastewater treatment plant with biological nutrient removal. J Environ Chem Eng 10(3). https://doi.org/10.1016/j.jece.2022.107430
85. Zhang QF, Li H (2007) MOEA/D: a multiobjective evolutionary algorithm based on decomposition. IEEE Trans Evol Comput 11(6):712–731. https://doi.org/10.1109/Tevc.2007.892759
86. Zhang R, Xie WM, Yu HQ, Li WW (2014) Optimizing municipal wastewater treatment plants using an improved multi-objective optimization method. Bioresource Technol 157:161–165. https://doi.org/10.1016/j.biortech.2014.01.103
87. Zhang S, Jin Y, Chen W, Wang J, Wang Y, Ren H (2023) Artificial intelligence in wastewater treatment: a data-driven analysis of status and trends. Chemosphere 336:139163. https://doi.org/10.1016/j.chemosphere.2023.139163
88. Zhou CP, Wang D, Pan XY, Shen HB (2021) Protein structure refinement using multi-objective particle swarm optimization with decomposition strategy. Int J Mol Sci 22(9). https://doi.org/10.3390/ijms22094408
89. Zhou P, Wang X, Chai TY (2022) Multiobjective operation optimization of wastewater treatment process based on reinforcement self-learning and knowledge guidance. IEEE Trans Cybernetics. https://doi.org/10.1109/Tcyb.2022.3164476
90. Zitzler E, Thiele L (1999) Multiobjective evolutionary algorithms: a comparative case study and the strength Pareto approach. IEEE Trans Evol Comput 3(4):257–271. https://doi.org/10.1109/4235.797969

Chapter 3
Overcoming Challenges in Traditional Waste Water Treatment Through AI-Driven Innovation

Shanmugapriya Sakkaravarthy, Negasso Argeta Jano, and Ajitha Vijayakumar

Abstract This chapter delves into the profound impact of artificial intelligence (AI) on wastewater treatment, addressing long-standing challenges and showcasing its transformative potential. AI emerges as a potent force of innovation, poised to overcome traditional challenges and user in an era of efficiency, sustainability, and accountability in wastewater treatment, minimizing environmental impact and optimizing resource usage. It becomes a solution to common issues in traditional wastewater treatment methods, such as pollutant removal inefficiencies, high operating costs, and inflexibility in adapting to changing conditions. The multifaceted role of AI in wastewater treatment is explored, focusing on real-time monitoring, predictive maintenance, and treatment process optimization. Cutting-edge technologies like sensor integration and the Internet of Things (IoT) are introduced for data collection and monitoring. Real-world case studies demonstrating the successful implementation of these innovations, providing valuable insights are also discussed. This chapter also highlights the power of predictive modeling driven by machine learning algorithms. AI predicts influential features and treatment performance, leading to increased efficiency. Automated control systems are emerging, with AI making critical decisions and monitoring equipment, reducing human intervention and errors. AI-driven energy optimization and intelligent dosing enhance resource efficiency and cost reduction while minimizing environmental impact as AI is adaptable to variability, seasonal changes, fluctuations in influent quality, and building resilience against changing conditions. AI streamlines regulatory compliance through real-time monitoring, automated reporting, and adherence to environmental standards.

S. Sakkaravarthy (✉)
Assistant Professor, Department of Biotechnology, Kongunadu Arts and Science college, Coimbatore 641029, India
e-mail: shanmuga.7@rediffmail.com

N. A. Jano · A. Vijayakumar
Department of Biotechnology, Kongunadu Arts and Science college, Coimbatore 641029, India

Department of Biotechnology, Rathinam College of Arts and Science, Coimbatore, India

A. Vijayakumar
e-mail: ajithav.bbt22@rathinam.in

M. C. Garg (ed.), *The AI Cleanse: Transforming Wastewater Treatment Through Artificial Intelligence*, Springer Water, https://doi.org/10.1007/978-3-031-67237-8_3

Ethical considerations surrounding AI deployment are thoroughly examined, along with strategies for responsible AI use in wastewater treatment.

Keywords Waste water treatment · Artificial intelligence · Internet of things

3.1 Introduction

Wastewater treatment is the process of cleaning and purifying water that has been used in houses, businesses, and industries before it can be safely returned to the environment. However, this process comes with several challenges. One of the major challenges is the increasing demand and scarcity of clean water, which puts pressure on wastewater treatment plants to treat more water with limited resources [1]. Another challenge is the aging infrastructure of treatment plants, which can lead to inefficiencies and breakdowns in the system. Moreover, emerging contaminants such as microplastics and pharmaceuticals pose a significant threat to the environment and human health [2]. Energy consumption and costs associated with wastewater treatment are also significant challenges that need to be addressed. Treatment plants require a considerable amount of energy to operate, making them a significant contributor to greenhouse gas emissions [3]. Limited funding and resources are another challenge faced by many wastewater treatment facilities, particularly in low-income areas where there is a lack of investment in infrastructure upgrades and maintenance. Finally, public awareness and education are crucial in promoting responsible behavior regarding wastewater disposal and reducing the number of harmful substances entering the system. Despite these challenges, advancements in technology and innovative approaches offer hope for more efficient and sustainable wastewater treatment systems in the future. Traditional methods of treating wastewater can be challenging and often require a lot of resources. However, with the help of AI-driven innovation, these challenges can be addressed more efficiently [4]. Artificial intelligence (AI) technology can assist in monitoring water quality, predicting potential issues and optimizing treatment processes to ensure that wastewater is properly treated before being released into the environment. This chapter serves as a comprehensive exploration of the transformative role of AIin wastewater treatment. It delves into the traditional challenges faced by wastewater treatment methods and outlines how AI-driven innovations offer practical solutions to these issues. It begins by introducing the challenges that have historically hindered wastewater treatment processes, including inefficiencies, high operational costs, and limited adaptability. It then pivots to the potential benefits of AI in overcoming these challenges, such as real-time monitoring, predictive maintenance, and process optimization. The chapter proceeds to explore AI-driven data collection and monitoring, shedding light on sensor technology, IoT integration, and their role in enhancing data analysis. Real-world case studies will be presented to demonstrate successful AI-driven data monitoring implementations. Another critical aspect covered is predictive modeling for process optimization. Machine learning algorithms are discussed in the context of predicting influent characteristics and optimizing treatment processes. The practical applications of AI in these areas will be illustrated through case examples.

3.2 Traditional/Conventional Wastewater Treatment Methods

Conventional wastewater treatment processes are methods used to clean and purify water that has been contaminated with waste materials. These processes typically involve several stages, each designed to remove different types of pollutants from the water. The stages are designed for either chemical, physical or biological but the whole process contain all [5]. The first stage is usually a physical process called screening, which involves passing the wastewater through a series of screens or filters to remove large objects such as sticks, leaves, and plastics. The second stage is known as primary treatment, where the water undergoes sedimentation in large tanks. During this stage, heavier particles settle at the bottom while lighter one's float on top and are skimmed off. Next comes secondary treatment which utilizes biological processes to break down organic matter using microorganisms. This process can be done aerobically (with oxygen) or anaerobically (without oxygen). Finally, tertiary treatment removes any remaining impurities that may still be present in the treated water using advanced filtration techniques such as reverse osmosis or ultraviolet disinfection. Contemporary wastewater treatment methods often fall short of completely eliminating pollutants; instead, they tend to concentrate or transform pollutants into different phases. Among a plethora of available techniques, adsorption stands out as one of the most efficient methods for wastewater treatment. This distinction arises from its several unique characteristics, which make it more effective when compared to alternative methods [5].

3.2.1 Activated Sludge Process

The Activated Sludge Process is a method used in wastewater treatment plants to clean up dirty water. It involves mixing the wastewater with microorganisms and oxygen in large tanks called reactors. The microorganisms break down the organic matter in the water, turning it into harmless substances like carbon dioxide and water (Fig. 3.1). This process takes several hours to complete, after which the treated water is separated from the solid waste material (sludge) that has formed during the process [6]. The sludge can then be further treated or disposed of safely without harming the environment. The treated water can also be discharged into nearby bodies of water or reused for other purposes such as irrigation or industrial processes.

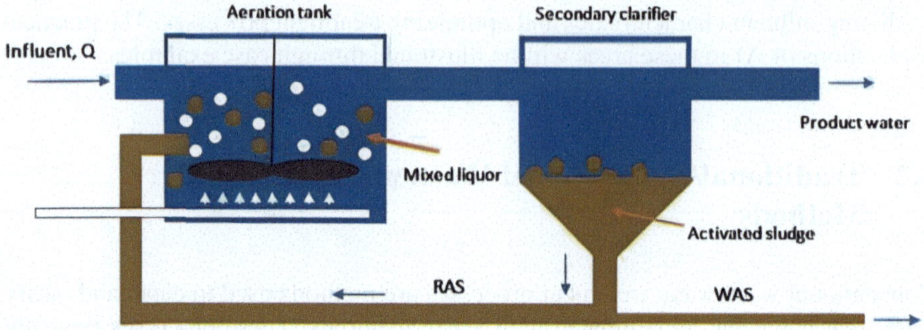

Fig. 3.1 Activated sludge process

3.2.2 Trickling Filter Process

The Trickling Filter Process is a way to treat wastewater that involves letting the water trickle over a bed of rocks or other materials. The rocks are covered in bacteria that break down the harmful substances in the wastewater, making it safe to release back into the environment. This process is often used in small communities or rural areas where there may not be access to large-scale treatment plants (Fig. 3.2). It's a simple and effective way to clean up wastewater and protect our waterways from pollution.

In the treatment process, influent wastewater is introduced into the TF (Trickling Filter) reactor through a distribution system that disperses it evenly across the top surface of the biofilm. As the wastewater percolates downward through the reactor, it interacts with the biofilm, where microorganisms actively break down organic substances and nitrogen compounds present in the liquid phase. This biological degradation process transforms these contaminants into more environmentally benign forms. Subsequently, the treated effluent drains out of the reactor, leaving behind a cleaner and more purified liquid stream [7].

3.2.3 Rotating Biological Contactor Process

The Rotating Biological Contactor Process is a method used to treat wastewater. It involves using a rotating cylinder with plastic discs attached to it, which are partially submerged in the water. The discs provide a surface for microorganisms to grow on, and as the cylinder rotates through the water, these microorganisms consume organic matter and pollutants in the water. The discs used in this system are typically constructed from corrosion-resistant materials like plastic or other similarly durable materials. As these discs rotate, a crucial process unfolds: microorganisms affix themselves to the surface of the discs, forming-biofilm play a pivotal role in consuming and breaking down the organic matter present in the wastewater [8]. It ensures that

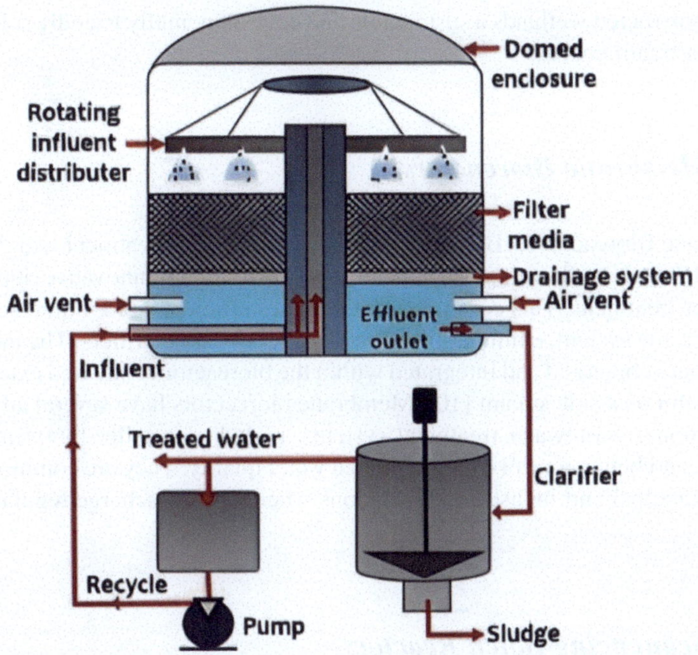

Fig. 3.2 Trickling filter process

the biofilm remains in constant motion, exposing it continually to the oxygen present in the surrounding environment that promotes the growth and metabolic activity of the microorganisms within the biofilm, allowing them to efficiently and effectively break down the organic substances contained in the wastewater. This process is an effective way to remove contaminants from wastewater because it allows for large amounts of microorganisms to be present in a small space. Additionally, the rotation of the cylinder ensures that all parts of the disc come into contact with oxygen-rich air, which helps facilitate biological reactions.

3.2.4 Constructed Wetlands

Constructed wetlands are man-made areas that mimic the functions of natural wetlands. These constructed wetlands are designed to treat wastewater, stormwater runoff, and other types of polluted water by using plants, soil, and other natural materials to filter out pollutants.

Constructed wetlands play a vital role in the treatment of wastewater by employing a combination of physical, chemical, and biological processes. These processes collectively contribute to the effective purification of incoming wastewater, making constructed wetlands a sustainable and environmentally friendly solution for wastewater treatment [9].

3.2.5 Membrane Bioreactors

A Membrane Bioreactor (MBR) integrates activated sludge treatment with Microfiltration (MF) or Ultrafiltration (UF) membranes, offering an innovative approach to wastewater treatment. This combination facilitates efficient solid–liquid separation within the same system, eliminating the need for secondary clarifiers. The membrane can be either submerged and integrated within the bioreactor or situated externally to the bioreactor as a side stream [10]. Membrane bioreactors have several advantages over traditional wastewater treatment systems, including smaller footprint, higher efficiency, and better control over the treated water quality. They are commonly used in both municipal and industrial applications where strict discharge regulations are in place.

3.2.6 Sequencing Batch Reactors

SBR, or Sequencing Batch Reactor, represents a specific variant of activated sludge technology utilized in wastewater treatment [11]. The Sequencing Batch Reactor (SBR) is an advanced wastewater processing unit that combines both biological and secondary treatment processes within a single tank. It accomplishes this by employing a controlled timed sequence, effectively integrating functions that would typically require multiple separate units in traditional Activated Sludge (AS) systems.

The operation of an SBR involves a well-defined sequence of five fundamental procedures, namely: Fill, React, Settle, Draw, and Idle [12]. These sequential steps enable efficient and comprehensive wastewater treatment in a single vessel, making SBR technology a versatile and space-efficient solution for various wastewater treatment applications. The process starts with filling up the reactor tank with wastewater and adding some bacteria to it.

Overall, SBRs are an effective way to treat large volumes of wastewater while minimizing environmental impact. They are commonly used in industrial settings such as factories and power plants but can also be found in smaller-scale applications like municipal sewage treatment facilities or even individual households with septic systems.

3.2.7 *Anaerobic Digestion*

Anaerobic digestion is a process that breaks down organic matter in the absence of oxygen. This means that microorganisms break down waste materials like food scraps, manure, and sewage into biogas and fertilizer without using any air. The biogas produced during anaerobic digestion can be used as a renewable energy source for heating or electricity generation. Additionally, the leftover fertilizer can be used to enrich soil and promote plant growth. Other benefits include a reduction of total biosolids volume of up to 50–80% and a final waste sludge that is biologically stable can serve as a rich humus for agriculture [13].

3.2.8 *Chlorination and Ultraviolet Disinfection*

Chlorination and ultraviolet disinfection are two methods used to clean water. Chlorination is when chlorine, a chemical substance, is added to the water to kill harmful bacteria and viruses. This method has been used for many years and is effective in removing most types of germs from the water. UV light and chlorination represent two of the most commonly employed disinfection methods for treating swimming pools and recreational water facilities. Interestingly, the combination of these techniques not only enhances the efficiency of microorganism elimination but also promotes safety for the pool users. This combination notably reduces the formation of chlorination by-products, further contributing to the overall well-being of individuals enjoying the pool [14].

Ultraviolet disinfection, on the other hand, uses UV light to destroy bacteria and viruses in the water. This method does not involve adding any chemicals to the water which makes it more environmentally friendly. However, this process requires electricity which means that it can be more expensive than chlorination. Both methods have their advantages and disadvantages but they are both important ways of ensuring that our drinking water is safe for consumption.

3.3 Common Challenges Faced by Traditional Waste Treatment Methods Methods

3.3.1 *Inefficiency in Pollutant Removal*

Traditional wastewater treatment methods often struggle to efficiently remove various pollutants from wastewater, including organic matter, nutrients (such as nitrogen and phosphorus), heavy metals, and emerging contaminants (e.g., pharmaceuticals and micro-plastics). Inefficient pollutant removal can lead to the discharge of inadequately treated wastewater into the environment, causing water pollution, harm to aquatic ecosystems, and potential health risks to humans.

The limited removal efficiency can result from outdated treatment technologies and processes that are less effective in treating emerging or complex contaminants. According to Abuhasel et al. [15], conventional methods for treating oily wastewater have proven to be inefficient in achieving desired results. In this study, Dissolved Air Flotation (DAF) technique, struggles to effectively separate oil droplets measuring approximately 20 μm or smaller from the wastewater. To address the challenge of smaller droplet sizes, additional pretreatment units are required prior to the flotation process. Unfortunately, this necessitates a larger physical footprint and higher capital investments. Furthermore, the formation of activated sludge as a byproduct of flocculation and demulsification in oily wastewater treatment processes poses a significant environmental concern when discharged into natural watercourses (Fig. 3.3). Another limitation is encountered when employing the Electrocoagulation (ET) method, as it is susceptible to electrode material corrosion. Conventional treatment methods have demonstrated their ineffectiveness in addressing emerging contaminants like pharmaceuticals. Implementing changes in treatment design, processes, and infrastructure to tackle these new challenges often entails a significant increase in costs [1].

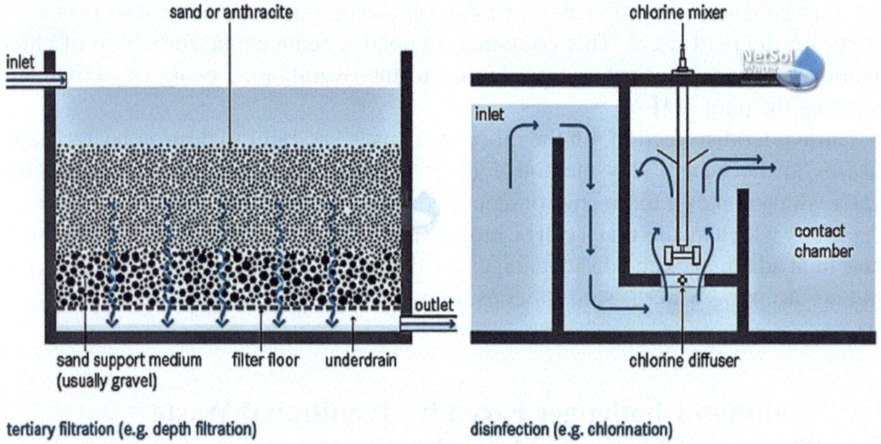

Fig. 3.3 Tertiary filtration and disinfection method

3.3.2 High Operational Costs

Traditional wastewater treatment plants require significant resources and operational costs to maintain and operate effectively. These costs include labor, energy, chemicals, maintenance, and infrastructure upkeep. High energy consumption is a notable cost factor, especially for aeration in biological treatment processes, which require constant air supply. The purchase and handling of chemicals for coagulation, flocculation, and disinfection also contribute to operational expenses. Frequent maintenance and repairs are necessary to keep aging infrastructure functioning, further increasing costs over time. For example, when utilizing plant-wide modeling approaches, the exploration of altering operational strategies to enhance bio-methane recovery through chemically enhanced primary treatment has been assessed. The findings indicate that while there are favorable outcomes such as elevated methane production for renewable energy generation and reductions in CO_2 emissions, subsequently decreasing the plant's carbon footprint, these advantages can be significantly counterbalanced by heightened chemical consumption. This, in turn, results in elevated operational expenses and adverse life cycle assessment (LCA) impacts, including the depletion of abiotic elements and fossil fuel resources [16].

3.3.3 Limited Adaptability to Changing Conditions

Traditional wastewater treatment systems are typically designed for specific influent characteristics and environmental conditions. They may struggle to adapt to variations in influent quality, flow rates, or seasonal changes. Climate change can exacerbate these challenges by causing unpredictable shifts in weather patterns, which can affect the quality and quantity of wastewater entering treatment facilities. Rapid urbanization and population growth can also strain existing treatment systems, making it difficult for them to cope with increased demand. Limited adaptability can lead to suboptimal treatment performance and regulatory compliance issues. Addressing these challenges is where AI-driven innovation in wastewater treatment can play a crucial role by enhancing efficiency, reducing costs, and improving adaptability to changing conditions.

3.4 The Role of Artificial Intelligence in Wastewater Treatment

AI has emerged as a transformative force across various industries, and the field of wastewater treatment is no exception. AI, a branch of computer science that focuses on creating intelligent machines capable of learning and decision-making, has found numerous applications in wastewater treatment, revolutionizing the way

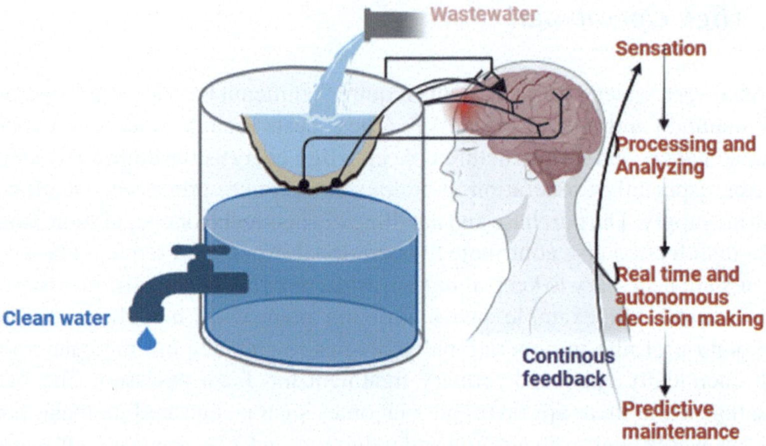

Fig. 3.4 Schematic representation of artificial intelligence integration in methods for wastewater treatment

we approach the management of water resources (Fig. 3.4). AI encompasses a range of technologies and techniques that enable machines to perform tasks that typically require human intelligence. At its core, AI leverages data-driven algorithms and computational power to analyze, learn from, and make predictions or decisions based on large datasets. Some key components of AI relevant to wastewater treatment include:

Machine Learning (ML): ML is a subset of AI that focuses on the development of algorithms that can learn and improve from experience. This technology is particularly valuable for pattern recognition, classification, and predictive modeling.

Deep Learning: Deep learning is a specialized form of ML that involves artificial neural networks inspired by the human brain's structure. It excels in handling complex, unstructured data and is well-suited for tasks like image and speech recognition.

Natural Language Processing (NLP): NLP enables machines to understand, interpret, and generate human language. This technology is useful for text-based data analysis, such as regulatory compliance documentation.

AI-based wastewater treatment systems rely on a structured process to produce safe water. Data is continuously collected from sensors measuring various parameters, which is then preprocessed and analyzed using AI algorithms. These algorithms make real-time decisions to optimize the treatment process, adjusting chemical dosages, aeration rates, and more as needed. Predictive maintenance ensures equipment reliability, while adaptive control systems adapt to changing conditions. A constant feedback loop maintains water quality, and safety protocols prevent harmful actions. Compliance with water quality standards and reporting are prioritized, and resource efficiency and cost savings are achieved through precise control and optimization. Overall, AI-based wastewater treatment ensures the production of safe water while enhancing efficiency and reducing operational costs. In doing so, they address the limitations of conventional methods by adapting to changing conditions, reducing operational costs, and minimizing environmental impact.

3.4.1 Applications of AI in Wastewater Treatment

AI-driven innovations have the potential to address various challenges in traditional wastewater treatment methods. Conventional wastewater treatment is crucial for reducing pollutants and improving water quality. However, natural conditions, influent variations, and treatment technology complexity introduce uncertainty and variability. These factors lead to fluctuations in treated water quality, operational costs, and environmental risks to receiving waters. Artificial intelligence offers a potent solution to address these challenges in wastewater treatment, streamlining processes and reducing complexities [17].

a. **Real-Time Monitoring and Data Analytics**

AI-powered sensors and data analytics platforms can continuously monitor water quality parameters and treatment processes in real time. Predictive analytics identify potential issues and enable proactive maintenance, reducing downtime.

b. **Predictive Modeling for Process Optimization**

AI algorithms can predict influent characteristics, allowing treatment plants to adjust their processes in advance to optimize treatment efficiency. Optimization algorithms can adjust aeration rates, chemical dosing, and other variables to minimize energy consumption and chemical usage.

c. **Autonomous Control Systems**

AI can automate the control of pumps, valves, and other equipment to maintain optimal treatment conditions. These systems adapt to changing conditions and operate efficiently without constant human intervention.

d. **Resource Efficiency and Cost Reduction**

AI-driven energy optimization reduces electricity consumption, contributing to cost savings and environmental sustainability. Smart dosing systems reduce the use of expensive chemicals while maintaining treatment performance.

e. **Addressing Variability and Adaptation**

AI helps treatment plants adapt to changing influent quality due to weather, seasonal variations, or industrial discharges. It enhances resilience by enabling rapid adjustments in response to unexpected events.

f. **Regulatory Compliance and Reporting**

Table 3.1 Comparative analysis of traditional and ai-enhanced wastewater treatment approaches across various aspects

Aspects	Conventional wastewater treatment methods	AI-based wastewater treatment method
Process complexity	Often involves basic physical, chemical, and biological processes	Utilizes advanced algorithms and machine learning for optimization
Efficiency	May have limitations in adapting to varying conditions and optimizing processes	Adapts to changing conditions in real-time, optimizing treatment processes for efficiency
Data utilization	Limited use of historical data and manual adjustments	Analyzes extensive data in real-time to make precise decisions
Maintenance	Routine maintenance based on fixed schedules	Predictive maintenance based on real-time monitoring and data analysis
Resource consumption	Can be energy and resource-intensive due to fixed processes	Can reduce resource consumption and energy usage through optimization
Environmental impact	May release pollutants during process inefficiencies	Minimizes environmental impact by optimizing treatment processes
Cost	Initial setup costs may be lower, but operational costs can be high	Higher initial investment in technology, but potential long-term cost savings
Adaptability	Limited adaptability to changing wastewater composition	Adapts to changing influent characteristics for consistent treatment quality

AI systems ensure real-time regulatory compliance monitoring and automated reporting, reducing the risk of non-compliance and associated penalties.

Table 3.1 illustrates the comparison of conventional and AI-based wastewater treatment methods in terms of different aspects.

3.4.2 AI-Driven Data Collection and Monitoring

AIhas revolutionized data collection and monitoring in wastewater treatment, offering a dynamic and efficient approach to managing water quality.

a. **Sensor Technology and IoT Integration**

AI-driven data collection and monitoring rely heavily on advanced sensor technology and seamless integration with the Internet of Things (IoT). Modern sensors are capable of measuring various water quality parameters, such as pH, turbidity, dissolved oxygen, and nutrient levels, with high precision [18]. They can also detect contaminants and pollutants. IoT devices and networks connect these sensors to centralized data platforms. This connectivity enables real-time data transmission, allowing treatment facilities to monitor conditions continuously. Sensor data is transmitted to a central server or cloud-based platform, where AI algorithms process and analyze the information.

b. **Data Collection and Analysis for Improved Insights**

AI-driven data collection and analysis offer several advantages for wastewater treatment plants. AI has the potential to enhance water treatment in several ways. It can enable real-time monitoring and automation of treatment systems, leading to increased efficiency and reduced energy consumption. AI-driven decision support systems assist operators in making informed decisions regarding treatment strategies, resource allocation, and maintenance scheduling. Moreover, AI techniques can optimize processes like evaporation by analyzing weather data and humidity levels, ultimately reducing water loss and improving overall efficiency [19]. This real-time data is crucial for rapid decision-making. AI can also integrate data from various sensors and sources, providing a holistic view of the treatment process. This enables a deeper understanding of complex interactions. Hence, AI algorithms predicts changes in water quality, allowing operators to take proactive measures to maintain treatment efficiency.

3.5 Predictive Modeling for Process Optimization

Predictive modeling powered by machine learning algorithms is a cornerstone of AI-driven innovation in wastewater treatment. This section explores the core components of predictive modeling and presents case examples demonstrating how it enhances efficiency in wastewater treatment processes.

3.5.1 Machine Learning Algorithms for Predictive Modeling

Regression Analysis: Linear and nonlinear regression models are employed to predict treatment performance based on historical data. These models establish relationships between various input parameters and treatment outcomes.

Classification Models: Classification algorithms, such as decision trees and support vector machines, can predict the quality of influent and identify the potential presence of contaminants or anomalies.

Time Series Analysis: Time series models, like ARIMA (AutoRegressive Integrated Moving Average), are used to forecast influent characteristics and treatment plant performance over time.

Artificial Neural Networks (ANNs): Deep learning models, including ANNs and convolutional neural networks (CNNs), excel in handling complex, high-dimensional data and are used for predicting influent variations and optimizing treatment processes.

3.5.2 Predicting Influent Characteristics and Treatment Performance

Influent Characterization: Predictive modeling can anticipate influent characteristics, including flow rates, pollutant concentrations, and seasonal variations. This allows treatment plants to adjust processes in advance.

Optimizing Chemical Dosing: AI models predict the optimal dosage of chemicals like coagulants and flocculants required for effective pollutant removal, reducing both costs and environmental impact.

Treatment Efficiency: Predictive models assess treatment plant efficiency by forecasting key performance indicators such as removal rates for pollutants, turbidity levels, and water quality parameters.

Energy Consumption: Machine learning models optimize energy usage by predicting fluctuations in energy demand, allowing treatment plants to schedule energy-intensive processes during off-peak hours.

3.6 Autonomous Control Systems

AI-driven autonomous control systems represent a significant advancement in wastewater treatment, offering enhanced efficiency, reliability, and reduced operational costs. In this section, we explore the core components of autonomous control systems and their applications, specifically in controlling pumps, valves, and chemical dosing, as well as their role in reducing human intervention and error.

3.6.1 AI-Driven Autonomous Control and Decision-Making

Advanced Control Algorithms: AI employs sophisticated control algorithms, including PID (Proportional-Integral-Derivative), fuzzy logic, and reinforcement learning, to make real-time decisions based on sensor data and historical patterns.

Predictive Control: Autonomous systems utilize predictive models to anticipate influent variations and optimize treatment processes. This proactive approach reduces the need for manual intervention.

Adaptive Control: AI-driven systems adapt to changing conditions, such as influent quality fluctuations or equipment malfunctions, ensuring treatment plant operations remain efficient and effective.

3.6.2 Applications in Controlling Pumps, Valves, and Chemical Dosing

AI-based wastewater treatment methods system utilizes predictive information to program control mechanisms, including designated control valves and pumps. In comparison to conventional setpoint control systems, this AI-driven control system effectively reduces water losses, leading to cost savings in production. Furthermore, the AI model estimates the potential for membrane fouling in real-time and recommends targeted backwashing to minimize fouling, offering operational flexibility within the wastewater treatment process [20].

Pumps: Autonomous control systems adjust pump speeds and flow rates based on real-time demand, optimizing energy usage and preventing overflows or underflows.

Valves: AI can regulate valve positions to control flow rates and pressure within treatment processes, ensuring consistent performance and reducing the risk of equipment damage.

Chemical Dosing: AI-driven dosing systems precisely calculate and adjust chemical dosages based on influent characteristics, improving treatment efficiency and reducing chemical costs.

Asset Management: Autonomous control systems monitor the condition of equipment and assets, scheduling maintenance or repairs when necessary to prevent unexpected breakdowns.

3.6.3 Reducing Human Intervention and Error

Minimizing Manual Adjustments: AI automates many aspects of wastewater treatment, reducing the need for manual adjustments by operators. This minimizes the potential for human error.

Enhancing Reliability: Autonomous systems operate consistently and reliably, ensuring treatment processes meet or exceed performance standards.

24/7 Operation: AI-driven control systems operate around the clock, monitoring and optimizing treatment processes even during non-business hours or adverse weather conditions.

Safety Improvements: By automating critical safety protocols and responses to emergencies, autonomous control systems enhance overall safety in wastewater treatment facilities.

Data-Driven Decision-Making: AI uses real-time data and predictive analytics to make informed decisions, reducing the risk of human bias and enhancing the accuracy of process control.

The integration of autonomous control systems in wastewater treatment not only improves operational efficiency but also enhances safety and environmental compliance. These systems are instrumental in meeting the increasing demands for high-quality water treatment while minimizing human intervention and the associated potential for error.

3.7 Resource Efficiency and Cost Reduction

Resource efficiency and cost reduction are critical objectives in wastewater treatment. Artificial Intelligence (AI) plays a pivotal role in achieving these goals by optimizing energy consumption, reducing chemical usage through intelligent dosing, and ultimately lowering operational costs while simultaneously reducing the environmental impact.

3.7.1 AI-Driven Energy Optimization

AI plays a pivotal role in the realm of smart energy management within wastewater treatment facilities. These systems utilize AI to analyze real-time data from sensors and historical energy consumption patterns, enabling the optimization of crucial equipment such as pumps, blowers, and aeration systems. By fine-tuning the operation of these components, AI ensures that energy is used with precision and efficiency, resulting in reduced operational costs and a decreased environmental impact. Additionally, AI facilitates load shifting by scheduling energy-intensive processes during off-peak hours when energy costs are lower, leading to substantial overall energy savings. Furthermore, AI-driven predictive maintenance plays a crucial role in identifying equipment issues before they become critical, preventing costly breakdowns, and minimizing energy loss due to inefficient equipment operation, thus enhancing both financial savings and equipment longevity. Lastly, AI recognizes opportunities for energy recovery within treatment processes, such as harnessing biogas produced during anaerobic digestion to generate electricity or heat. By capturing and utilizing energy that would otherwise go to waste, AI elevates energy efficiency and sustainability in wastewater treatment operations, ultimately contributing to economic and environmental sustainability.

3.7.2 Reducing Chemical Usage Through Intelligent Dosing

Artificial intelligence (AI) has become a potent tool for addressing real-world challenges and has garnered significant attention for its applications across various domains. In recent times, AI techniques have found utility in the realms of water treatment and desalination, enabling the optimization of processes and providing viable solutions to combat water pollution and scarcity. The application of AI is poised to reduce operational costs in water treatment by enhancing cost-effectiveness and optimizing the utilization of chemicals [21].

The integration of AI into chemical dosing processes within wastewater treatment systems introduces precision and efficiency to this critical aspect of treatment. AI algorithms perform precise dosage calculations based on real-time influent characteristics, ensuring that chemicals are administered in exact quantities necessary to achieve treatment objectives effectively. What sets AI apart is its ability to establish a dynamic feedback loop through autonomous control systems. These systems continually adapt dosages as influent conditions change, maintaining consistent treatment performance while minimizing chemical waste. This adaptability is particularly valuable in responding to the variability of influent quality, ensuring that resources are used judiciously.

Furthermore, AI-driven systems specialize in optimizing the dosing of coagulants and flocculants, reducing the need for excess chemicals. This optimization not only contributes to cost savings but also prevents overdosing, which can necessitate additional treatment steps and result in increased operational expenses. By leveraging AI to fine-tune chemical dosing, wastewater treatment facilities can achieve optimal treatment outcomes with efficiency and cost-effectiveness, ensuring the reliable removal of contaminants from wastewater streams.

3.7.3 Lowering Operational Costs and Environmental Impact

AI applications have proven effective in predicting and reducing operational costs in water treatment processes by optimizing chemical usage. Multiple AI models have demonstrated success and accuracy in assessing the efficiency of different adsorbents employed for removing various contaminants from water [22]. The integration of AI into wastewater treatment processes brings about a multitude of benefits, with a central focus on resource efficiency and cost reduction. Firstly, AI significantly improves overall treatment efficiency, mitigating the operational costs linked to extended treatment times, excessive energy consumption, and increased chemical usage. This enhancement in efficiency directly translates to reduced operational expenditures.

Furthermore, AI's predictive maintenance capabilities play a pivotal role in minimizing unplanned downtime, thereby curbing emergency repair costs and extending the lifespan of equipment. These preventive measures result in substantial long-term savings.

Another noteworthy advantage is the optimization of chemical dosing through AI, leading to cost reductions in chemical procurement and disposal. This optimization fosters cost-effectiveness and environmental sustainability by reducing chemical waste and associated expenses.

The environmental benefits of AI-driven resource optimization are substantial. Decreased energy consumption and reduced chemical usage result in fewer greenhouse gas emissions and less chemical waste, aligning wastewater treatment with broader sustainability objectives.

Finally, the consistent efficiency achieved through AI-driven processes ensures compliance with stringent environmental regulations. This compliance prevents potential fines and legal expenses that could arise from non-compliance, contributing to further cost savings.

Generally, the integration of AI into wastewater treatment processes offers a **win–win scenario**, delivering cost reductions while simultaneously advancing environmental sustainability. By fine-tuning operations and optimizing resource usage, wastewater treatment facilities can achieve significant savings, effectively balancing economic and ecological priorities.

3.8 Addressing Variability and Adaptation

Addressing variability and adaptation challenges in wastewater treatment is essential to maintain consistent treatment performance and comply with regulatory requirements. Artificial Intelligence (AI) offers solutions to these challenges by enabling treatment plants to handle seasonal and weather-related changes, adapt to fluctuations in influent quality, and enhance overall flexibility and resilience.

3.8.1 Handling Seasonal and Weather-Related Changes

AI plays a pivotal role in enhancing weather-informed wastewater treatment by harnessing predictive weather modeling, dynamic process adjustments, and reservoir management. Through predictive weather modeling, AI utilizes weather data and forecasting models to anticipate seasonal variations and extreme weather events. Study conducted by Weyn et al. [23] discusses the role of AI in handling seasonal and weather-related changes through a Deep Learning Weather Prediction (DLWP) model. This AI system efficiently predicts atmospheric variables with global coverage, creating ensemble forecasts and improving subseasonal-to-seasonal (S2S) forecasting, particularly for lead times of two to six weeks. It shows promise in handling complex weather patterns, including tropical cyclones, and competes favorably with established forecasting systems like ECMWF at longer lead times.

This advanced forecasting capability empowers treatment plants to proactively prepare for increased flow rates and potential contamination risks associated with adverse weather conditions. AI-driven control systems further bolster the response to weather-related challenges by automatically adjusting crucial process parameters, such as aeration rates and chemical dosing, based on real-time weather data and forecasts. This adaptability ensures the consistent efficiency of treatment processes. Additionally, AI aids in reservoir management during heavy rainfall periods, preventing overflow incidents and mitigating the risk of combined sewer overflows (CSOs) or sanitary sewer overflows (SSOs). By integrating AI, wastewater treatment facilities can effectively navigate the impact of weather on their processes, maintaining both treatment quality and environmental compliance.

3.8.2 Adapting to Changes in Influent Quality

AI brings valuable capabilities to influent monitoring in wastewater treatment processes. Firstly, it offers real-time monitoring of influent quality parameters, allowing for continuous surveillance. AI systems excel at swiftly detecting sudden changes or pollutant spikes in the influent, triggering immediate adjustments in treatment processes to maintain efficiency, even when influent conditions fluctuate unexpectedly. According an article [24] that focuses on designing an adaptive control system for dissolved oxygen concentration in wastewater treatment, particularly in response to fluctuating influent quality, While it's challenging to precisely predict influent changes during rain or storms, the approach leverages the predictability of influent variations in dry weather conditions. Through a nonlinear optimization process based on dry weather data, the system determines an optimal setpoint and weighting gain. During weather events, this algorithm dynamically adjusts the setpoint using the optimization results and ammonium predictions. A constrained nonlinear neural-network model predictive control system is employed to track these setpoint changes. Simulations comparing various approaches with a fixed-setpoint PI control demonstrate improved effluent quality, reduced energy consumption, or both, highlighting the effectiveness of AI in adapting to influent quality changes in wastewater treatment.

Moreover, AI-driven predictive models are employed to anticipate influent quality changes. These models leverage historical data and weather forecasts to proactively optimize treatment processes based on expected variations. This proactive approach reduces the risk of non-compliance events and helps treatment plants adapt to influent fluctuations more effectively.

Additionally, AI systems are adept at identifying emerging contaminants and unusual influent compositions. Early detection of these anomalies enables treatment facilities to take immediate mitigation actions, preventing potential harm to the environment and ensuring that treated water consistently meets regulatory standards. In essence, AI's role in influent monitoring not only enhances the efficiency of wastewater treatment but also contributes significantly to the early detection and management of influent-related challenges, ultimately improving the reliability and environmental responsibility of treatment processes.

3.8.3 AI-Driven Flexibility and Resilience

The implementation of AI in wastewater treatment provides valuable solutions for addressing challenges related to variability and adaptation. Autonomous control systems driven by AI continuously adapt treatment processes in real time, ensuring efficient operation under varying conditions and unexpected events. This adaptability not only maximizes treatment efficiency but also minimizes disruptions. AI also plays a pivotal role in facilitating swift responses to emergencies, such as chemical spills or equipment failures, by automatically triggering mitigation actions and alerting operators [25]. This proactive approach helps prevent and mitigate potential environmental incidents. Additionally, AI optimizes resource allocation and load balancing, ensuring consistent treatment performance across multiple units or treatment stages, even as conditions change. This guarantees that the treatment process remains effective and efficient throughout its course. Furthermore, AI-driven simulations enable treatment plants to create and evaluate scenarios for potential challenges, aiding in the development of resilient response strategies. By anticipating and planning for various scenarios, facilities can be better prepared to handle unexpected events. Overall, AI-driven solutions enhance the resilience, efficiency, and environmental responsibility of wastewater treatment processes, contributing to both environmental protection and operational effectiveness.

3.9 Regulatory Compliance and Reporting

Regulatory compliance is a fundamental requirement for wastewater treatment facilities. Artificial Intelligence (AI) offers powerful tools for real-time compliance monitoring, automated reporting, and documentation, ensuring that treatment plants adhere to environmental standards effectively and efficiently.

3.9.1 AI for Real-Time Regulatory Compliance Monitoring

Continuous monitoring is a critical component of AI-driven wastewater treatment systems. These systems leverage sensor networks and data analytics to keep a watchful eye on influent and effluent parameters around the clock. These sensors measure a wide range of critical parameters, including pH levels, turbidity, chemical concentrations, flow rates, and temperature, among others. The real-time data collected is then compared against stringent regulatory standards to ensure that the treated water consistently meets quality requirements.

One of the key strengths of AI in this context is its ability to perform anomaly detection. AI algorithms are adept at identifying deviations from regulatory limits or expected patterns within the data [26]. When any such deviations are detected, the system instantly triggers alerts to notify operators and plant personnel. This rapid response capability is invaluable in addressing potential compliance issues promptly, preventing any violations, and safeguarding the environment and public health.

Moreover, AI adds a predictive dimension to wastewater treatment compliance. By analyzing historical data and accounting for influent variations, AI can forecast compliance trends. This predictive compliance approach allows treatment plants to take proactive measures to maintain compliance. It enables operators to anticipate potential issues and adjust treatment processes, chemical dosages, and other parameters in advance, thereby minimizing the risk of non-compliance events. In essence, AI empowers wastewater treatment facilities with the tools to not only meet regulatory standards but to do so efficiently, effectively, and with a proactive mindset.

3.9.2 Automated Reporting and Documentation

Recent advancements in smart infrastructure and the rise of the Internet of Things (IoT) offer a valuable opportunity in the wastewater sector. This opportunity involves the capability to collect, document and report real-time data on sewer conditions, which can be utilized for effective operation management and maintenance [27]. Data integration is a foundational aspect of AI-driven wastewater treatment systems. These systems excel at aggregating data from diverse sources and sensors into a centralized platform. This centralized approach simplifies data management and reporting, as all relevant information is readily available within a unified framework. This integration not only streamlines data handling but also enhances the overall efficiency of wastewater treatment operations.

One of the notable benefits of AI in this context is its ability to automate reporting processes. AI-driven systems are capable of generating compliance reports automatically. This automation significantly reduces the time and effort previously required for manual reporting, ensuring that reports are produced promptly and accurately. This is especially critical for adhering to regulatory requirements, as timely reporting is essential for demonstrating compliance.

In addition to real-time data management and automated reporting, AI systems also excel at data logging. These systems maintain comprehensive logs of all relevant data points, capturing both current and historical information. This feature proves invaluable for auditing purposes and regulatory reviews. When regulatory authorities or auditors need to examine past data, AI systems provide easy and efficient access to the historical records, simplifying the compliance verification process. In essence, AI-driven wastewater treatment systems offer a comprehensive data management solution that not only enhances operational efficiency but also facilitates compliance and regulatory transparency.

3.9.3 Ensuring Adherence to Environmental Standards

AI plays a pivotal role in ensuring environmental compliance within wastewater treatment facilities. By harnessing the power of AI, treatment plants can consistently meet stringent environmental standards and avoid violations that might lead to fines or penalties [28]. AI achieves this through several key mechanisms.

Firstly, AI enables trend analysis by scrutinizing historical data, identifying long-term trends, and patterns in influent and effluent parameters. This valuable insight helps treatment facilities develop strategies to proactively maintain compliance, even as regulations evolve over time. It empowers them to anticipate changes and adapt their processes accordingly.

Moreover, AI introduces adaptive control to wastewater treatment processes. Autonomous control systems, driven by AI, have the capability to make real-time adjustments to treatment processes. These adjustments are vital for ensuring compliance when influent conditions unexpectedly change. AI-driven control systems can fine-tune parameters such as chemical dosages, aeration rates, and flow rates to swiftly respond to variations in water quality, helping to prevent compliance lapses.

Additionally, AI platforms can be programmed to stay current with changing regulations. This feature ensures that treatment processes remain aligned with the latest environmental standards. By automating the incorporation of regulatory updates, AI mitigates the risk of inadvertent non-compliance due to outdated procedures or standards.

3.10 Challenges and Ethical Considerations

Machine learning, deep learning and smart technologies are used in wastewater treatment for generation, prediction enhancement, and classification tasks, providing a guide for future water resources challenges. These models can be used to make decisions in water resources management and governance, but ethics and future directions need to be addressed and focused [29]. The integration of Artificial Intelligence (AI) in wastewater treatment, while promising, is not without its challenges and ethical considerations. This section explores the potential limitations and challenges of AI in this context, highlights ethical concerns, and outlines strategies to mitigate risks and ensure responsible AI use.

3.10.1 Potential Limitations and Challenges of AI in Wastewater Treatment

The successful implementation of AI in wastewater treatment comes with its own set of challenges and considerations:

Data Quality: AI systems heavily rely on high-quality data. Inaccurate or incomplete data can lead to incorrect predictions and suboptimal decisions. Ensuring the accuracy and reliability of data collected from sensors and other sources is crucial for the effectiveness of AI-driven wastewater treatment.

Data Privacy: Collecting and sharing sensitive wastewater treatment data may raise privacy concerns, especially when it comes to regulatory compliance and public health. Protecting the confidentiality and security of this data is paramount, and appropriate measures must be in place to address privacy issues.

Initial Investment: The adoption of AI in wastewater treatment may require a significant initial investment in technology, training, and infrastructure upgrades. While AI can lead to long-term cost savings and efficiency improvements, the upfront costs can be a barrier for some facilities.

Technological Maturity: AI technologies are rapidly advancing, but not all AI solutions are equally mature. Some AI technologies are still evolving, and their effectiveness in wastewater treatment applications may vary. Facilities need to carefully evaluate the maturity and suitability of AI solutions for their specific needs.

Interoperability: Integrating AI systems with existing wastewater treatment infrastructure and control systems can be complex. It may require custom solutions and careful planning to ensure seamless interoperability. Compatibility issues between AI components and legacy systems need to be addressed to avoid disruptions in operations.

Addressing these challenges requires a thoughtful and strategic approach to implementing AI in wastewater treatment. Facilities must prioritize data quality and privacy, consider the long-term benefits of AI, and carefully assess the suitability of AI technologies for their unique operational requirements. Additionally, collaboration with experts and technology providers can help navigate the complexities of integrating AI into existing treatment processes.

3.10.2 Ethical Considerations in AI Implementation

Intelligent technologies are recognized as key drivers of innovation and knowledge for innovation. Consequently, numerous regional, national, and even supranational strategies and policies have emerged to harness the potential benefits of AI for their respective populations. These strategies span countries like Canada, Mexico, Japan, and India, with regional initiatives in places like Bavaria and the Northern Baltic countries. Additionally, there are collaborative efforts at the supranational level, including within organizations such as the OECD and the United Nations International Telecommunication Union (ITU) [30].

Transparency: Ensuring transparency in AI decision-making processes is crucial. Stakeholders, including operators, regulators, and the public, should have a clear understanding of how AI algorithms arrive at conclusions and recommendations. Transparent AI systems help build trust and facilitate informed decision-making.

Bias and Fairness: AI systems may inherit biases present in historical data, potentially leading to unfair or discriminatory outcomes. Efforts must be made to address and mitigate bias in AI models. Regular audits and fairness assessments can help identify and rectify bias in decision-making processes.

Accountability: Establishing clear lines of accountability for AI-driven decisions and actions is essential. Defining who is responsible for the outcomes of AI systems helps prevent potential ethical breaches. Accountability frameworks should encompass both human operators and the AI itself.

Job Displacement: The automation of certain tasks through AI could lead to job displacement for human workers. Ethical considerations include providing retraining opportunities and alternative employment options to mitigate the impact on the workforce. Balancing efficiency gains with social responsibility is essential.

Environmental Impact: While AI can improve resource efficiency in wastewater treatment, it also consumes energy for computations. Ensuring that the environmental benefits of AI outweigh its energy footprint is a critical ethical consideration. Facilities should aim to minimize energy consumption and explore sustainable AI solutions.

The use of AI-based solutions in public administrations is expected to generate a positive societal and environmental impact. It should empower citizens in their technology usage, while also providing the means for end users to comprehend and engage with AI. A significant concern posed by AI revolves around its effects on employment and the economy [31].

3.10.3 Strategies to Mitigate Risks and Ensure Responsible AI Use

To effectively implement AI in wastewater treatment, facilities must prioritize data governance, algorithmic fairness, and human oversight to ensure responsible and ethical AI usage. This includes establishing ethical guidelines, engaging with stakeholders, and providing continuous training for staff. Additionally, sustainability assessments should be conducted to minimize environmental impact, and rigorous compliance with regulations must be maintained, emphasizing data privacy and environmental protection throughout the AI integration process.

3.11 Future Trends and Outlook

As AI continues its rapid evolution, its integration into wastewater treatment processes holds great promise for shaping the industry's sustainable practices. This essay explores the emerging AI technologies, the pivotal role AI plays in advancing sustainable wastewater management, and the opportunities for further research and development.

3.11.1 Emerging AI Technologies and Their Potential Impact

Predictive algorithms are becoming increasingly complex, making the need for explanations of their decisions more crucial. Balancing accuracy and interpretability is a key challenge. Complex models like Recurrent Neural Networks (RNNs) offer high accuracy but are often less interpretable, while traditional models like logistic regression are more interpretable but may sacrifice accuracy. Achieving the highest accuracy with large datasets often involves complex models that are difficult to explain. Various approaches exist to enhance understanding of complex model predictions, but their relationships and optimal use cases remain unclear [32]. One of the most noteworthy emerging AI technologies is Explainable AI (XAI). XAI techniques enhance transparency in AI decision-making, making it easier for operators and stakeholders to understand and trust AI-driven processes. This transparency is crucial for the widespread acceptance of AI in wastewater treatment.

AI also holds the key to advancing the circular economy in wastewater treatment. By optimizing resource recovery from wastewater, such as extracting valuable nutrients and energy, AI contributes significantly to sustainability efforts. The potential benefits are immense, as wastewater treatment facilities become hubs for resource regeneration rather than simply waste disposal sites.

Quantum computing, with its incredible processing power, promises to revolutionize AI applications in wastewater treatment. It can solve complex problems and optimize processes at unprecedented speeds, leading to more efficient and effective water treatment.

Additionally, continuous advancements in sensor technology and the Internet of Things (IoT) integration provide richer and more accurate data streams. This enhances AI's ability to optimize treatment processes, resulting in improved efficiency and resource utilization.

3.11.2 The Role of AI in Advancing Sustainable Wastewater Management

The United Nations' Sustainable Development Goals (SDGs) aim to enhance the quality of life for people across developed, emerging, and developing nations. These goals encompass social and economic dimensions, with a primary emphasis on environmental sustainability. Simultaneously, data-driven technologies have had a profound impact on various aspects of our lives, triggering significant economic and societal transformations [33]. AI's influence on sustainability metrics is profound. It enables the development of comprehensive sustainability metrics, allowing treatment plants to assess their environmental impact and make data-driven decisions to reduce their carbon footprint. This data-driven approach is essential for aligning wastewater treatment with broader sustainability goals.

Smart grid integration is another significant role that AI will play in the industry. AI-driven wastewater treatment facilities will integrate with smart grids to optimize energy usage and participate in demand response programs. This contributes to a more resilient and sustainable energy ecosystem, benefiting both the treatment plants and the broader community.

Enhanced resource recovery is a crucial aspect of sustainability, and AI facilitates the recovery of valuable resources from wastewater, such as phosphorus and biogas. This supports a sustainable and circular approach to resource management, reducing waste and environmental impact.

Furthermore, AI's predictive capabilities can assist treatment plants in adapting to the effects of climate change. It can predict extreme weather events, optimize responses, and ensure continued operation during adverse conditions, contributing to climate resilience.

3.11.3 Opportunities for Further Research and Development

Numerous opportunities for further research and development exist in the realm of AI-driven wastewater treatment.

Water Reuse and Desalination: Research into AI-driven optimization of water reuse and desalination processes will be crucial for addressing water scarcity issues globally.

Ecosystem Health Monitoring: AI can be further developed for monitoring the health of aquatic ecosystems impacted by wastewater discharges, helping to protect biodiversity.

Integration with Smart Cities: Research opportunities exist in integrating AI-driven wastewater treatment with broader smart city initiatives, enhancing urban sustainability and resilience.

Multi-Objective Optimization: Further research into multi-objective optimization algorithms will enable treatment plants to simultaneously consider energy efficiency, cost reduction, and environmental impact.

Human–Machine Collaboration: Exploring ways to optimize human–machine collaboration in wastewater treatment, where AI augments human expertise and decision-making, is an avenue for research and development.

3.12 Conclusion

In conclusion, our exploration of AI in wastewater treatment has revealed a landscape of transformative potential. Through AI, we unlock data-driven optimization, enhanced sustainability, and unwavering regulatory compliance. It has the power to redefine wastewater treatment, turning it into a force for efficiency and innovation. Despite ethical considerations and challenges, the opportunities AI presents are immense. To fully realize its benefits, we must continue to delve into its capabilities, invest in research and development, and uphold ethical AI practices. Embracing AI not only ensures regulatory compliance but also propels wastewater treatment toward an era marked by environmental responsibility, resource efficiency, and enduring sustainability. This journey serves not just communities but also the well-being of our planet.

References

1. Matheri AN, Mohamed B, Ntuli F, Nabadda E, Ngila JC (2022) Sustainable circularity and intelligent data-driven operations and control of the wastewater treatment plant. Phys Chem Earth Parts A/B/C 126:103152. https://doi.org/10.1016/j.pce.2022.103152
2. Vieira Y, Lima EC, Foletto EL, Dotto GL (2021) Microplastics physicochemical properties, specific adsorption modeling and their interaction with pharmaceuticals and other emerging contaminants. Sci Total Environ 753:141981. https://doi.org/10.1016/j.scitotenv.2020.141981

3. Capodaglio AG, Olsson G (2019) Energy issues in sustainable urban wastewater management: use, demand reduction and recovery in the urban water cycle. Sustainability 12(1):266. https://doi.org/10.3390/su12010266
4. Asthana AN (2023) Wastewater management through circular economy: a pathway towards sustainable business and environmental protection. Adv Water Sci 34(3):87–98
5. Rashid R, Shafiq I, Akhter P et al (2021) A state-of-the-art review on wastewater treatment techniques: the effectiveness of adsorption method. Environ Sci Pollut Res 28:9050–9066. https://doi.org/10.1007/s11356-021-12395-x
6. Jagaba AH, Kutty SRM, Baloo L, Birniwa AH, Lawal IM, Aliyu MK, Yaro NSA, Usman AK (2022) Combined treatment of domestic and pulp and paper industry wastewater in a rice straw embedded activated sludge bioreactor to achieve sustainable development goals. Case Stud Chem Environ Eng 6:100261. https://doi.org/10.1016/j.cscee.2022.100261
7. Abdelfattah A, Hossain MI, Cheng L (2020) High-strength wastewater treatment using microbial biofilm reactor: a critical review. World J Microbiol Biotechnol 36:75. https://doi.org/10.1007/s11274-020-02853-y
8. Waqas S, Harun NY, Sambudi NS, Bilad MR, Abioye KJ, Ali A, Abdulrahman A (2022) A review of rotating biological contactors for wastewater treatment. Water 15(10):1913. https://doi.org/10.3390/w15101913
9. Sharma P. Wetlands for wastewater treatment and management: a review. Adv Sci Technol 95
10. Alfonso-Muniozguren P, Serna-Galvis EA, Bussemaker M, Torres-Palma RA, Lee J (2021) A review on pharmaceuticals removal from waters by single and combined biological, membrane filtration and ultrasound systems. Ultrason Sonochem 76:105656. https://doi.org/10.1016/j.ultsonch.2021.105656
11. Al-dhawi BNS, Kutty SRM, Ghaleb AAS, Almahbashi NMY, Saeed AAH, Al-Mekhlafi AA, Alsaeedi YAA, Jagaba AH (2023) Pretreated palm oil clinker as an attached growth media for organic matter removal from synthetic domestic wastewater in a sequencing batch reactor. Case Stud Chem Environ Eng 7:100294. https://doi.org/10.1016/j.cscee.2022.100294
12. Azeez GK, Al-Jaberi F, Ahmed SA, Hussain AA (2023) Sequencing batch reactor (SBR) technology in wastewater treatment: a mini-review. In: AIP conference proceedings, vol 2806, no 1. AIP Publishing
13. Bdour A (2007) Perspectives on sustainable wastewater treatment technologies and reuse options in the urban areas of the Mediterranean region. In: World environmental and water resources congress 2007: Restoring Our Natural Habitat, pp 1–15
14. Arberas-Jiménez I, Sifaoui I, Reyes-Batlle M, Rizo-Liendo A, Sancho L, Urruticoechea A, Piñero JE, Lorenzo-Morales J (2022) Ultraviolet—chlorine combined treatment efficiency to eliminate Naegleria fowleri in artificial surf lagoons. Heliyon 8(11):e11625. https://doi.org/10.1016/j.heliyon.2022.e11625
15. Abuhasel K, Kchaou M, Alquraish M, Munusamy Y, Jeng YT (2020) Oily wastewater treatment: overview of conventional and modern methods, challenges, and future opportunities. Water 13(7):980. https://doi.org/10.3390/w13070980
16. Solon K, Volcke E, Sperandio M, van Loosdrecht M (2019) Resource recovery and wastewater treatment modelling. Environ Sci Water Res Technol. https://doi.org/10.1039/c8ew00765a
17. Zhao L, Dai T, Qiao Z, Sun P, Hao J, Yang Y (2019) Application of artificial intelligence to wastewater treatment: a bibliometric analysis and systematic review of technology, economy, management, and wastewater reuse. Process Saf Environ Prot 133:169–182. https://doi.org/10.1016/j.psep.2019.11.014
18. Park J, Kim KT, Lee WH (2020) Recent advances in information and communications technology (ICT) and sensor technology for monitoring water quality. Water 12(2):510. https://doi.org/10.3390/w12020510
19. Gunasekaran K, Boopathi S (2023) Artificial intelligence in water treatments and water resource assessments. In: Artificial intelligence applications in water treatment and water resource management. IGI Global, pp 71–98
20. Li L, Rong S, Wang R, Yu S (2021) Recent advances in artificial intelligence and machine learning for nonlinear relationship analysis and process control in drinking water treatment: a review. Chem Eng J 405:126673. https://doi.org/10.1016/j.cej.2020.126673

21. Alam G, Ihsanullah I, Naushad M, Sillanpää M (2021) Applications of artificial intelligence in water treatment for optimization and automation of adsorption processes: recent advances and prospects. Chem Eng J 427:130011. https://doi.org/10.1016/j.cej.2021.130011
22. Uwamungu JY, Kumar P, Alkhayyat A, Younas T, Capangpangan RY, Alguno AC, Ofori I (2022) Future of water/wastewater treatment and management by industry 4.0 integrated nanocomposite manufacturing. J Nanomater
23. Weyn JA, Durran DR, Caruana R, Cresswell-Clay N (2021) Sub-seasonal forecasting with a large ensemble of deep-learning weather prediction models. J Adv Model Earth Syst 13(7):e2021MS002502. https://doi.org/10.1029/2021MS002502
24. Sadeghassadi M, Macnab CJ, Gopaluni B, Westwick D (2018) Application of neural networks for optimal-setpoint design and MPC control in biological wastewater treatment. Comput Chem Eng 115:150–160. https://doi.org/10.1016/j.compchemeng.2018.04.007
25. Etxegarai M, Camps M, Echeverria L, Ribalta M, Bonada F, Domingo X (2022) Virtual sensors for smart data generation and processing in AI-driven industrial applications. In: Industry 4.0-perspectives and applications. IntechOpen
26. Dogo EM, Nwulu NI, Twala B, Aigbavboa C (2019) A survey of machine learning methods applied to anomaly detection on drinking-water quality data. Urban Water J 16(3):235–248
27. Edmondson V, Cerny M, Lim M, Gledson B, Lockley S, Woodward J (2018) A smart sewer asset information model to enable an 'Internet of Things' for operational wastewater management. Autom Constr 91:193–205. https://doi.org/10.1016/j.autcon.2018.03.003
28. Gill SS, Xu M, Ottaviani C, Patros P, Bahsoon R, Shaghaghi A, Golec M, Stankovski V, Wu H, Abraham A, Singh M, Mehta H, Ghosh SK, Baker T, Parlikad AK, Lutfiyya H, Kanhere SS, Sakellariou R, Dustdar S, Uhlig S et al (2022) AI for next generation computing: emerging trends and future directions. Internet Things 19:100514. https://doi.org/10.1016/j.iot.2022.100514
29. Zakur YA, Biswadip BM, Yousif AZ, Digvijay P (2023)"Survey on the artificial intelligence and machine learning techniques on the applications of wastewater treatment for sustainable environment. In: Handbook of research on safe disposal methods of municipal solid wastes for a sustainable environment. IGI Global, pp 241–248
30. Bartneck C, Lütge C; Wagner A, Welsh S (2021) SpringerBriefs in ethics. An introduction to ethics in robotics and AI. https://doi.org/10.1007/978-3-030-51110-4
31. Cortés U, Cortés A, Barrué C, Sánchez A, Moya-Sánchez EU, Garcia-Gasulla D (2021) To be fair or not to be: using ai for the good of citizens. IEEE Technol Soc Mag 40(1):55–70. https://doi.org/10.1109/MTS.2021.3056173
32. Wongburi P, Park JK (2021) Prediction of sludge volume index in a wastewater treatment plant using recurrent neural network. Sustainability 14(10):6276. https://doi.org/10.3390/su14106276
33. Bachmann N, Tripathi S, Brunner M, Jodlbauer H (2021) The contribution of data-driven technologies in achieving the sustainable development goals. Sustainability 14(5):2497. https://doi.org/10.3390/su14052497

Chapter 4
Advancing Nutrient Removal and Resource Recovery Through Artificial Intelligence: A Comprehensive Analysis and Future Perspectives

K. S. Hemanth, Tigulla Deepika, and Vidya Lakshmi

Abstract The increasing difficulties associated with effectively controlling wastewater treatment operations while simultaneously satisfying the imperatives of nutrient removal and resource recovery have necessitated the use of advanced technology. This book chapter provides a comprehensive analysis of the use of artificial intelligence (AI) methods within this complex context. Utilizing a vast array of scholarly investigations and real-world implementations, this study explores the intricate domain of wastewater treatment, providing a comprehensive understanding of how artificial intelligence algorithms are used to enhance the efficiency of nutrient removal procedures and expedite the recovery of valuable resources. This chapter presents a thorough examination of the impact of artificial intelligence (AI) on sustainable innovations in wastewater treatment facilities. It accomplishes this through a comprehensive analysis of relevant data and the inclusion of real-world case studies. The findings of this research highlight the transformative effect of AI on conventional approaches to wastewater treatment, enabling the adoption of environmentally friendly and resource-efficient practices. The integration of artificial intelligence (AI) with wastewater management offers a fascinating story that highlights the shifting paradigm in the field of environmental engineering and the efficient exploitation of resources.

Keywords Artificial intelligence (AI) · Wastewater treatment · Nutrient removal · Resource recovery · Sustainable innovations · Environmental engineering

K. S. Hemanth · T. Deepika · V. Lakshmi
School of Computer Science and Applications, REVA University, Bengaluru, India
e-mail: 22130622807@reva.edu.in

V. Lakshmi
e-mail: 22130620575@reva.edu.in

K. S. Hemanth (✉)
Department of Computer Science, Christ University, Bangalore, India
e-mail: hemanth.ks@christuniversity.in

M. C. Garg (ed.), *The AI Cleanse: Transforming Wastewater Treatment Through Artificial Intelligence*, Springer Water, https://doi.org/10.1007/978-3-031-67237-8_4

4.1 Introduction

4.1.1 Background and Context

Drainage systems were present in ancient towns to divert rainfall from pavements and roofs, such as the Cloaca Maxima in ancient Rome. However, urban drainage and sewage during the Middle Ages were limited, with waste often dumped into gutters and overflowed. This led to surface water pollution, as water closets were connected to storm sewers. Significant quantities of sewage were generated, even after dilution, therefore requiring the treatment or purification of wastewater before to disposal. In the late 19th and early twentieth centuries, centralized sewage treatment plants were built, combining physical, biological, and chemical processes to remove pollutants. Concerns about environmental quality in the middle of the twentieth century forced governments to regulate wastewater disposal methods more strictly and to demand higher treatment levels. Wastewater treatment plants grew to be big, intricate structures that consumed a lot of energy. Energy efficiency started to play a big role in the creation of modern pollution control systems, leading to increased attention to land and subsurface disposal of sewage [1]. Wastewater is a rich resource from which it is possible to recover clean water, energy, nutrients, and other resources. However, a staggering 80% of wastewater is discharged into the environment without undergoing any treatment on a global scale. Utilizing recycled wastewater is a viable approach to combatting pollution and water shortages, particularly in regions where 36% of the global population lives in areas with limited access to water. After undergoing the cleaning process, it may be used for purposes such as irrigation, industrial operations, and recreational activities as a substitute for freshwater. Additionally, it may be used to sustain the continuity of the ecosystem, and the resulting outputs from its processing can provide energy and nutrients. Organic wastes are viewed as potential resources in a bio-based circular economy, with the potential to be employed in the production of chemicals, fuels, and nutrients.

4.1.2 Emerging Challenges and Need for Advance Technology

Physical, chemical, and biological pollutants are among the many things found in wastewater. Organic materials, pathogenic organisms, synthetic and inorganic chemicals, microplastics, sediments, radioactive materials, oil, and heat are examples of common pollutants. Before wastewater enters natural water bodies including rivers, lakes, estuaries, and seas, it is treated to remove pollutants. Human activity is the main source of water pollution, and treating wastewater is an essential part of controlling water pollution.

Some of the challenges in wastewater treatment operations are:

i. **Energy Consumption**

Operating a wastewater treatment plant is costly due to energy consumption from equipment and pollutants. Biological processes require high oxygen volumes to maintain microbiological organisms and reduce BOD, Ammonia–Nitrogen, and Phosphorus. A developed country's wastewater treatment plants use two to three percent of its electrical power, with biological treatment accounting for 50–60% of plant usage.

This challenge can be dealt in many ways such that changes in wastewater treatment equipment can reduce energy consumption by 40%. Fine screens, membrane aeration technology, and direct treatment of high return stream concentration are some of the solutions.

New technologies like microfiltration, ultrafiltration, nanofiltration, and reverse osmosis are being used to remove pollutants from wastewater, producing potable and non-potable reuse waters. Anaerobic digesters are used in biological systems to reduce sludge and organic loadings, while solar energy is being implemented to reduce energy costs depending on the wastewater application or location [2].

ii. **Workforce**

Wastewater treatment systems require 24-h operation and 365-day maintenance, requiring qualified staff for operation to understand technicalities and operate machinery. Technicians must be on-call 24/7 to handle mishaps and maintain the plant, handling mechanical, electrical, and maintenance functions. Their workload increases during water quality degradation or seasonal variations.

So in order to deal with this kind of challenge new technologies, including semi-automated and fully automated systems, utilize sensors, probes, and analysers to control system can be used.

iii. **Chemical Treatment Cost**

Wastewater filtration processes can have environmental impacts, as organic matter removed from wastewater can end up in local waterways. Industrial wastewater treatment systems use chemicals to separate pollutants, but supply chain issues, product shortages, and price increases have led to the need for new technologies.

To deal with this the best implementation would be.

New green technology-powered systems treat wastewater with eco-friendly processes, eliminating environmental pollutants. As environmental regulations evolve, eradicating residual chemicals becomes increasingly crucial for safe wastewater treatment.

iv. **Sludge Production**

Sludge, generated during wastewater treatment, poses environmental challenges. Inorganic coagulants, low in cost but high in chemical dosages, can cause pH noncompliance and create large floc particles in wastewater. Proper dewatering can reduce sludge volumes, increasing disposal charges and operational costs.

To address this, the most effective method is Recycling sludge for agriculture is a sustainable solution, with modern treatment technologies potentially reducing production and promoting long-term sustainability in wastewater treatment plants.

v. **Footprint**

The space needed to process activated sludge is a significant concern. Constructing an activated sludge factory incurs significant expenses and requires a substantial land area. Extensive areas of land are required to accommodate large and expensive settling tanks and aeration basins for primary and secondary treatment operations. Municipal wastewater treatment plants must increase their capacity to accommodate the expanding population.

These challenges may be addressed by using advanced technologies that use smaller process basins and enhance biomass per unit volume by including biofilm attachment medium (such as IFAS and MABR) or boosting biomass concentration (such as MBR). These technologies are at the forefront of minimizing space requirements. A smaller footprint not only saves costs associated with land, but also decreases CAPEX by reducing steel, concrete, and equipment [2].

4.1.3 Purpose and Scope

In this chapter, the examination revolves around the impact of wastewater and the specialized role of AI in the recovery of essential nutrients and minerals. The investigation also encompasses an analysis of the distinct components within wastewater, originating from both industrial and municipal sources. The study's scope extends to the exploration of AI-driven methodologies for nutrient collection, with a particular emphasis on their applicability to cater to the distinctive needs of farmers, organizations, and government entities.

The central focus lies in harnessing artificial intelligence to enhance the process of nutrient removal. To ensure a comprehensive perspective, the chapter incorporates pertinent global research findings. This work elucidates the pivotal role of AI in the realm of resource recovery and nutrient removal, providing insights into its potential to revolutionize wastewater treatment. The compilation of existing material is aimed at fostering a better understanding of environmentally conscious approaches to wastewater treatment and resource management.

4.2 Wastewater Treatment Fundamentals

4.2.1 Overview of Wastewater Treatment Processes

The quality of the water depends on numerous variables that fall under the categories of physical, chemical, biological, and quantity-related aspects. Physical factors encompass temperature, which impacts the solubility of substances and biological processes, turbidity, which gauges the water's electrical conductivity and is influenced by sediment and particles, and conductivity, which measures the water's electrical conductivity and is associated with the presence of ions. Chemical variables include pH, which gauges the water's acidity or alkalinity, dissolved oxygen levels, which are essential for aquatic life, and oxidation–reduction potential, which shows how readily the water may react chemically. The presence of microorganisms like bacteria and viruses, as well as algae and pesticides, may have a substantial influence on water ecosystems. These are biological variables. Water level, which impacts the availability of habitat, flow pressure, which regulates the transit of substances, and velocity, which affects the mixing and dispersion of particles and pollutants, are all quantity factors. All these elements work together to determine the overall quality of water and whether it is suitable for different applications, such as drinking, leisure, and the support of aquatic life [3].

Wastewater is a byproduct of industrial or commercial operations. The terms "wastewater treatment" and "wastewater processing" refer to the methods and processes used to treat it. Before it may be safely released into the environment, this form of wastewater must be treated. The following document describes the numerous techniques involved in wastewater treatment. Figure 4.1 represent the wastewater treatment stages, first stage, referred to as preliminary treatment, includes actions such as screening and grit removal. Then comes primary treatment, which focuses on sedimentation to remove suspended particles. In the future, secondary treatment will use microorganisms to oxidize organic contaminants and remove nutrients like nitrogen and phosphorus. Finally, tertiary treatment entails removing any leftover organic and inorganic substances from the effluents, which is frequently followed by chemical or radiation disinfection, guaranteeing that the treated wastewater can be discharged into the environment [4].

4.2.2 Nutrient Removal as a Critical Component

Eliminating nutrients from wastewater is a very important task. The varied nutrient contents in the various wastewater types are shown in Table 4.1. All living things depend on nutrients like nitrogen and phosphorus for their growth and vigor. On the other hand, eutrophication occurs when waste streams with high nutrient levels are released into aquatic habitats. Effective nutrient removal from wastewater is necessary in order to meet strict laws pertaining to nutrient release. Nutrient recovery

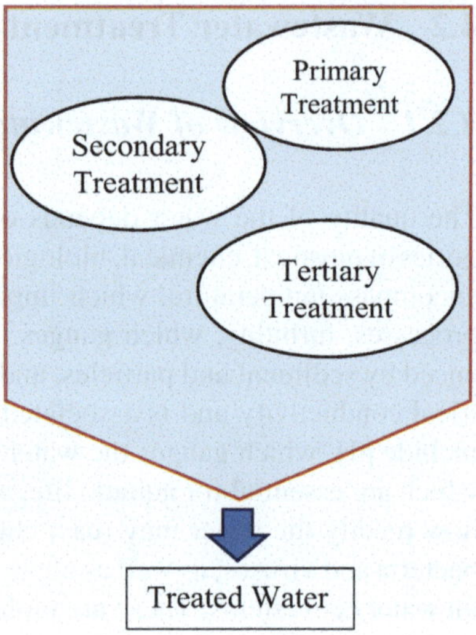

Fig. 4.1 Treatment of wastewater

from waste streams is important for promoting a circular economy and reducing the loss of limited resources.

A wide variety of nutrient sources, both point and non-point in nature, provide food for aquatic ecosystems [5]. There are many different sources of pollutants, including manure, aquaculture, and animal farming, as well as sewage from homes and businesses, combined sewer overflows, septic overflows, municipal wastewater discharges, stormwater discharges, industrial effluents from thermal power plants, mineral processing, and detergent manufacturing as show in Fig. 4.2.

4.2.3 Resource Recovery Opportunities in Wastewater

Water recycling facilities provide an extraordinary opportunity to decrease energy consumption, slash expenses, and enhance the generation of renewable energy by turning traditional waste streams into useful goods [7]. These resources are water, energy, nutrients (phosphorus and nitrogen), and biosolids in water resource recycling facilities [7]. As shown in the figure direct effects of sustainable wastewater treatment include improved water quality, reduced environmental pollution, and enhanced public health.

- **Biogas Production**: Anaerobic digestion of organic waste in wastewater results in the production of biogas, which is mostly made up of methane. This biogas may be collected and utilised to generate heat and power using renewable energy sources [8].

Table 4.1 Nutrient concentrations in various types of wastewaters [6]

	Units	Municipal wastewater influent	Household wastewater	Stormwater
Total nitrogen	mg Nl^{-1}	20–80	44–189	5.5
		20–85		
Total Kjeldhal nitrogen	mg Nl^{-1}	20–80	16–189	1.90–4.18
NH_3	mg Nl^{-1}	12–50	1.6–94	5–56
		12–50		
NO_3	mg Nl^{-1}	0.5	20–40	0.86–2.21
		0	BDL (0.2)–1.1	(NO_3/NO_2)
NO_2	mg Nl^{-1}	0.1		–
		0		
Organic nitrogen		8–30		
		8–35		
Total phosphorous	mg Pl^{-1}	4–14(6–23)	0.2–32	0.35
		4–15		
Orthophosphate	mg Pl^{-1}	3–10(4–14)	5–15(as PO_4)	
Polyphosphate	mg Pl^{-1}	0(1–5)		
Organic phosphate	mg Pl^{-1}	1–4(1–4)		

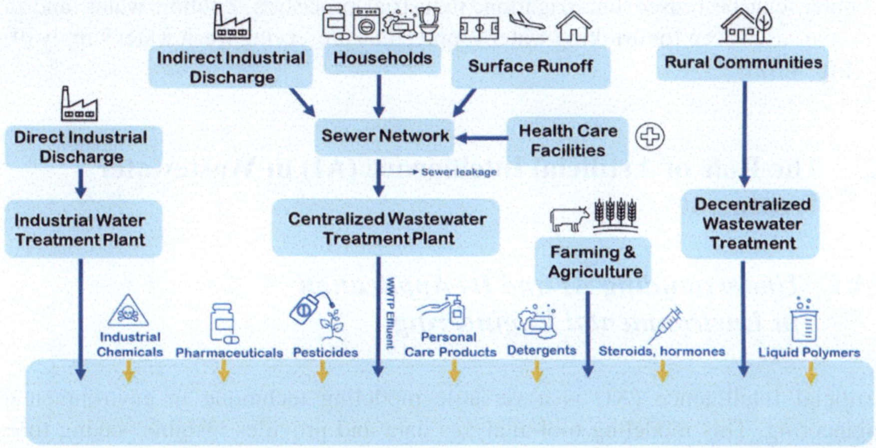

Fig. 4.2 Sources nutrients in wastewater

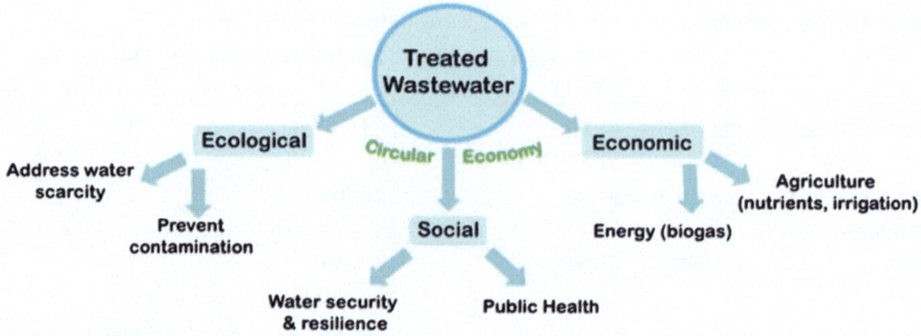

Fig. 4.3 Direct effect of sustainable wastewater treatment

- **Biodiesel**: Biodiesel is a renewable fuel for transportation that is created from lipids generated from wastewater. A class of organic molecules known as lipids are soluble in organic solvents like alcohol and ether but insoluble in water.
- **Hydrogen**: Water waste is electrolyzed to produce hydrogen, which is then utilised as a fuel for energy production and transportation.
- **Thermal energy**: Thermal energy may be utilised to heat and cool spaces and is recovered from wastewater using heat exchangers.
- **Electricity**: Produced using a variety of methods, including microbial fuel cells and turbine generators, it may be consumed locally or sold back to the grid.
- **Water Reuse**: Treated wastewater, often referred to as reclaimed or recycled water, can be reused for irrigation, industrial processes, cooling water, and in some cases even for drinking water supply, thus preserving fresh water supply [9] (Fig. 4.3).

4.3 The Role of Artificial Intelligence (AI) in Wastewater Treatment

4.3.1 Understanding AI and Its Application in Environmental Engineering

Artificial Intelligence (AI) is a versatile modeling technique in environmental engineering. This modeling tool analyzes data and provides insights, saving time over laborious research and calculations. Adaptive Neuro-Fuzzy Inference Systems (ANFIS), Genetic Algorithms (GAs), and Artificial Neural Networks (ANNs) are notable AI applications for environmental engineering. Due to their generalization, speed, and accuracy, Artificial Neural Networks (ANNs) are generally selected among these AI methods. ANNs can estimate CO_2 emissions, cost-effective soot

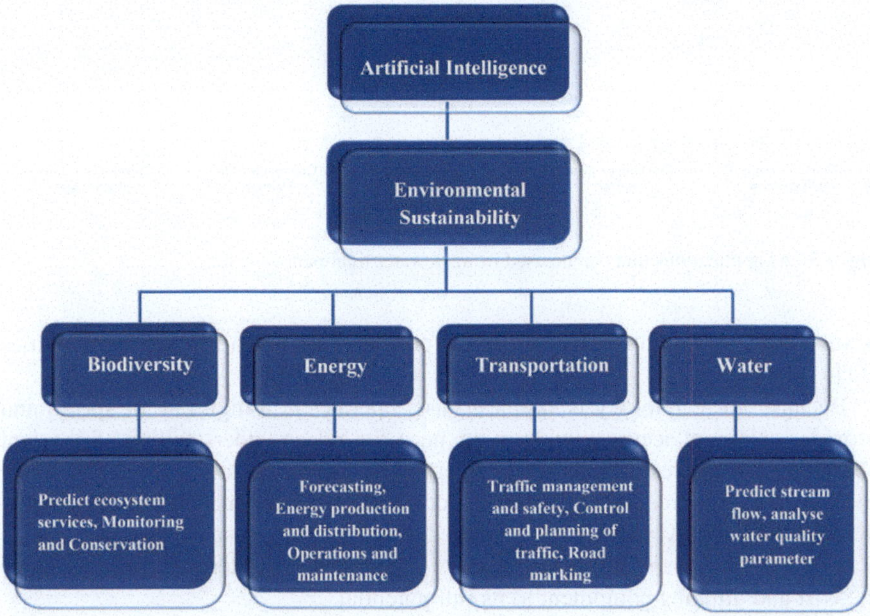

Fig. 4.4 Application of environmental engineering

emissions, and NOx emissions from solid fuel combustion in large-scale circulating fluidized bed boilers using air or oxyfuel. A variety of artificial intelligence applications in environmental engineering are shown in Fig. 4.4.

AI and IoT technology have automated tasks, benefiting enterprises. Using deep learning (DL), ANNs have modeled sorption well, and ANFIS has predicted thermodynamic parameters, energy, and exergy. Genetic Algorithms continue to model and optimize energy and environmental engineering systems. GEP allows mathematical expressions between dependent and independent variables. Automated machine learning (AutoML) is exciting for modeling home prosumer agents and analyzing worst-case energy usage [10].

4.3.2 *Rationale for Integrating AI into Wastewater Treatment*

Artificial intelligence (AI) has become a powerful tool in various fields, including medicine, finance, and environment. Its ease of use, high-speed operation, and accuracy make it useful in various applications. AI has also been applied in wastewater treatment, water environment quality improvement, river water quality modelling, water resource recycling and many more as presented in Fig. 4.5. Wastewater treatment is complex, influenced by chemical, physical, and microbiological factors, and

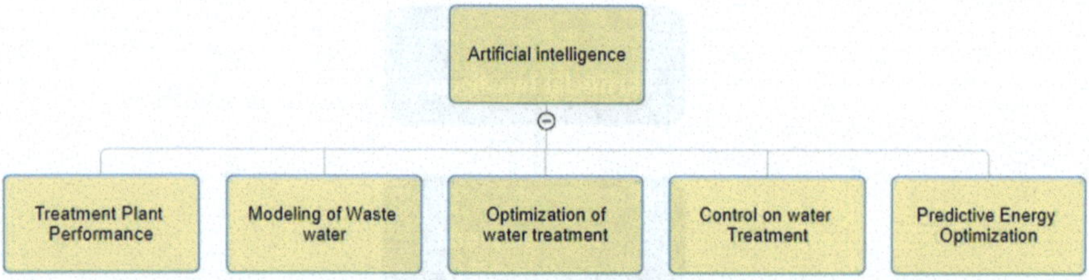

Fig. 4.5 AI applications that can be used in wastewater treatment

is subject to uncertainties due to natural phenomena and anthropogenic activities [11].

Because AI technology is simple to use, operates at a high rate of speed, and provides sufficient accuracy without the need for a deep understanding of physical phenomena, many researchers have implemented it in wastewater treatment plant for river water quality modelling, water resource recycling, and machine fault diagnostics. Additionally, AI technology is applied to prevent, diagnose, and treat disease in wastewater. AI is also used in the financial sector of wastewater treatment to reduce the cost and utilizing equipment to its full potential.

4.3.3 AI Techniques Commonly Used in Wastewater Treatment

TECHNIQUE 1: Using Artificial Neural Network

Hazardous contaminants such as hydrogen potential (pH), total suspended solids (TSS), total dissolved solids (TDS), oil and grease (O&G), biochemical oxygen demand (BOD), and chemical oxygen demand (COD) are present in the wastewater discharged by the detergent industry. It is imperative to eliminate these detrimental contaminants from the effluent prior to its ultimate release into the environment.

Two types of artificial neural network (ANN) have been used in this.

- Feed Forward Neural Network (FFNN) an example of 5 layer FFNN is shown in Fig. 4.6.

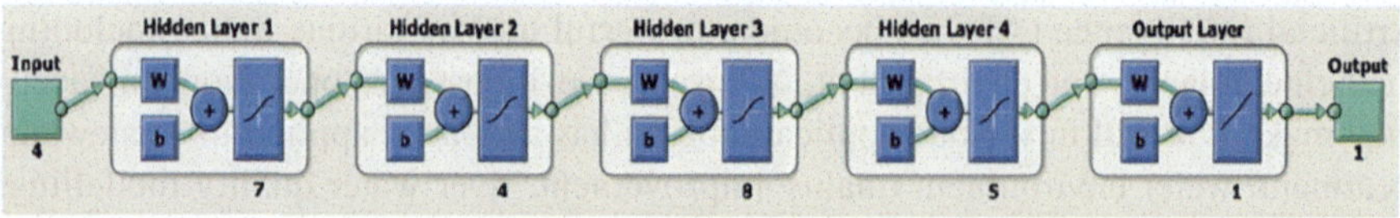

Fig. 4.6 SS5-layer feed forward neural network

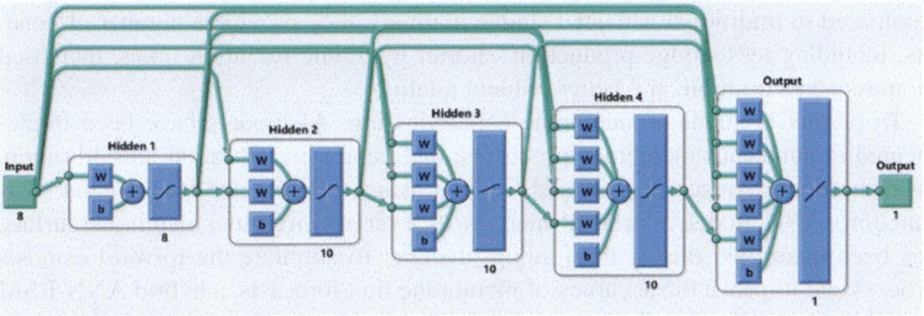

Fig. 4.7 5-layer cascade forward neural network

- Cascade Forward Neural Network (CFNN) an example of 5 layer CFNN is shown in Fig. 4.7.

SVR (support vector Regression) models were designed for better prediction results. It allows users to define acceptable error margins and tolerances for mistakes, identifying appropriate hyperplanes in higher dimensions and allowing them to adjust their tolerance for exceeding that margin. The Levenberg–Marquardt algorithm was used to train the models. Amongst the three the best result was produced by Feed forward neural networks.

TECHNIQUE 2: Feed Forward Neural Network

Improved intelligent approaches in wastewater treatment can be used in complex modelling systems for improved performance prediction. Key factors like chemical oxygen demand, biological oxygen demand, and total suspended substances are used in wastewater treatment plant models.

In this Feed forward neural network was used with 3 optimization algorithms.

(i) L-M algorithm- which is a combination of The Gauss–Newton (G-N) algorithm and descending gradient method
(ii) BR algorithm (Bayes's Regularization)
(iii) SCG algorithm

Three models had appropriate MSE (<0.001), but not optimized. SCG algorithm was fastest due to shortest training duration [12].

TECHNIQUE 3: Hybrid Artificial Neural Network-Response Surface

Methodology

Bioreactors (MBRs), or membrane processes, have been designed and optimized for wastewater treatment in laboratory-based research by means of artificial intelligence models. Using biological treatment in conjunction with membrane filtration, these MBRs offer a useful and effective method of treating wastewater.

Compared to traditional activated sludge methods, they provide a number of benefits, including less sludge production, shorter hydraulic retention times, increased biomass concentration, and better effluent quality.

To predict or manage membrane contamination, AI models have been implemented in numerous membrane processes, including microfiltration, ultrafiltration, nanofiltration, reverse osmosis, and forward osmosis. Using the RBF (radial basis function)-ANN model, interfacial energy with a variably irregular membrane surface has been quantified during the fouling process. To simulate the forward osmosis process and improve the accuracy of membrane flux forecasts, a hybrid ANN-RSM model has been developed.

AI models are being utilized to maximize the performance of microbiological treatment (MBR) by considering several characteristics like as pollutant load, pH, and temperature. With an average prediction accuracy of 95.7%, a five-stage machine learning model was created to simulate and forecast the behaviours of AGS reactors. The COD removal rates in wastewater treated by Chinese traditional medicine were simulated using a Backpropagation Neural Network model. The investigation also discovered that pH has a significant impact on the effectiveness of COD and TN removal in O-AMSMBR [13].

AI Models Employed in the Optimization of Process Parameters Include

A hybrid model that combines genetic algorithms (GA) and artificial neural networks (ANN) was used to forecast the best conditions for raising the biomass of green microalgae in an algal biorefinery within the framework of environmental research and treatment procedures. This approach assisted in the discovery of a highly effective decontaminant that works quickly and affordably in mesoporous nanohybrids. Notably, throughout the decontamination phase, the ANN-PSO model showed greater effectiveness. Furthermore, a self-organizing map (SOM) was utilized to identify critical qualities inside the framework of a municipal wastewater treatment facility. Using the predictive power of ANN and ANFIS models, the main goal here was to improve the extraction of MB from commercially viable agricultural waste sources like as sugarcane bagasse and peanut hulls. Fourier Transform Infrared (FTIR) spectroscopy proved an excellent means of validating the biosorption and prediction capabilities of these AI models.

FTIR techniques were successfully used to identify and analyze pollutants, notably nitrates and microplastics. By recognizing functional groups, Convolutional Neural Network (CNN) models were created to speed up the analysis of gas-phase FTIR spectra. Moreover, combining AI models that use image data with spectroscopic techniques like Raman, SEM, and FTIR has a lot of potential to improve the identification of pollutants, especially micropollutants in water-related sectors. This adaptable method highlights the various uses of artificial intelligence in the field of environmental treatment and monitoring systems.

AI Used in Process of Performance Prediction

When it comes to AI modeling, the Neural Networks (NN) ensemble model performs better than other ensemble models in terms of prediction, but the ANFIS model stands out due to its greater performance when compared to other individual AI models. In addition, real-time forecasts of effluent water quality were obtained by seamlessly integrating an improved Feedforward Neural Network (IFFNN) with Genetic Algorithms (GA). Deep cascade-forward backpropagation (DCB) and DL time series forecasting (DLTSF) models were used to forecast effluent water quality at the El-Berka WWTP. These models provided important information on the performance of the treatment process.

Using both Artificial Neural Network (ANN) and ANFIS models, empirical validation verified the adsorption efficacy of Trapa bispinosa peel (UFBC) in the chlorpyrifos (CPS) adsorption process. Interestingly, a dynamic PCA-NARX model was presented for the purpose of predicting effluent quality, highlighting its benefits over static ANN models for effluent quality simulation. Additionally, a thorough evaluation was carried out utilizing Response Surface Methodology (RSM), ANN, and ANFIS models to ascertain the removal efficiencies of Chemical Oxygen Demand (COD) and Total Suspended Solids (CTSS) in textile effluent using a complex salt-Luffa cylindrica seed extract (CS-LCSE) as a coagulant.

An Artificial Neural Network (ANN) model in conjunction with the fabrication of Fe/Cu nanoparticles under various conditions made it easier to estimate the removal efficiency of COD from home wastewater. As a result of this investigation, the ideal process parameters for getting the greatest outcomes were determined. Results from RSM analyses, adsorption isotherm evaluations, and kinetic investigations highlight the efficacy of Fe/Cu nanoparticles as very efficient COD adsorbents. The procedure of ascertaining the ideal circumstances for effective eradication is further streamlined by the ANN model.

4.4 Enhancing Nutrient Removal with AI

4.4.1 AI-Driven Optimization of Biological Nutrient Removal Processes

Wastewater treatment plants may leverage Industry 4.0 to increase their sustainability and operational effectiveness. The optimisation of the treatment procedure, the reduction of energy use, and the improvement of effluent quality may all be achieved by integrating Industry 4.0 technology into wastewater treatment plants.

A cutting-edge strategy that uses artificial intelligence (AI) and machine learning techniques to optimise biological nutrient removal (BNR) processes in wastewater treatment is designed to increase the efficacy and efficiency of nutrient removal systems.

4.4.1.1 Case Study: AI in Activated Sludge Systems

Mechanistic models of activated sludge are used in wastewater treatment facilities (WWTPs) to simulate and forecast the operation of the process. To enhance predictions and comprehension of WWTP performance, it is suggested that biological data from gene sequencing technology be analysed using AI techniques like deep learning and combined with process engineering models.

Let's look at a hypothetical WWTP that is having trouble effectively forecasting and managing N_2O emissions, a greenhouse gas that has a big influence on climate change. The biological processes in the WWTP have traditionally been modelled using ASMs, but their ability to adequately capture N_2O dynamics has limits. To get around these constraints, ASMs and AI are being integrated.

The researchers suggest using graph-based feature extractions from metagenomics data of microorganisms present in the activated sludge system to combine AI with ASMs. Convolutional neural networks, which have already demonstrated effectiveness in categorising metagenomics data based on phylogenetic tree distances, may be employed with these featurizations as input.

The metagenomics data may be analysed by AI models to determine the genetic sequences of important enzymes involved in the processes that produce N2O. The accuracy of predictions relating to N_2O emissions can therefore be improved by formulating special characteristics for graph neural network models.

In this case study, the combination of AI with ASMs enables a more thorough comprehension of the influence of the microbial population on N_2O dynamics in the WWTP. Researchers are able to create sophisticated prediction models that take into account the cellular-level metabolic activities of microorganisms and their influence on the operation of the process plant as a whole by fusing domain-specific information from activated sludge microbiology with AI approaches. The activities of AI approaches are present in the flowchart as shown in Fig. 4.8.

The case study demonstrates how AI may advance process modelling in WWTPs. Researchers can create more accurate and long-lasting models for WWTP operation by using AI and molecular data, which will enhance the management of N2O emissions and other performance metrics.

4.4.2 Predictive Modelling for Nutrient Removal

To optimise the procedure and get better outcomes in terms of effluent quality and energy savings, predictive modelling is required for the removal of nutrients from wastewater. Traditional control techniques might not be enough to handle the complicated interactions between the many variables involved in wastewater treatment facilities since these processes are extremely nonlinear and time-varying. Model predictive control (MPC), for example, uses predictive modelling approaches to forecast

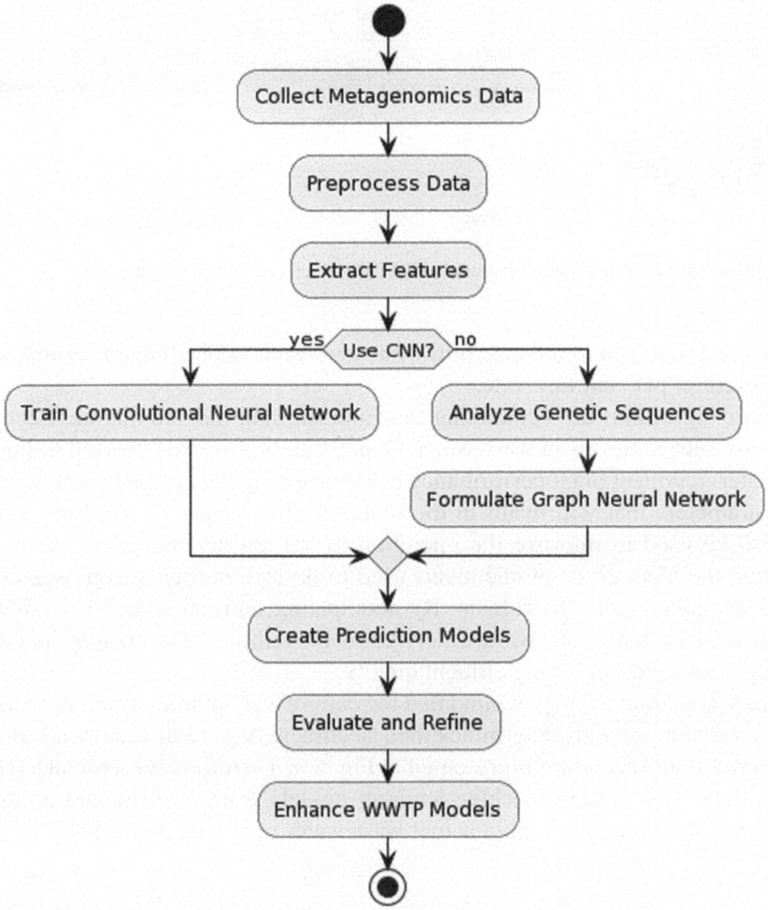

Fig. 4.8 A flowchart of AI in activated sludge systems

future system behaviour and optimise control actions based on real-time measurements and historical data. As a result, resources like energy and chemicals may be used more effectively, and legal standards for effluent quality may be better met.

4.4.2.1 Application of Machine Learning in Predicting Nutrient Levels

Machine learning can anticipate nutrient levels in wastewater by analysing vast volumes of data from many sources, including sensors, meteorological data, and historical data from wastewater treatment facilities.

Artificial neural networks (ANN) are one of the most often utilised machine learning algorithms for forecasting nutrient levels in wastewater. ANN models may

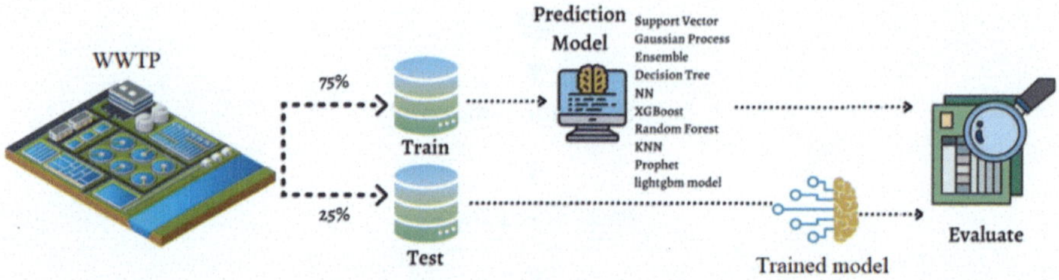

Fig. 4.9 Illustration of the general framework of ML-based prediction procedure

be trained on past data to forecast future nutrient levels depending on variables such as temperature, pH, and flow rate.

Genetic algorithms are yet another machine learning method that has been used to forecast nutrient levels in wastewater. Genetic algorithms can be used to improve wastewater treatment plant performance by determining the optimal combination of input parameters that will result in the required nutrient levels. Machine learning may also be used to improve the operation of wastewater treatment facilities by predicting the ideal doses of chemicals used to remove nutrients from wastewater, such as coagulants and flocculants. By anticipating nutrient levels in wastewater, machine learning may assist wastewater treatment facilities in optimising operations, lowering costs, and improving effluent quality.

But it's important to keep in mind that the calibre and volume of data used to train machine learning models determines their accuracy. A general framework of ML-based predictions procedure is presented in Fig. 4.9. Therefore, it is crucial to ensure that the data used to train machine learning models is trustworthy and accurately reflects the environment found in actual wastewater treatment plants [4].

4.4.3 Real-Time Monitoring and Control of Nutrient Removal

Real-time nutrient removal monitoring and management are crucial in wastewater treatment facilities to detect inadmissible inflow wastewater, reduce environmental harm, and lower treatment costs. It optimizes the nutrient removal process, increasing plant efficiency and saving operating costs. Real-time monitoring ensures the plant meets regulatory nutrient removal standards, avoiding fines and penalties. It also helps prevent nutrient overloading downstream habitats, causing hazardous algal blooms and environmental issues. Operators can quickly identify any issues or deviations from the required process conditions through real-time monitoring, allowing for necessary adjustments to ensure maximum performance.

4.4.3.1 IoT Sensors and AI for Process Optimization

In the wastewater treatment plant, IoT sensors may be utilised to collect real-time data on characteristics such as flow rate, pH, temperature, and dissolved oxygen levels. This information may be used to track the plant's performance and suggest areas for improvement.

AI algorithms may be used to analyse data received by IoT sensors and detect patterns and trends that human operators may not notice. AI algorithms, for example, may be used to detect correlations between multiple parameters and estimate the impact of changes in one parameter on plant performance.

Wastewater treatment facilities may optimise their processes in real time by utilising IoT sensors and AI. For example, if the AI algorithm detects that the dissolved oxygen levels in the aeration tank are too low, it may automatically raise the air flow rate.

AI and IoT sensors may also be utilised to forecast equipment breakdowns and maintenance requirements. AI algorithms can forecast when equipment is likely to fail and warn maintenance people to undertake preventative maintenance by analysing data from sensors on equipment such as pumps and blowers.

The study details numerous **sensors** and tools that may be used to gauge the many elements that influence water quality.

- Spectro::lyser: Turbidity, temperature, pressure, colour, dissolved ions, UV254.
- SmartCoast: pH, dissolved oxygen (DO), conductivity, temperature, turbidity, phosphate, water level.
- Kapta 3000 AC4: Chlorine, temperature, pressure, conductivity.
- Smart water (Libelium): pH, DO, conductivity, temperature, oxidation–reduction potential (ORP), turbidity, dissolved ions.
- I::scan: Colour, turbidity, UV254.
- Lab-on-chip: Any specific bio-chemical [3].

4.5 Expedited Resource Recovery Through AI

4.5.1 Resource Identification and Extraction

4.5.1.1 AI Applications in Identifying Valuable Resources in Wastewater

The research titled "Microalgal cultures for the remediation of wastewaters with different nitrogen to phosphorus ratios: Process modelling using artificial neural networks" showed that C. vulgaris can readily adjust to various N:P molar ratios in synthetic wastewaters. This adaptability is essential for ensuring consistency in the process when dealing with actual wastewaters. In addition, microalgae demonstrated a remarkable ability to eliminate nitrogen and phosphorus from wastewater, achieving an impressive removal rate and efficiency of 92.0% for nitrogen and 98.2% for phosphorus (0.2%/0.92 0.03 mgP L^{-1} d^{-1}). Varied quantities of nitrogen and phosphorus were used in the investigation, yielding significant consequences on nutritional consumption while exerting little influence on the growth of microalgae. At low N:P ratios of 2 and 3, nitrogen impeded the absorption of phosphorus by microalgae, whereas at high ratios of 66 and 67, phosphorus had the same inhibitory effect. Therefore, to support the proliferation of cells in response to changes in the environment, C. vulgaris probably adjusted its nutrient absorption rates and, as a result, the composition of its biomass. The GA-ANN models had a strong fitting capacity for nitrogen and phosphorus removal rates and biomass production, as shown by their respective R2 values of 0.951, 0.800, and 0.793. The results were consistent with the experimental data, and the models accurately depicted the changes in these parameters caused by diverse input inputs. The beginning biomass concentration was chosen as the threshold variable to distinguish between two productivity regimes: one occurring during the stationary phase and the other during the exponential growth phase. The starting concentrations of nitrogen and phosphorus were used as threshold variables to determine the rates at which nitrogen and phosphorus were removed, respectively. Furthermore, these findings demonstrate that the concentration of nutrients significantly influences the process of nutrient removal. The study's results suggest that microalgae-based wastewater treatment methods are reliable, efficient, and durable. Furthermore, GA-ANN models have the potential to be advantageous in the simulation and regulation of microalgal systems [14].

4.5.2 *Sustainable Practices Enabled by AI*

4.5.2.1 Balancing Resource Recovery with Environmental Sustainability

The Fourth Industrial Revolution (4IR) is transforming the treatment of water and wastewater by utilizing advanced technologies such as online multi-sensor data collection, wireless communication, the Internet of Things, data pre-processing, big data storage, analytics, human–machine interface, data comprehension, data mining, blockchain, evolutionary computational, modeling-simulation, metadata, data pipeline, faulty detection algorithms, and digital twins. The effectiveness of traditional treatment approaches in managing emerging pollutants has been shown, leading to increased costs and decreased flow rates in wastewater treatment plants.

Artificial intelligence (AI)-enabled processes and monitoring in water digitization can facilitate decision-making in public policy and water management using cyber-physical systems (CPS). In water management, digital twins can help with decision-making processes, reduce operating expenses, stop untreated wastewater from leaking, and enhance data gathering. The COVID-19 pandemic may be resolved by wastewater-based epidemiology (WBE), which provides an early warning and digital intelligent monitoring system. Wastewater treatment facilities that are both decentralized and centrally located can support industry 4.0 technologies, the sustainable development goals, and the circular economy (CE).

Advancements in machine learning, deep learning, and AI-based data analysis have shown the ability to tackle complex problems in drinking water treatment. This has led to the creation of predictive model platforms and universal process analysis, enabling more effective solutions. AI infrastructure monitoring enables the cost-effective use of satellite imagery, leak detection, and data mining capabilities. Plant troubleshooting is facilitated by the use of digital twins, which are virtual replicas of water infrastructure assets that can be examined and supervised.

Industry 4.0 depends on industrial cyber-physical systems (ICPS), which connect the virtual and physical realms. With the help of digital twins, decision-making is made holistically and remotely in real time, allowing for infrastructure and parameter control [11].

4.6 Transformative Impact of AI on Wastewater Treatment

Traditional wastewater treatment technologies have relied on statistical and multivariate data analysis approaches to evaluate pollutant removal efficiency and estimate uncertainty. However, without the use of complex mathematical formulae and extensive knowledge about the relationship between the input and output variables, AI technologies can deliver better accuracy in the prediction of regression coefficients and extremely low average error [15].

AI technology may be utilised to forecast the effluent quality of the treatment process and take appropriate measures to prevent quality deterioration. Automated sampling assists the operator in analysing trends and patterns, and quick intervention can avoid a tragedy. Continuous monitoring of systems using sensors aids in enhancing operational efficiency, reducing energy usage, and lowering costs [15]. AI may be used to improve the efficiency of wastewater treatment facilities by predicting the ideal doses of chemicals used to remove contaminants from wastewater, such as coagulants and flocculants. This can assist to minimise chemical consumption and enhance treatment process efficiency.

In wastewater treatment facilities, AI may also be used to forecast equipment failures and maintenance requirements. AI algorithms can forecast when equipment is likely to fail and warn maintenance people to undertake preventative maintenance by analysing data from sensors on equipment such as pumps and blowers. This can aid in lowering downtime and maintenance expenses [15].

Artificial intelligence can be used to optimise the energy consumption of wastewater treatment systems. AI algorithms can find chances to minimise usage of energy and boost energy production by analysing data on energy consumption and production, such as solar and wind power. AI can be utilised to increase the prediction accuracy of wastewater treatment models. AI systems may uncover patterns and trends in wastewater treatment operations by analysing enormous volumes of data. These patterns and trends are not immediately evident to human operators. This can assist to increase the accuracy of prediction models and optimise wastewater treatment plant performance.

4.6.1 Environmental Benefits

4.6.1.1 Reduced Energy Consumption and Carbon Footprint

One of the most significant environmental benefits of AI in wastewater treatment facilities is lower energy use and carbon footprint. A wastewater treatment plant must use energy to function, and the increasing number of new contaminants and other variables that make significant energy consumption necessary have become vital in the planning and administration of these facilities [16].

By giving real-time data on the performance of treatment procedures, AI may assist in optimising the usage of resources such as energy and chemicals. Using data-driven soft sensors to precisely estimate energy use, which may enable daily energy management and assist in optimising the wastewater treatment process and operation. This can result in lower energy consumption and chemical use, as well as fewer greenhouse gas emissions and environmental effect.

Reducing energy usage and carbon footprint in wastewater treatment facilities can provide several environmental benefits, including reduced greenhouse gas emissions and climate change mitigation, resource conservation, and improved air and water quality. A fascinating path towards sustainable water treatment that is in line with the SDGs, the circular bioeconomy, and industry 4.0 is to include life cycle environmental or cost management technologies into optimisation models.

Other ways to reduce energy consumption and carbon footprint in wastewater treatment plants include optimising plant design and operation, using energy-efficient equipment and technologies, implementing energy management systems, and utilising renewable energy sources such as solar, wind, and biogas [16, 17].

AI may be used to forecast equipment failures and maintenance requirements, allowing for proactive maintenance and lowering downtime, both of which can assist to reduce the environmental effect of equipment failures. Overall, the application of artificial intelligence (AI) in wastewater treatment facilities can assist to enhance environmental performance and lower the environmental effect of wastewater treatment operations [16].

4.6.1.2 Lower Chemical Usage and Waste Generation

AI can help to optimize the use of chemicals by providing real-time data on the performance of treatment processes. Various strategies may be used to reduce chemical consumption and waste generation in wastewater treatment facilities, including the use of data mining and optimisation techniques to minimise energy-material saving management.

Data mining entails gathering and pre-processing data from diverse sources, employing algorithms and statistical techniques to extract patterns and insights from the data, and then applying these insights to optimise the design and operation of the treatment process. Alternative treatment technologies like as advanced oxidation, membrane filtration, and activated carbon adsorption can be employed to successfully remove emerging pollutants while reducing the requirement for harmful chemicals.

Lowering chemical consumption and waste creation in wastewater treatment facilities can have several environmental benefits, including lowering the quantity of harmful chemicals discharged into the environment, preserving natural resources, and lowering trash generation. Furthermore, the use of circular bioeconomy and resource recovery may alter the wastewater treatment sector by converting trash into a valuable resource and minimising the requirement for virgin resources.

4.6.2 Economic Advantages

4.6.2.1 Cost Savings Through Efficient Operations

The application of AI in wastewater treatment facilities can result in several environmental advantages, as well as financial savings through more efficient operations.

With a high degree of efficiency and accuracy, data-driven optimisation strategies may lower the operating cost of energy and material under desired effluent parameters in real-time. This implies that wastewater treatment facilities may optimise their operations by employing data mining and optimisation approaches to minimise energy and material usage, resulting in considerable cost savings. Cost saving can also be done by various AI technologies such as predictive maintenance and process optimization [16].

AI technology can aid in optimising the energy usage of wastewater treatment facilities, lowering the plant's operational expenses. AI can find locations where energy usage may be lowered without affecting plant efficiency by continually monitoring the systems using sensors. This can result in considerable financial savings for the facility as well as a reduction in its carbon impact. The circular bioeconomy and resource recovery may alter wastewater commercialization by adding value through resource recovery, creating new income streams, and reducing the demand for virgin resources. This can lead to cost savings for the plant. Furthermore, AI may aid in system automation, reducing the requirement for manual labour and increasing plant efficiency. This may result in cost savings for the plant.

4.6.2.2 Revenue Generation from Recovered Resources

Applying AI to wastewater treatment plants could result in money being made from resources that are recovered. These kinds of facilities can recover useful elements from wastewater, such as water, energy, and nutrients. Through the identification of the most efficient treatment techniques and operational parameters, artificial intelligence (AI) technology enhances resource recovery. As a result, wastewater treatment plants can increase the value of their operations and lessen their reliance on expensive inputs. It has a dual benefit in that it lowers facility costs and brings in new sources of income that support the plant's long-term viability. For example, biogas can be recovered by AI optimization of the anaerobic digestion process, and this can then be used as a renewable energy source [17].

AI can aid in the identification of the optimal treatment methods for recovering nutrients such as nitrogen and phosphorus from wastewater. These nutrients may be utilised in agriculture as fertilisers, generating money for the plant.

4.6.3 Social Implications

4.6.3.1 Enhanced Public Health Through Improved Wastewater Treatment

By streamlining the wastewater treatment process, the use of AI in wastewater treatment plants has the potential to improve public health. Since these facilities oversee eliminating dangerous substances and bacteria from wastewater before it is released into the environment, they are vital to the protection of public health. Treatment processes can be optimized to determine the most effective treatment protocols and operating factors by leveraging AI technology. As a result, pollutants and viruses are removed more successfully, lowering the related health risks.

AI can aid in the early detection of irregularities in the treatment process, preventing the discharge of untreated or partially treated wastewater into the environment. This can help to lower the risk of waterborne infections and other problems [15].

4.6.3.2 Job Creation and Skill Development in AI-Assisted Wastewater Management

The incorporation of AI and big data in the water business is providing new job possibilities and necessitating the acquisition of new skill sets. The growth of big data in the water business is creating a demand for data analysts, data scientists, and other data management and analysis experts. The water industry's digital revolution is opening new opportunities for skill development and capacity growth. Wastewater treatment facilities may enhance their operations and generate new income streams by implementing digital technology and AI-assisted wastewater management. This can open up new chances for skill development and capacity building for water sector operators and other professionals. Furthermore, the use of AI in wastewater management is increasing the demand for individuals with AI and machine learning skills [17].

4.7 Real-World Case Studies

4.7.1 Case 1: AI-Driven Nutrient Removal in a Large Municipal Wastewater Treatment Plant

Sewage treatment plant operation and maintenance issues can result in serious ecological and public health problems, including water-borne illnesses. High levels of organics suspended and dissolved particles, phosphorus (P), inorganic nitrogen (N), and organics are found in municipal wastewater. These nutrients encourage the growth of algae, which contaminates groundwater and causes eutrophication in bodies of water. By 2030, there could be a 40% water deficit worldwide, hence it's critical to develop efficient municipal wastewater treatment systems. Figure 4.10 presents the AI-driven nutrient removal in a large municipal wastewater treatment plant.

The disadvantages of traditional wastewater treatment methods include high energy consumption, instability, carbon emissions, and resource waste. Examples of these methods are chemical phosphorus removal, coagulation-denitrification, and aerobic-activated sludge. Higher efficiency, lower costs, and less energy usage are possible with microalgae-based wastewater treatment; however, performance is dependent on several factors.

To forecast the nitrogen and phosphorus removal efficiency in microalgae-based municipal wastewater treatment, this work employs statistical modelling approaches such as RSM, ANN, and SVR. Global circumstances are optimized using a multi-objective optimization technique and a Bayesian optimization algorithm, which maximizes efficiency and offers a workable method for treating wastewater.

Here, the use of microalgae (*Chlorella kessleri* sp.) to remove N and P from municipal wastewater simultaneously was investigated. Microalgae have the potential to effectively remove N and P from household wastewater, according to an early trial.

In this investigation, three different methodologies were used. Novel approaches were created, employing computational techniques like Response Surface Methodology (RSM), Artificial Neural Networks (ANN), and Support Vector Regression (SVR), to evaluate the efficacy of removing N and P microalgal systems in aqueous settings. Real-world experimental data obtained from the experiment's design, notably the Box-Behnken Design (BBD) matrix, was examined using an ANOVA as part of the study to determine the importance of each operational variable.

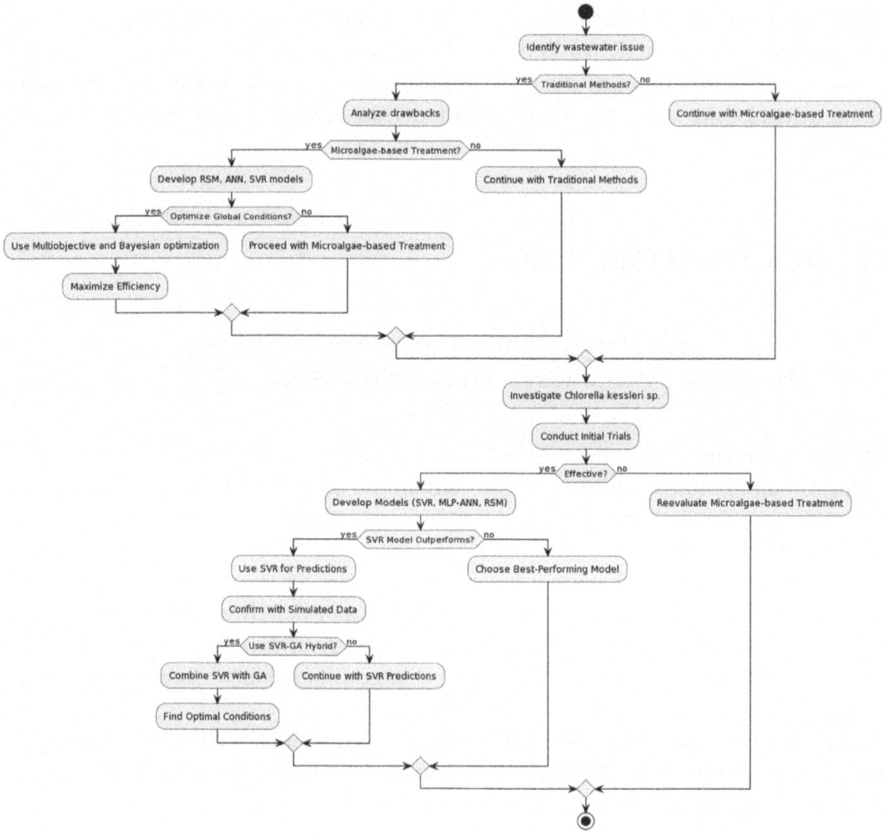

Fig. 4.10 Flowchart of AI-driven nutrient removal in a large municipal wastewater treatment plant

Out of all the computational models, the SVR model outperformed the Multilayer Perceptron Artificial Neural Network (MLP-ANN) and RSM models, yielding predictions that were more accurate. The outstanding R-squared (R2) values for N removal (99.96%) and P removal (95.92%) demonstrate that the SVR model produced results that were remarkably accurate. Notably low were the Root Mean Square Error (RMSE) values, which were 5.3534 for P removal and 0.7819 for N removal. With a mean absolute error (MAE) of 1.8936 for P and 0.5079 for N removal, the error levels stayed low.

To validate the Support Vector Regression (SVR) model and determine its effectiveness, more synthetic data was included. SVR-GA is a hybrid platform that was developed by combining the proven SVR models with Genetic Algorithm (GA) to find the optimal global conditions for increasing the efficiency of nutrient removal. As a result, the ideal configuration exceeded 93% in terms of nutrient removal efficiency [18].

4.7.2 Case 2: Resource Recovery Success in an Industrial Wastewater Facility

Prior to being returned to the environment or utilized, wastewater treatment systems aim to reduce the nutrient content. It has been demonstrated that microalgae are effective at extracting nutrients from a range of wastewaters, including industrial and municipal wastewater. Furthermore, they possess the capacity to generate by-products such as biodiesel, which finds applications in the treatment of effluent, production of biofuels, and reduction of carbon dioxide emissions. The composition of wastewater significantly affects carbon, nitrogen, and phosphorus, which are the three essential elements required for microalgal treatments. Differing concentrations of carbon, nitrogen, and phosphorus in distinct wastewater streams may have an impact on the efficacy of microalgal removal.

For instance, the Chlorella vulgaris strain displayed varying levels of treatment efficacy in various wastewater stream types. It is important to create an environment that favors microalgae development over that of bacteria and fungi, as the native microbial population in wastewater may compete with imported microalgae when nutrients become scarce. The effectiveness of nitrogen removal and microalgae productivity are likewise highly dependent on temperature and light.

An investigation was conducted to assess the capability of Chlorella protothecoides, a kind of freshwater microalgae, in treating urban wastewater using a small-scale photobioreactor that is illuminated externally. The highest levels of COD, total nitrogen, and total phosphorus elimination were seen under optimal conditions. When choosing microalgae strains for wastewater treatment, it is important to consider factors such as their ability to produce a large amount of biomass, develop quickly, withstand the local environment, and efficiently remove nutrients [19] (Fig. 4.11).

4.7.3 Case 3: AI-Powered Small-Scale Wastewater Treatment in Developing Regions

Current water quality monitoring (WQM) methods are limited in the setting of sewage treatment plants (STPs) in terms of their temporal and spatial coverage, operational expenses, and the availability of near-real-time data. The incorporation of Internet of Things (IoT) technology has been crucial in monitoring wastewater treatment plants' equipment and operations to overcome these limitations. Systems for treating wastewater that are Internet of Things (IoT) based are essential for cutting down on emergency reaction times, streamlining administrative procedures, saving energy, and improving overall operational efficiency.

A direct result of cutting-edge technology like cloud computing, artificial intelligence, machine learning algorithms, and service-oriented architecture is the creation of Smart Water Quality Monitoring Systems (SWQMS). To provide near-real-time awareness and support well-informed decision-making, these systems leverage data

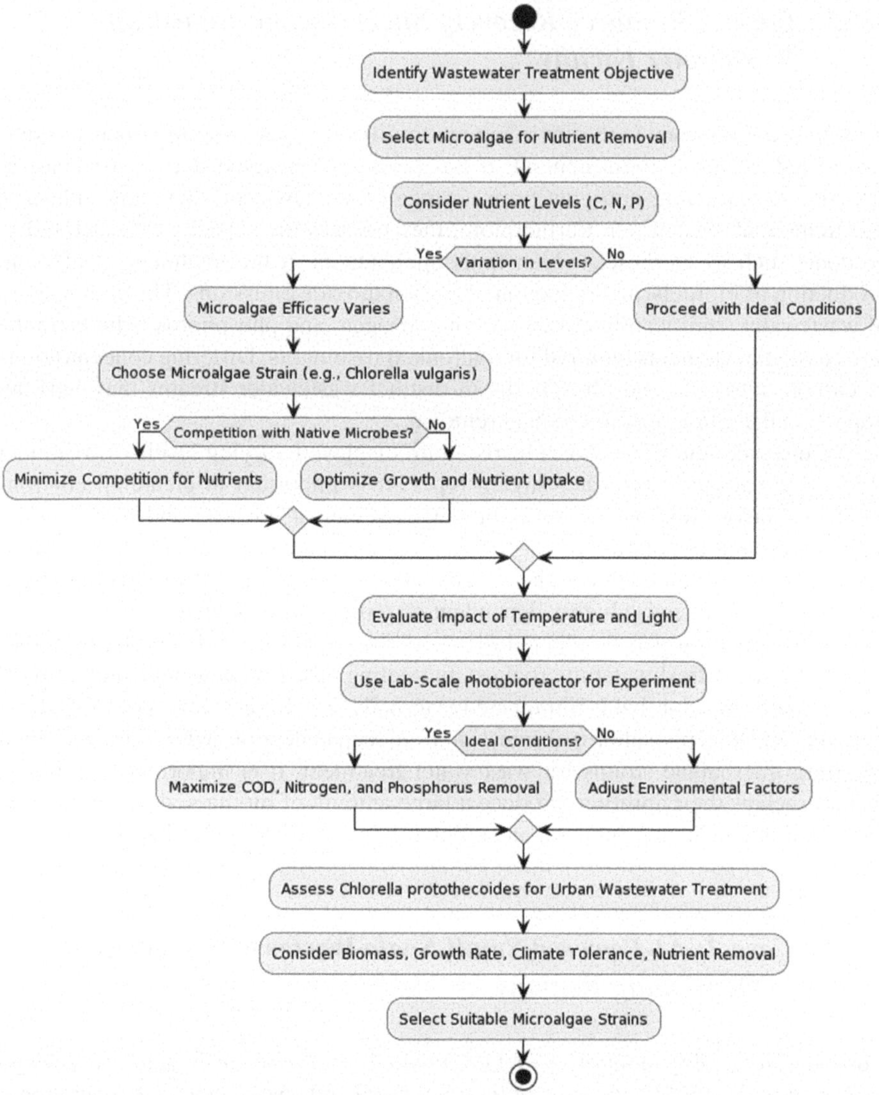

Fig. 4.11 A flowchart of resource recovery success in an industrial wastewater facility

from a variety of sources, including sensors, cameras, news, satellite imaging, and crowdsourcing.

Wireless Sensor Networks (WSNs) have shown promise in the field of environmental monitoring applications, enabling remote, near-real-time process monitoring with little to no human interaction. This development opens the door to more research into the development of decentralized SWQMSs, especially in the context of smart cities.

A noteworthy example of this advancement is provided by Martínez and associates, who utilized a Wireless Sensor Network (WSN) within a wastewater treatment facility to create an affordable water quality monitoring apparatus as shown in Fig. 4.12. This gadget detects nitrate and nitrite in natural waters using ion chromatography detection. It easily interfaces with an Internet of Things software platform for wastewater quality monitoring and management as well as a decentralized SWQMS. By processing and interpreting sensor data, the EcoSens Aquamonitrix System provides nearly real-time data processing and monitoring. Its usefulness has been demonstrated through testing at four wastewater treatment facilities in Spain [20].

4.8 Challenges and Limitations

4.8.1 Ethical Concerns and Data Privacy in AI Applications

Data security: To build models and make predictions, AI uses a lot of data. However, this information could include private or confidential information about people or businesses, such trade secrets. To avoid unauthorised access or misuse, it is crucial to guarantee the privacy and security of this data.

Discrimination and bias: Artificial intelligence models may exhibit bias if their underlying algorithms are biased or if the training data fails to accurately represent the entire population. This may lead to prejudice against specific individuals or groups, which may have ethical consequences.

Since they are exposed to extreme external conditions and subpar maintenance, it is necessary to maintain and check the accuracy of signals from online sensors. Transparency and accountability: AI models can be complicated and tough to comprehend, making it challenging to discover biases or errors. For AI models to be utilised ethically and to avoid unforeseen consequences, transparency and accountability must be ensured during their creation and usage.

4.8.2 Initial Investment and Infrastructure Requirements

The use of AI in wastewater treatment plants has several challenges. To begin with, integrating AI is expensive and requires spending money on personnel training, software purchases, and other infrastructure changes. This initial monetary outlay may be a major obstacle for smaller firms or those with limited means.

Second, the infrastructure needs of some AI applications are constrained. These technologies typically require high-speed internet and strong cloud computing abilities, yet these resources might not be generally available everywhere. As a result,

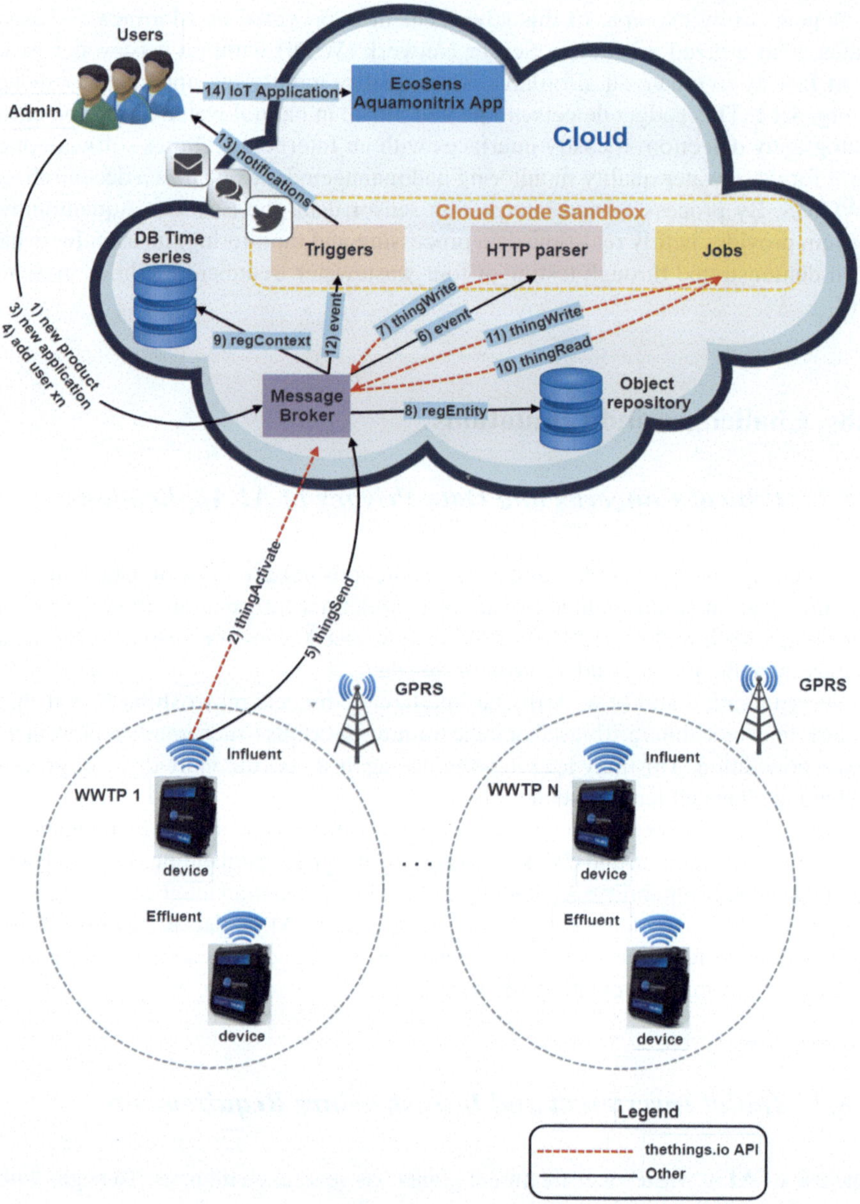

Fig. 4.12 Smart wastewater quality monitoring system

several businesses may find it challenging to appropriately integrate AI into their wastewater treatment systems owing to infrastructure limitations.

Technical expertise is also necessary for the creation and use of AI models. It is vital to be proficient in disciplines like data science and computer programming, but not all organisations have these skills. The broad deployment of AI in wastewater treatment systems may be constrained by this competency gap.

Data management and collecting also provide a significant difficulty. Large datasets are necessary for training and developing AI models, but their availability and quality might be troublesome. Assuring the sufficiency of data for AI applications may be challenging if data collection methods are not well-established or if data management procedures are insufficient. Additionally, managing the many formats and types of data gathered during wastewater treatment can be difficult and could call for extra resources.

4.8.3 Technical Challenges and System Reliability

Systems for treating wastewater have a number of technical obstacles when integrating AI. First and foremost, creating and putting into practice AI models can be a technically challenging task, especially when dealing with complex systems like wastewater treatment plants. The accuracy and dependability of these models depend on crucial elements including data quality, model selection, and algorithm design, all of which require careful consideration.

The reliability of the system is another crucial issue. AI models may show sensitivity to modifications in the monitored systems' surroundings. This makes ongoing maintenance and monitoring necessary to identify and resolve any problems as soon as they occur, guaranteeing the long-term dependability of these models.

It can be difficult to integrate AI with current wastewater treatment systems, particularly if AI was not taken into consideration during system design. To guarantee smooth compatibility, a successful integration process necessitates intensive planning and coordination. Inadequate data management procedures and a dearth of academic transparency exacerbate these problems. It is difficult to evaluate the performance of different AI models since the data and source code used in wastewater treatment experiments are not publicly available, which hinders academic openness and repeatability of results.

Additionally, certain AI models—like artificial neural networks (ANN)—have poor reproducibility and interpretability. They frequently function as "black boxes," mapping connections without disclosing the underlying mechanics. This might cause randomness to enter the process and impede the acquisition of optimal network parameters. The lack of physical meaning in many AI models' parameters makes it more difficult for them to explain the fundamental workings of wastewater treatment systems.

The adoption of AI in this sector is further complicated by the lack of standards, benchmarks, and fair comparison techniques within the field, which makes it difficult for researchers to determine which AI models perform best when applied to certain real-world wastewater treatment problems.

4.9 Future Directions and Emerging Trends

4.9.1 Advancements in AI and Machine Learning for Wastewater Treatment

- To effectively apply AI models to water treatment technology (WWT), collaboration between mathematicians, biologists, engineers, and computer experts is needed. Additional research is required to comprehend important issues and intricate WWT systems. In domains relevant to WWT, hybrid AI models—which blend data-driven and knowledge-driven models—are becoming more and more popular.
- Artificial Neural Networks (ANNs) are popular for wastewater treatment applications due to their ease of training and understanding. Deep neural networks, such as convolutional and recurrent neural networks, have gained attention for their versatility and flexibility. Hybrid models, coupled with dynamic optimization, are being used to predict outputs and increase biodiesel and lipid production.
- Analytics for wastewater Implementing machine learning.

Data generated by sensor nodes can be secured using blockchain technology, and the data can be leveraged to get important insights into wastewater management. The effectiveness of wastewater management can be increased by using different wastewater datasets protected by blockchain to train machine learning algorithms. Artificial intelligence can be used to find problems in wastewater management, like typical leakage points and defective sensors. Analyses allow for the consideration of appropriate steps to enhance the therapeutic process [21].

- AI models are predominantly data-driven and opaque, although, like ANN, they may represent complex nonlinear interactions between inputs and outputs. While the precise nature of their implementation and underlying mechanisms remains obscure, these technologies exhibit promise for use in wireless sensor network (WWT) engineering applications. The integration of AI models with conventional mathematical models has the potential to decrease data demands while yielding significant outcomes. A few AI models have limited interpretability because of their parameters, weak reproducibility, and the challenge of training NN, particularly DNN. Determining optimal network architectures through trial and error is impractical, potentially resulting in overfitting or underfitting. Additional research is necessary to support the development of AI applications in WWT by addressing the challenges of NN training, parameter optimization, poor repeatability, and limited interpretation, as well as constructing AI models. AI models heavily depend on data, and the lack of open-source code and data presents a significant challenge when attempting to compare the performance of different models. To reducing experiment costs, ensuring equitable comparisons, and promoting the widespread adoption of AI models in WWT, standards and benchmarks should be established and unedited data and source code should be made public [13].

- Costs may be cut, and efficiency raised in wastewater treatment plants (WWTPs) through data-driven process control using machine learning algorithms and computational advances. But having the correct personnel with experience in control engineering and computer science is essential. Using IoT, real-time monitoring, and sophisticated models and sensors may lower greenhouse gas emissions and increase energy efficiency [22].

4.9.2 Integration of AI with Other Emerging Technologies (E.g., Blockchain, Nanotechnology)

A highly promising technological future is presented by the combination of Artificial Intelligence (AI) with new technologies, such blockchain and nanotechnology, in wastewater treatment. In the future, the breadth of this integration presents a variety of new developments that have the potential to drastically alter the industry. It is anticipated that blockchain will become more widely used in wastewater treatment systems, providing secure data management and facilitating the development of decentralized wastewater networks. It is envisaged that multi-stakeholder collaboration will be fostered by the secure recording and sharing of data on resource recovery, treatment, and water quality using blockchain. This will develop an ecosystem for wastewater management that is more accountable, transparent, and efficient.

AI and nanotechnology together are expected to be crucial in the creation of cutting-edge materials and treatment techniques. A new development in this field is the creation of AI-powered smart nanomaterials that are designed to improve the accuracy and efficiency of wastewater pollution and contaminant removal. By focusing on certain contaminants, these intelligent materials can improve treatment effectiveness and lessen the need for traditional chemical additives. Moreover, real-time monitoring and control features are anticipated when AI is included into nanomaterial-based therapeutic systems. These systems are designed to adjust to sudden variations in the quality of the water, making treatment procedures more efficient and meeting strict requirements.

Regarding emerging developments, it is predicted that the combination of AI, sensors, and the Internet of Things (IoT) would bring about a new age in wastewater treatment. A network of linked sensors and devices will be created as a result of this integration, continuously gathering information on energy usage, equipment condition, and water quality. This enormous dataset will be processed and analyzed mostly by AI algorithms, which will make predictive maintenance, anomaly detection, and system optimization easier. The future looks to fully autonomous wastewater treatment plants that require little human intervention as AI algorithms advance.

Increased energy efficiency, financial savings, and the adoption of ecologically friendly treatment techniques will result from this autonomy. In conclusion, the fusion of blockchain, nanotechnology, and the Internet of Things to produce smarter, more effective, and environmentally sustainable wastewater treatment systems portends a technologically sophisticated future. This is evidenced by the integration of AI with new technologies in wastewater treatment.

4.9.3 Regulatory Changes and Policies Promoting AI Adoption in Wastewater Treatment

The part wastewater management plays in accomplishing the Sustainable Development Goals (SDGs) of the UN. It demonstrates that 11 of the 17 SDGs may be aided by wastewater treatment.

AI has been a potential solution in water and wastewater management for over half a century. Because of things like population aging, climate change, and changing socioeconomic needs, traditional software decision support systems have limits that can be solved by AI-based techniques like pattern recognition. These technologies can combine data on influent characteristics, system operations, and water supply and demand that was previously inaccessible. Simple chatbots for customer service to sophisticated systems that advise or assist with the operation and maintenance of water and wastewater treatment are just a few of the important current and future uses of AI in the water supply and wastewater management industries.

The incorporation of themes into the constitution, private law, criminal law, administrative law, and the Intellectual Property Basic Act are among the legislative concerns pertaining to AI governance. The Local Autonomy Act, which may be relevant to AI-based decision-making, sets the maximization of impact from minimal cost for governmental services [23, 24].

There is a rising trend in the development of data analytics, cloud-based AI tools, smart sensors, and machine learning systems that are specifically designed to monitor and manage the Earth's most valuable natural resource. The advent of Water 4.0, sometimes referred to as Smart Water, is rapidly becoming a practical solution that will assist governments, enterprises, and other organizations in attaining their Sustainable Development Goals. This novel technique employs cutting-edge technology to guarantee the effective and sustainable administration of water resources.

4.10 Conclusion

This article has taken a comprehensive look at the treatment of wastewater, discussing its procedures, analysing its nutrients, and using artificial intelligence (AI) in order to make the treatment more effective. The emphasis placed on the part that AI plays in enhancing biological nutrient removal processes and making accurate predictions of nutrient levels highlights the potential that AI has to revolutionize wastewater treatment.

In addition, the paradigm shift that is occurring in environmental engineering, notably in the treatment of wastewater, is drawing attention to the growing use of AI models. In order to harness modern technologies and overcome issues in data management, interpretability, and system stability, a strong emphasis is placed on the joint efforts of mathematicians, biologists, engineers, and computer experts.

It has been determined that one of the most important aspects of sustainable wastewater treatment is the promotion of procedures that are both kind to the environment and economical with resources. In its conclusion, the report highlights the revolutionary influence that AI models may have in wastewater treatment facilities by lowering energy consumption, enhancing overall environmental sustainability, and maximizing resource usage. This combination of AI and conventional methods of treating wastewater offers tremendous improvements and resource recovery, which will contribute to the creation of a wastewater management system that is both more effective and less harmful to the environment.

References

1. Wastewater treatment. Encyclopædia Britannica. Published 05 Oct 2023. https://www.britannica.com/technology/wastewater-treatment. Accessed 10 Oct 2023
2. What are the major challenges when operating an industrial wastewater treatment system? Ellis. Published 30 Mar 2023. https://www.elliscorp.com/what-are-the-major-challenges-when-operating-an-industrial-wastewater-treatment-system/. Accessed 10 Oct 2023
3. Radhakrishnan V, Wu W (2018) IoT Technology for Smart Water System. In: 2018 IEEE 20th international conference on high performance computing and communications; IEEE 16th international conference on smart city; IEEE 4th international conference on data science and systems (HPCC/SmartCity/DSS). Exeter, UK, pp 1491–1496. https://doi.org/10.1109/HPCC/SmartCity/DSS.2018.00246
4. Pandey S, Twala B, Singh R, Gehlot A, Singh A, Montero EC, Priyadarshi N (2022) Wastewater treatment with technical intervention inclination towards smart cities. Sustainability 14:11563. https://doi.org/10.3390/su141811563
5. Nutrient removal from domestic wastewater: a comprehensive review on conventional and advanced technologies. J Environ Manag. Published 13 July 2021. https://www.sciencedirect.com/science/article/abs/pii/S0301479721013086. Accessed 10 Oct 2023
6. Drexler ILC, Prieto AL, Yeh D (2014) Wastewater constituents. In: Ahuja S (ed) Comprehensive water quality and purification. Elsevier, pp 7–29. ISBN 9780123821836. https://doi.org/10.1016/B978-0-12-382182-9.00043-8

7. Wastewater treatment and water resource recovery facilities (WRRFs). WBDG. Published 09 Oct 2020. https://www.wbdg.org/resources/wastewater-treatment-and-water-resource-recovery-facilities-wrrfs. Accessed 10 Oct 2023
8. Tang L, Xu L, Li X, Li Y, Wang X, Li S (2021) Fate of microplastics in wastewater treatment plants. Sustain Water Resour Manage 7(2):19. https://doi.org/10.1007/s40899-021-00509-5
9. Jahan MB, Sumon MTZ, Shahin MMK, Rahman MH (2018) Water and sanitation: a case study of Rohingya Camps in Cox's Bazar. In: IOP conference series: materials science and engineering, vol 365, no 2, p 022047. https://doi.org/10.1088/1757-899X/365/2/022047
10. Zhao L, Dai T, Qiao Z, Sun P, Hao J, Yang Y (2020) Application of artificial intelligence to wastewater treatment: a bibliometric analysis and systematic review of technology, economy, management, and wastewater reuse. Process Saf Environ Prot 133:169–182
11. Hamood WA et al (2022) The role of conventional methods and artificial intelligence in wastewater treatment: a comprehensive review. Processes 10(9):1832
12. Kshirsagar PR, Tirth V, Islam S, Qaiyum S, Al Duhayyim M, Waji YA (2022) IOT based smart wastewater treatment model for industry 4.0 using artificial intelligence. Sci Program
13. Wang Y, Cheng Y, Liu H, Guo Q, Dai C, Zhao M, Liu D (2023) A review on applications of artificial intelligence in wastewater treatment. Sustainability 15(18):13557
14. Salgado EM, Esteves AF, Gonçalves AL, Pires JC (2023) Microalgal cultures for the remediation of wastewaters with different nitrogen to phosphorus ratios: process modelling using artificial neural networks. Environ Res 231:116076
15. Malviya, Jaspal D (2021) Artificial intelligence as an upcoming technology in wastewater treatment: a comprehensive review. Environ Technol Rev 10(1):177–187
16. Zhang S, Jin Y, Chen W, Wang J, Wang Y, Ren H (2023) Artificial intelligence in wastewater treatment: a data-driven analysis of status and trends. Chemosphere 139–163
17. Matheri N, Mohamed B, Ntuli F, Nabadda E, Ngila JC (2022) Sustainable circularity and intelligent data-driven operations and control of the wastewater treatment plant. Phys Chem Earth Parts A/B/C 126:103152
18. Hossain SZ, Sultana N, Jassim MS, Coskuner G, Hazin LM, Razzak SA, Hossain MM (2022) Soft-computing modeling and multiresponse optimization for nutrient removal process from municipal wastewater using microalgae. J Water Process Eng 45:102490
19. Fathima J, Chatterjee P (2022) A techno-economic assessment of nutrient recovery from wastewater using microalgae: scenario in India collected from published literature. Water Sci Technol 86(6):1325–1341
20. Atal Innovation Mission (AIM). https://www.aim.gov.in/images/Waste-Water-ver2_18102022.pdf. Accessed 10 Oct 2023
21. Hakak S, Khan WZ, Gilkar GA, Haider N, Imran M, Alkatheiri MS (2020) Industrial wastewater management using blockchain technology: architecture, requirements, and future directions. IEEE Internet Things Mag 3(2):38–43
22. Sundui B, Ramirez Calderon OA, Abdeldayem OM, Lázaro-Gil J, Rene ER, Sambuu U (2021) Applications of machine learning algorithms for biological wastewater treatment: updates and perspectives. Clean Technol Environ Policy 23:127–143
23. Sakiewicz P, Piotrowski K, Ober J, Karwot J (2020) Innovative artificial neural network approach for integrated biogas–wastewater treatment system modelling: effect of plant operating parameters on process intensification. Renew Sustain Energy Rev 124:109784
24. Wastewater: a resource that can pay dividends for people, the environment, and economies, says World Bank. World Bank. Published 20 Mar 2020. https://www.worldbank.org/en/news/press-release/2020/03/19/wastewater-a-resource-that-can-pay-dividends-for-people-the-environment-and-economies-says-world-bank. Accessed 10 Oct 2023

Chapter 5
Intelligent Data Acquisition and Analytics for Enhanced Wastewater Treatment Efficiency

J. Anuradha, Prashantkumar B. Sathvara, Sandeep Tripathi, and R. Sanjeevi

Abstract Intelligent data acquisition and analytics (IDA&A) is the use of sensors, data loggers, and other technologies to collect data from wastewater treatment plants (WWTPs) and then use advanced analytics to improve the effectiveness of the treatment procedure. IDA&A can be used to monitor a variety of parameters in a WWTP, like Flow rate, Temperature, pH, Turbidity, Biological oxygen demand, Volatile suspended solids, Nutrients (nitrogen and phosphorus), Chemical oxygen demand. By collecting data on these parameters, IDA&A can help operators to identify problems early on and take corrective action before they cause major disruptions to the treatment process. For example, if the flow rate suddenly increases, IDA&A can alert operators to the potential for overflows or spills. Or, if the pH level drops too low, IDA&A can warn operators that the treatment process is not working properly and corrective action is needed. In addition to monitoring parameters, IDA&A can also be used to optimize the treatment process. For example, IDA&A can be used to determine the optimal dosage of chemicals to be used in the treatment process. Or, IDA&A can be used to identify opportunities to reduce energy consumption or water usage. IDA&A is a controlling tool that can help to improve the efficiency and sustainability of WWTPs. By collecting and analyzing data, IDA&A can help operators to identify problems early on, optimize the treatment process, and reduce costs. The Public Utilities Board is using IDA&A to monitor the performance of its WWTPs. The data collected is used to recognize parts where the treatment process can be enhanced, such as by optimizing the dosage of chemicals or reducing energy consumption. The metropolitan city is using IDA&A to monitor the quality of its recycled water. The data collected is used to ensure that the water meets the standards for drinking water. The National Environmental Engineering Research Institute is using IDA&A to develop new wastewater treatment technologies. The data collected is used to optimize the performance of these technologies and to ensure that they are

Present Address:
J. Anuradha · P. B. Sathvara · S. Tripathi · R. Sanjeevi (✉)
Department of Environmental Science, Nims Institute of Allied Medical Science and Technology, NIET, Nims University Rajasthan, Jaipur, India
e-mail: r.sanjeevi@nimsuniversity.org

M. C. Garg (ed.), *The AI Cleanse: Transforming Wastewater Treatment Through Artificial Intelligence*, Springer Water, https://doi.org/10.1007/978-3-031-67237-8_5

effective in removing pollutants from wastewater. IDA&A is a rapidly developing field, and new applications for this technology are being developed all the time. As IDA&A becomes more sophisticated, it is projected to play a growing role in enhancing wastewater treatment efficiency and sustainability.

Keywords Artificial intelligence · Wastewater treatment plant · Decision support systems · Process control

5.1 Introduction

Water resources, which are critical to human survival and manufacture, are now under risk from severe contamination caused by anthropological actions and natural practices. Every day, a significant quantity of wastewater is created; the majority of it includes hazardous chemicals, and it is dumped into the environment untreated or without being recycled [1]. Untreated sewage has a high concentration of organic materials, nutrients, organic micro pollutants, suspended solids, pathogenic and nonpathogenic microorganisms, and other contaminants. Wastewaters may be divided into six groups based on their source: municipal, residential, Medical, industrial, agricultural, and nuclear applications are all possible. Metropolitan and native wastewaters are the most common, and most WWT study is focused on these two types of sewage [2]. It is crucial to evaluate the characteristics of the wastewaters from various sources before deciding on the best treatment method since the physical and chemical parameters of the wastewaters might vary greatly. It is critical to research wastewater treatment and reuse as a resource in order to meet growing water demand while safeguarding finite the environment, water resources, and human health. Wastewater treatment (WWT) uses a mix of physical, biological and chemical processes to eliminate toxins from sewage and manufacture clean water that may be safely returned to the environment [3]. How to effectively decrease water contamination to an acceptable level though creating fewer undesirable environmental consequences and lowering energy consumption are the main concerns of WWT. Overall, WWT has been extensively utilized in the industrial and agricultural sectors and is a crucial procedure for the reuse of water resources and sustainable development. It also protects the environment and the public health. Technologies that are efficient and cutting-edge are urgently required to increase WWT's productivity, lower costs, and minimize energy usage [4–6].

Artificial intelligence (AI) is a computer program's ability to replicate human intelligence and accomplish autonomous learning, reasoning, judgment, and decision-making. AI, one of the most spectacular innovations of our century, is advancing quickly and is being used widely in numerous fields, including autopilot, computer vision, and natural language processing (NLP) [7]. Because of its tremendous efficiency, AI can be used to classify and analyse regression patterns in vast volumes of data that are produced everywhere and at any time, boosting businesses and considerably advancing the growth of all spheres of life. Machine learning (ML), a key area of artificial intelligence (AI), Data, methods, statistics, and mathematical optimization are used to simulate how persons learn, progressively enhancing its correctness and approaching artificial intelligence. With the introduction of the large dataset age and the advancement of supercomputing capacity, ML has been increasingly popular and successful in an extensive range of manufacturing, including manufacturing, cultivation, health, preservation of the environment, and scientific research [8, 9].

Major components of the WWT process are process design, laboratory-scale research, and water quality monitoring. Particularly in recent years, AI models are suitable more widespread in wastewater connected fields, where they are employed for estimate and optimization of the Wastewater system. In previous WWT study, AI simulations were successfully employed for WWT process development, water standard monitoring, WWT procedure parameter optimization, and WWT process efficiency prediction [10]. These models have demonstrated extremely strong prediction and optimization performances. This research have demonstrated that a model based on artificial intelligence, as a powerful tool, has been tremendously effective in domain-relevant WWT applications. However, the bulk of review publications on AI uses for WWT application on specific methodologies or procedure design features, such as adsorbent processes, membrane-based, membranes bioreactors procedures, and WWTP. Until now, there hasn't been an in-depth investigation of artificial intelligence presentation in the WWT process, which involves water-related observing, lab tests, and actual process patent. Furthermore, there aren't many reviews that fully show the most common AI models in wastewater-treatment and identify their compensations, drawbacks, and suggestions [11, 12].

5.1.1 AI Models

Figure 5.1 depicts the most often used artificial intelligence (AI) models for the WWT in the source material. Artificial Neural Networks (ANN), Machine Learning (ML), and Search Algorithms (SA) constitute the three major categories that may be used to categorize these prediction and optimization models.

5.1.1.1 Artificial Neural Networks

Modern machine learning and artificial intelligence systems rely heavily on artificial neural networks (ANNs). The organization and operation of organic neural networks, such as the human intelligence, serve as inspiration for these computational models. Interconnected nodes, or neurons, are arranged into three layers in ANNs: input, hidden, and output. In order to process information, each neuron in an ANN performs weighted sums of its efforts and applies an stimulation purpose [13]. The network learns to modify these weights and biases during training in order to provide precise predictions or classifications depending on the input data. Back

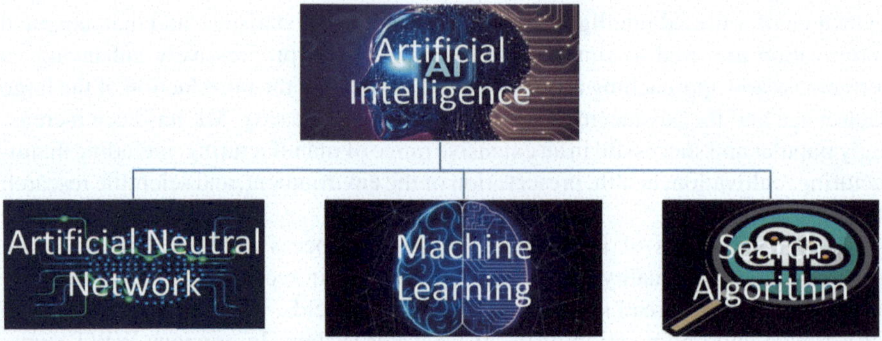

Fig. 5.1 AI model classification for WWT

propagation is a common method for achieving this learning process, in which the network iteratively adjusts its parameters by minimizing a predetermined loss function. Neural networks based on artificial intelligence have shown to be highly effective in a range of applications, including recognition of images, processing natural languages, recommendation systems, autonomous vehicles, and others. Computer vision and speech recognition are only two examples of the domains that Deep Learning, a subclass of ANNs with many hidden layers, has revolutionized. Recurrent Neural Networks (RNNs) thrive with chronological data, such as time sequence or language sequences, whereas Convolutional Neural Networks (CNNs) specialize in jobs involving grid-like data, such as pictures. With continued research, network structures, training methods, and applications for ANNs are getting better [14, 15]. They are an essential component of AI technology due to their capacity to understand intricate patterns and create data-driven predictions (Fig. 5.2).

Fig. 5.2 Artificial neural networks

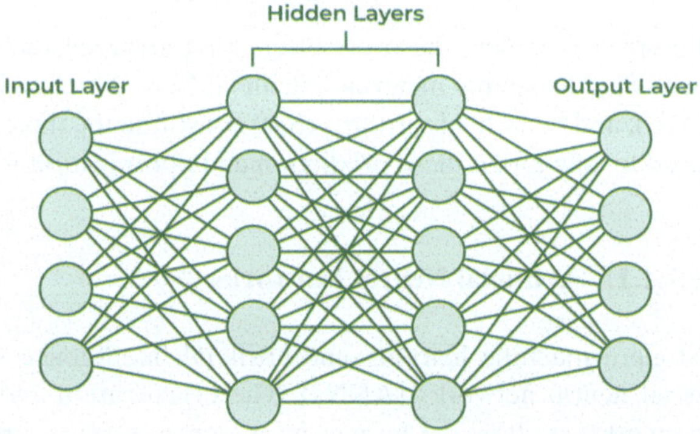

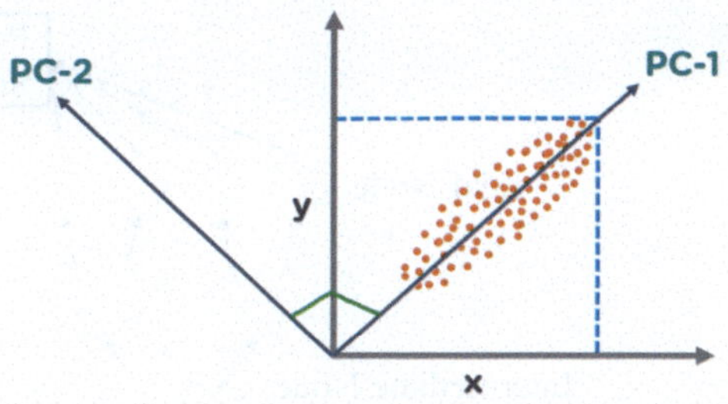

Fig. 5.3 Principal component analysis

5.1.2 ML Models

5.1.2.1 PCA (Principal Component Analysis)

PCA, or principal component analysis, is a common predictive model and reduction of dimensionality approach. Its primary purpose is to minimize data loss while extracting the most critical information from large datasets. PCA is very useful in a variety of disciplines, including as image processing, biology, finance, and more. When using PCA, the original data is converted into a new coordinate system, with the first principle component (PC) capturing the majority of the variation. Increasingly less variation is captured by subsequent components, which maintain their orthogonal (uncorrelated) relationships with the preceding ones. The underlying structure and patterns in the data are effectively represented by these primary components [16]. PCA is frequently used to reduce data dimensionality, which may be used to streamline complicated datasets and enhance the efficiency of machine learning algorithms. Researchers and data scientists can simplify computations and eliminate noise or redundant data by keeping just the top main components that account for the majority of variation. PCA is essential for data visualization, too. Data exploration and understanding are aided by the visualization of data clusters, patterns, or outliers, which is made possible by projecting high-dimensional data into a lower-dimensional environment (such as 2D or 3D). In conclusion, PCA is an effective method for lowering the number of dimensions in data, improving data visualization, and spotting crucial trends in datasets. It has applications in many different industries where effective machine learning and data-driven decision-making depend on data analysis, visualization, and simplification [17, 18] (Fig. 5.3).

5.1.2.2 Decision Tree (DT)

Decision Trees are useful in both basic machine learning and complicated problem-solving contexts because of their famed simplicity and interpretability. Here is how they function: Construction of trees: Choosing the optimal feature to divide the data into subsets is the first step in the process, which is often done by optimizing a measure of impurity (for example, Gina impurity for classification or mean squared error for

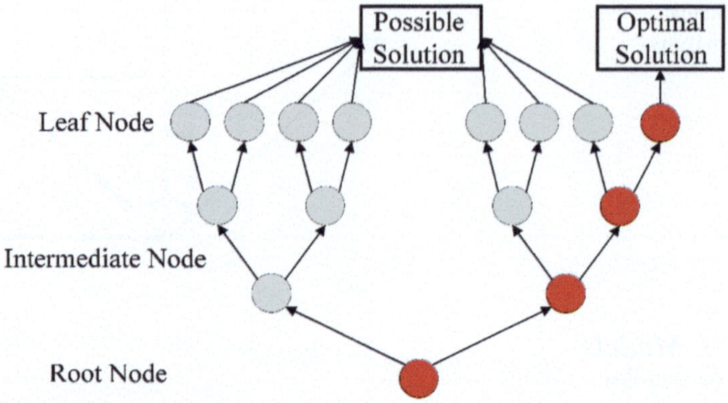

Fig. 5.4 Decision tree

regression). This characteristic turns into the tree's trunk. The values of the selected feature are used to create child nodes that are connected to the root by constructing subsets of the data. A branching structure results from this procedure repeating for each child node. A preventing requirement, such as a maximum seriousness, a minimum amount of samples per leaf, or the point at which additional splitting does not appreciably reduce impurity, is then reached. Based on the dominant class or the mean value of the training samples in each leaf node, a class tag (in classification) or a statistical value (in regression) is assigned to each leaf node [19, 20]. Decision because it's simple to follow the pathway from the origin to a leaf to understand why a specific prediction was made, trees are interpretable. However, if the tree is too deep and begins to capture data noise, they may be vulnerable to over fitting. Techniques like pruning and the use of ensemble methods like Random Forests are frequently used to alleviate this. Decision trees may be used for a variety of activities, such as fraud detection, consumer segmentation, and illness diagnosis, in a variation of manufacturing, including medical, economics, and marketing (Fig. 5.4).

5.1.2.3 SVM (Support Vector Machine)

An effective and adaptable machine learning model this is a classification and regression issue Support Vector Machine (SVM). It works especially effectively when there are issues with distinct class borders. Finding the appropriate hyper plane to divide data points into discrete classes while maximizing the margin between them is the goal of SVMs. The boundary is the parting among each class's closest data opinions and the hyper plane. A greater margin often results in higher generalization and enhanced classification performance, hence SVMs try to maximize this margin. Both linearly separable and non-linearly separable data may be handled by SVMs. This is accomplished by the use of kernel purposes such as radial and polynomial foundation functions, to convert the dataset into a higher-dimensional space and

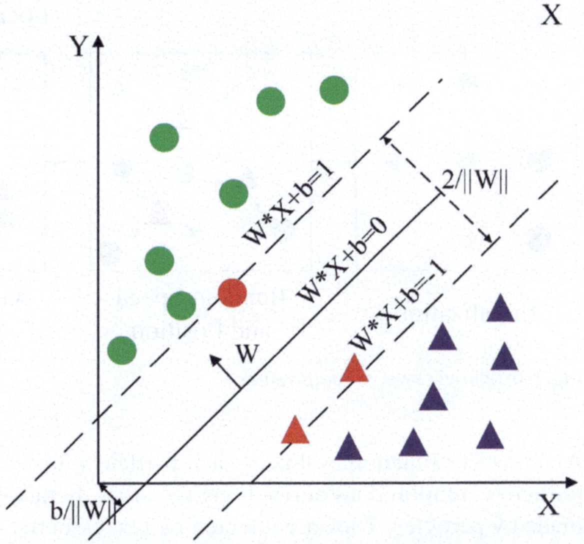

Fig. 5.5 Support vector machine

enable the discovery of a hyper plane that can divide classes that are not separable linearly. Support vectors are the dataset points neighboring to the point of conclusion boundary (the hyper plane). They are very important in figuring out where and how the hyper plane is oriented. To manage the trade-off amongst exploiting the margin and reducing classification errors, SVMs integrate regularization parameters (C in the case of soft-margin SVMs). This makes it possible to modify the behavior of the model. Due to its capacity to manage high-dimensional data and intricate decision boundaries, SVMs are widely employed in a variability of applications, including text organization, picture grouping, bioinformatics, and financial prediction. They are renowned for their dependability and efficiency in settings with few information. Even though SVMs have many advantages, choosing the right kernel functions and regularization parameters may be difficult and computationally expensive for very big datasets. However, SVMs continue to be a useful tool in the machine learning arsenal, particularly when trying to identify a distinct separation border in the data (Fig. 5.5).

5.1.2.4 PSO (Particle Swarm Optimization)

A population-based optimization technique called particle swarm optimization (PSO), is stimulated by the swarm behavior of fish or birds. It is a flexible and effective technique used in machine learning and optimization to locate the best answers in challenging search areas.

To tackle the optimizations problem, PSO mimics a population of particle moving around the search space. This is how it goes: Initialization: The search space contains a people of units, each of which characterizes a potential answer to the problem.

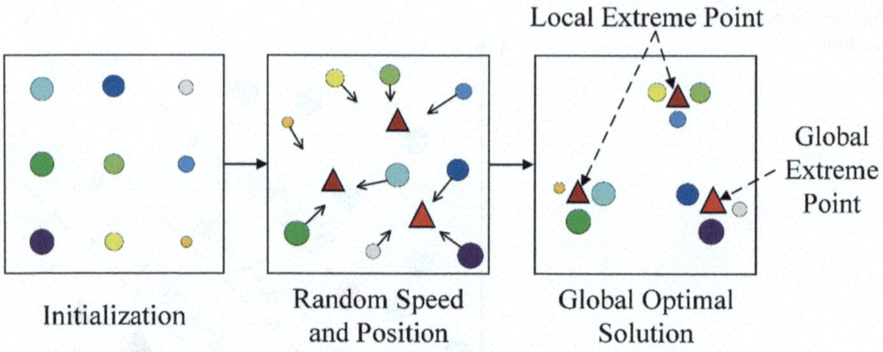

Fig. 5.6 Particle swarm optimization

An objective function evaluates each particle's fitness by quantifying how well the particle's solution does at resolving the issue. Iterative positioning adjustments are made by particles. They are directed by two essential bits of knowledge, much like how each particle keeps track of the best answer it has so far discovered [21, 22]. The top response obtained by any particle over the whole population. Particles change their velocities after each iteration according to their pBest, gBest, and current location. They move closer to potential search space areas thanks to this modification. Until a convergence requirement, when a condition, such as obtaining an ideal degree of fitness or a maximum amount of iterations, is met, the procedure runs for a particular amount of iterations. The control of PSO rests in its volume to strike a balance between exploitation (improving on promising answers) and exploration (finding new areas of the search space). It is very useful in situations where gradient-based approaches are either inapplicable or inefficient for continuous and multi-modal optimization issues. Function optimization, neural network training, feature selection, and hyper parameter tweaking are just a few of the many domains in which PSO has applications. It is an important tool in the toolbox of machine learning practitioners and academics due to its simplicity, versatility, and efficiency (Fig. 5.6).

5.1.3 Search Algorithm (SA)

In computer science and artificial intelligence, a search algorithm (SA) is a methodical methodology used to locate a particular target among a huge dataset or search area. When trying to obtain information on the internet, find your way through a maze, or optimize a complicated system, search algorithms come in handy. A search algorithm's principal goal is to swiftly identify an item of interest or solution, frequently by meticulously examining resources while lowering the number of assessments or steps required. The following are important search algorithm traits and categories: If the target is there in the search space, deterministic search algorithms are assured to discover it since they adhere to a predetermined set of criteria.

On the other hand, heuristic search algorithms make use of informed assumptions or heuristics to direct the search, frequently compromising optimality for speed in complicated, expansive search environments. Uninformed search algorithms scour the search space without any prior knowledge of the target's location. In order to make better judgments throughout the search, informed search algorithms incorporate extra information like estimations or problem-specific heuristics. Breadth-first search (BFS), Depth-first search (DFS), Binary search, linear or sequential search, and others are examples of common search types. Each kind has benefits and is appropriate for various problem domains. Time and space complexity, optimality, and completeness are common metrics used to assess a search algorithm's effectiveness. These elements aid in assessing the algorithm's appropriateness for a certain job [23, 24]. Numerous fields, including information retrieval, data mining, path finding in robots and video games, natural language processing, and optimization issues, use search methods. The kind of the problem, the scope of the search, and the computational resources at hand are a few more factors that should be taken into account when deciding which search method to utilize.

5.1.4 Genetic Algorithm (GA)

A strong optimization and search method influenced by the ideas of natural selection and genetics is known as a genetic algorithm (GA). When standard optimization techniques may not be as successful, GAs are employed to identify approximations of solutions to complicated optimization and search issues. A genetic algorithm functions as follows: The method begins with a population of individual possible solutions, which are frequently produced randomly or using some sort of heuristic. Every person stands for a potential answer to the optimization issue. Each person's fitness is assessed by considering how effectively they are able to tackle the task. A better solution is indicated by a higher fitness score. To create a new generation, people are chosen from the present population, with a larger likelihood of selection for those who are more fit. This procedure simulates how advantageous features are naturally selected. Through methods like one-point or two-point crossover, chosen people (parents) are united to generate children [25]. In order to produce possibly better solutions, this integrates the genetic information of the parents. In order to add variety and explore new areas of the solution space, random changes (mutations) are occasionally introduced into the genetic code of the progeny. The preceding generation is replaced by the current generation of people (offspring). Some tactics can entail retaining some of the top performers from the previous generation. The method iteratively repeats the selection, crossover, mutation, and replacement processes until a termination requirement is satisfied (such as a good solution is identified), or for a predetermined number of generations. When there are several local optima in the search space or when there are many huge solution spaces, GAs are very successful at solving difficult optimization issues. They are used for a variety of tasks in machine

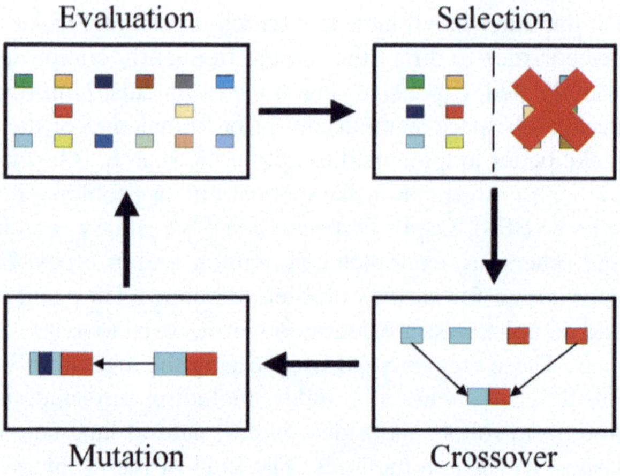

Fig. 5.7 Genetic algorithm

learning, such as feature selection, hyper parameter tuning, and neural network architectural search, as well as engineering design, financial portfolio optimization, evolutionary biology, and other fields. In order to effectively explore and utilize the solution space in search of nearly optimum solutions, GAs harness the power of evolution (Fig. 5.7).

5.1.5 Genetic Programming (GP)

A potent evolutionary technique used for automatic development and evolution of computer programmes or mathematical expressions is known as genetic programming (GP). It expands on the ideas of genetic algorithms (GAs) to grow populations of computer programmes in search of methods to solve complicated problems. A population of randomly chosen or predetermined computer programmes, frequently represented as syntax trees, is the basis of GP [26]. Every programme stands in for a prospective fix for the issue. Running each programme on a set of test inputs and comparing the output to the expected outcomes allows for the evaluation of its fitness. Programmes with superior results earn higher fitness ratings. To act as parents for the coming generation, programmes are chosen from the existing population. Selection usually favors programmes that deliver higher results and is inversely correlated with fitness. Crossover procedures are used to merge pairs of chosen programmes. In order to generate new programmes containing components from both parents, this entails switching the sub trees of the syntax trees. The syntax trees of programmes occasionally undergo random modifications or mutations, adding genetic variety to the population. The old generation of programmes is replaced by the current generation, which contains hybrid offspring and potentially altered programmes. The GP process continues until a termination requirement is satisfied, such as discovering a workable solution or hitting a computational limit, or for a predetermined number of

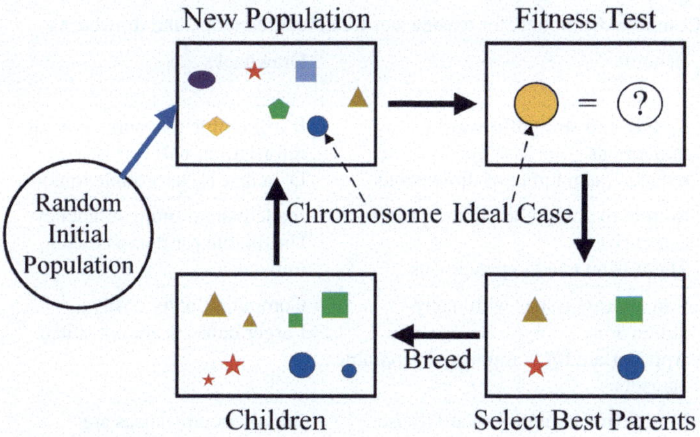

Fig. 5.8 Genetic programming

generations. In issues like symbolic regression, developing mathematical models, or evolving control techniques for autonomous systems, where the structure and form of the answer are not known in advance, genetic programming is very helpful. It is utilized in areas like symbolic regression, autonomous code creation, machine learning, and evolutionary robotics and is capable of producing very complicated and optimized solutions [27]. GP uses evolutionary concepts to develop programmes through generations, enabling it to find fresh approaches to difficult issues (Fig. 5.8 and Table 5.1).

5.1.6 Artificial Intelligence Model Applications in Wastewater Treatment

Designing smart WWT systems and reusing sewage benefits since the use of AI models, the Internet of Things, also known as the IoT framework, and conventional techniques. An AI model is a controlling applications for simulation, forecast, and optimization of the wastewater treatment process [31]. It has a wide range of applications in wastewater treatment, including removing of colors, heavy metals, vitamins and minerals, organic matter, solids, microbiological contaminants, medications, and insecticides from water. From a research standpoint, the primary presentations of AI models are in design of processes and laboratory-scale research. In practice, process design frequently comprises the optimization of development factors as well as the forecasting of procedure performance [32–34]. Water quality monitoring may be used to collect the massive amounts of data necessary to build and train AI models for optimization and prediction.

Table 5.1 Common AI models for wastewater treatment, benefits, and drawbacks

AI models	Benefits	Drawbacks	Ref.
PCA	Simple and straightforward to implement reduces the number of dimensions	It is possible that some critical information will be lost Data that is susceptible to noise	[28]
DT	Simple to comprehend, interpret, and categorise There is no need to pre-process	Inadequate training efficiency Unsuitable for datasets with imbalances	[28]
SVM	Can address issues with many dimensions Appropriate for complicated separable datasets	Computationally costly Larger datasets are not suited	[29]
PSO	Simple and straightforward to use Outstanding computational efficiency Broad applicability	Initial circumstances are important Not appropriate for discrete tasks	[30]
GA	Appropriate for difficult nonlinear issues Multi-objective optimisation is supported Flexible and efficient	It is difficult to train Inadequate local search capability High dimensions are not recommended	[30]
GP	Appropriate for complicated optimisation tasks Automatic search optimisation	There are several control factors Slowly come together High dimensions are not recommended	[29]

5.1.7 Monitoring Water Quality for Gathering Information

Monitoring the water's quality is an important method for gathering water quality measures or information for the use of artificial intelligence (AI) models in WWT [27]. When sensors are used to continually check the quality of influent and effluent water, a lot of data is accessible. Although various research use varied quantities of datasets as inputs, the amount of data used to train and evaluate AI models nearly always stays the same. The majority of studies employ 60–80% of the training information and the rest to be evaluated. Researchers attempted to develop a range of sensors that would enable rapid and accurate real-time monitoring, as well as the automation of WWT operations via real-time sensing, data processing, and online controls [35–37]. While it is simple to obtain and use certain Water quality parameters for monitoring influent water, including DO, COD, BOD, flow-rate, temperatures, and preliminary pollutant quantity, are usually employed to assess the effectiveness of wastewater treatment plants, while waste product, Chemical Oxygen Demand, pH, Dissolve Oxygen, and levels of pollutants are generally used to assess the achievement of wastewater management plants or the impact of WWT [38, 39].

Dataset-driven demonstrating and real-time prediction have become more appealing as a result of the proliferation of measurable data, AI models, and multivariate statistical techniques. With a correlation coefficient of $R^2 = 0.74$ for all samples, Post employed a CNN model in conjunction with laser-induced Raman and fluorescent spectroscopy (LIRFS) to offer present monitoring of WWTP micro pollutants [28]. The findings demonstrate that this approach may produce measurements with high accuracy, achieve detection limits, and identify micro pollutants that cannot be tracked using the observing techniques used by WWTPs. The grouping of highly complex fluorescent measurements with highly accurate Raman measures enhanced with AI is a capable continuous observing instrument for acknowledgment of images in WWT-related fields such as microbiology in order water quantity tests, micro pollutants identifiers, and even device-specific modifications in WWTP administration. To observer functioning stations and provide essential evidence wanted for successful operation from specialists, a present remote surveillance structure for the wastewater treatment plant based on PCA, ANN, plus multidimensional statistical control techniques was created [40, 41]. Examined high-accuracy uses of internet of things and artificial intelligence algorithms for water quality surveillance and prediction may offer customers services for safe water quality and serve as a crucial foundation for government decision-making about water quality management. Conducted a literature research on the utilization of data derived soft sensor technology for biological WWTP monitoring and real-time forecasting in order to compile the applications [42, 43]. A comprehensive assessment of advances in easy devices for online observing of a wastewater treatment plant was also carried out. They carried out an experiment in order to find sensors to monitor on-site the wastewater treatment plant without sensor maintenance for a year, and they showed that vigorous soft sensors can be logically built to happen immediate surveillance demands while drastically reducing maintenance frequency [44].

5.1.8 Laboratory-Scale Research

Because of its inexpensive price tag, short experiment length, security management, simplicity of functioning, and high repeatable, laboratory-scale study has proven a critical apparatus for developed novel methods and wastewater treatment technologies. We arrange the most prevalent applications of AI in small-scale investigation in laboratories. AI models in WWT are generally utilized in laboratory-scale research to build and optimize membrane processes or bioreactors [43, 45, 46]. Recent studies have employed membrane technologies or bioreactors with membranes (MBRs), and some scientists have created comprehensive assessments of the usage of artificial intelligence (AI) models for WWT using these approaches. MBR is an effective and cost-effective wastewater treatment technique that combines membrane filtering with biological (microbial) treatment. It's a mixture system that combines a classic organic therapy system with acute physical aqueous-solid separation activities performed by membrane filter technology. Because it associations the technological benefits of membrane-based segregation of matter with biodegradation through microorganism, MBR offers several benefits over the standard activated sludge method [47]. These benefits include higher biomass concentration, reduced sludge generation, a shorter time for hydraulic retention (HRT), the elimination of the need for additional clarifiers, a less plant area need, and enhanced effluent value are all benefits of this technology [48, 49]. As a consequence, MBR, among the most important recent developments in WWT, is a valuable instrument for long-term wastewater management and is currently broadly used to the process a varied range of industrial and municipal wastes [50]. Have reviewed the development of MBR technology historically towards sustainable wastewater treatment. The wastewater treatment Anaerobic MBR (AnMBR) technology was described by Tomczak and Gryta, who also showed that AnMBRs require less energy than conventional WWTPs [51].

Many different aerobic and aerobic MBRs have been advanced and utilized in treatment plant, however one of the major barriers to broad usage of membrane-based processes or MBRs is fouling of the membranes, which affects membrane performance and lifetime while increasing maintenance and operating expenses. Membrane fouling, a complex and inevitable phenomenon caused by the gathering and adsorption of contaminants in sewage on the filter membrane's surface as well as inside the membrane openings, reduces membrane permeability [52, 53]. Researchers have been working to create effective and long-lasting fouling management solutions since, Overall, fouling of the membrane is the most significant issue limiting membrane process performance and cost [54, 55]. Using 1% NaOH solutions, Tomczak, Gryta, and Grubecki developed a technique for preventing membrane catching in an MBR and showed that it is successful in recovering the initial membrane performance [56]. Two of the most pressing concerns in membrane technology or MBR-related application research are how to minimize membrane fouling and adjust MBR efficiency for large-scale uses in WWT in an efficient and cost-effective manner, as well as how to achieve rapid commercialization [57].

The most often explored membrane types for usage of AI models to anticipate or regulate membrane fouling include forward and reverse osmosis, ultrafiltration, micro filtration and Nano filtration [58]. A centrifugal basis functions (RBF) ANN model was used to assess interfacial energy with an arbitrarily uneven membrane area throughout the membrane ensnaring process [59]. They demonstrated how the RBF-ANN model is capable of accurately capturing the complicated relationships between surface energy and important influencing factors. In order to forecast With high precision, permeate flux can be measured for a laboratory-scale forward osmosis process, $R^2 = 0.973$, Jawad utilized an ANN model [60]. The findings show demonstrates the ANN model beats the MLP approach, and that adding fewer hidden layers along with additional neurons improves the ANN model's accuracy. They later developed an integrated ANN-RSM model to anticipate membrane flow and simulate the forward osmotic process. The RSM model is used for optimizations in the mixture model, while the ANN model for forecasting membrane flow is employed for investigational design [61]. RSM and ANN models' respective prediction performances are $R^2 = 0.98036$ and 0.9408.

5.1.9 Process Parameters Optimization

Process parameter optimization's primary goal is to save costs and boost WWT effectiveness [62]. The uses of AI models for process parameter optimization are compiled. The effects show that the ANN-PSO system overtakes the added models in determining the optimal conditions for maximum decontamination efficacy. The mesoporous nano hybrids might be employed to remediate other pollutants or organic contaminants in wastewater at a cheap cost and in a short amount of time. By tracking Based on the Oxidation–Reduction Potential (ORP) and DO 3200 aeration-non-aeration cycles, Jaramillo-Morán and Martn de la Vega were able to discover crucial features of a municipal wastewater treatment plant. This technique can increase the effectiveness of nutrient removal in WWTP [30, 63].

Optimized the process parameters by utilizing an ANN-GA model to forecast efficacy of the electro-oxidation (EO) process for treating sulphate wastewaters containing bromophenol blue dye. They discovered that the AI model is a potent instrument for planning and managing the WWT procedures. ANN and ANFIS models were used to determine the best environments for MB elimination using less-cost crop residue (sugarcane the bagasse and peanut hulls) for MB WWT. They also implemented Fourier Transform Infra-Red (FTIR) spectroscopy to validate the biosorption and remarkable forecast abilities of these biosorption in AI models. In reality, FTIR is a valuable spectroscopic method that has been broadly used to identify and characterize contaminants such as table salt, nitrate from agricultural fertilizers, and microplastics in water. Although FTIR spectra interpretation takes time, it may be used to evaluate the element structure of pollutants and to recognize functional sets in chemicals, complex chemicals, and bio-sorbent surfaces, providing critical information about them. In an effort to quicken up the examination of functional networks and simplify the understanding of FTIR spectra, I established the first generalizable approach utilizing CNN to detect functional categories in gas-phase FTIR bands. The outcomes suggest that CNN model with a high number of spectrum examples may be used to new tiny particles or chemical detection application areas and are successful at recognizing spectral characteristics. In general, spectroscopic methods like FTIR, SEM, and Raman spectroscopy combined with image-based AI models are useful for identifying and detecting contaminants, particularly micro pollutants in water-related industries. A CNN, LSTM, and AM-based hybrid deep learning CLSTMA models for a WWT system to track and evaluate the water's quality of an industry WWT system for papers for environmentally conscious manufacture. The CLSTMA model outperforms the CNN, LSTM, and CLSTM models in In terms of water quality monitoring and modelling. Based on artificial intelligence (AI) models and sensors, we developed an intelligent the WWT system for regulating sewage processing. A Gated Recurrent Division, model outperforms SVR and LSTM models, and the sophisticated wastewater processing network may be broadened to include a smaller measure sewage enterprises in supportable communities. Furthermore, five machine learning (AI) models were examined for simulating and foreseeing carbohydrate biomass production in WWT networks while taking into consideration the relationships among vitamins and minerals, carbon, development of biomass, and populations. The outcomes reveal that the CNN-1D model outclasses other models and can approximate system dynamics. Employed DNN and LSTM for the research of WWTP to forecast the N_2O emission rate and pinpoint the critical factors influencing the features of N_2O high emission. They demonstrated that a mixed model that blends mechanical and DL models may be used to quantitatively describe and grasp complex N_2O emission patterns from waste water treatment plants. They create that the LSTM model outperforms the DNN model. Zhu, Jiang, and Feng proposed a revised fed-forward NN with a least squares SVM (FFNN-LSSVM) approach to forecast the wastewater BOD/NH_3-N of a wastewater treatment plan [64]. Because of its high prediction accuracy, fast computing time, and simple calculation technique, the proposed model surpasses existing approaches for forecasting wastewater quality.

5.1.10 Process Performance Prediction

The WWT process' crucial component of process performance has drawn academics' attention. The effectiveness of pollutants being removed, the performance of WWTPs, the ideal process condition, and effluent quality are among the usual process performance forecasts [65]. AI models for process performance forecasting. Used single AI and ensemble models to analyse Nicosia WWTP performance in order to forecast WWTP performance. According to the findings, Amongst the ensemble models, the NN model of ensembles has the best prediction performance, while the ANFIS model outperforms other single AI models [66]. In order to anticipate a WWTP's real-time effluent water quality in also coupled enhanced Feed forward Neural Network (IFFNN) with GA. They discovered that this model's prediction performance was increased when compared to the traditional FFNN model, by 52.3% (COD) and 72.6% (TN) [67]. Deep cascade forward propagation (DCB) and DL time series foretelling (DLTSF) algorithms were obtainable to predict the El-Berka WWTP's effluent water quality and assess its treatment efficacy. Trapa bispinosa peel was used to duplicate the adsorption procedure of the chemical (CPS) for the removal of contaminants, and the performance of adsorption was tested experimentally. According to the findings, UFBC has a low optimum time and an excellent capacity for adsorption for CPS removal, making it a practical and effective adsorbent. Suggested an evolving PCA-NARX model for waste quality prediction and suggested potential present modifications for WWTP tasks. The results display that the dynamical model beats static artificial neural network (ANN) models in terms of modelling effluent quality [68, 69]. Estimated chemical oxygen demand and CTSS removal effectiveness from fabric wastewater using multifaceted salt-Luffa cylindrica abstract of seeds (CS-LCSE) as a coagulant based on ANFIS, RSM, and ANN models. The ANFIS model provides the best predictive performance, with a greater R^2 value (0.9997 and 0.9996 for CTSS and COD removal) and a smaller MSE score (0.0002643 and 0.0038472 for CTSS and COD removals). To identify the optimal process conditions, an ANN model was used to prophesy the COD effectiveness of elimination from residential wastewater by producing Fe/Cu NPs under changed operational parameters [62]. According to the adsorbed isotherm, RSM and kinetic studies results, Fe/Cu Np are a highly effective adsorbent materials for COD elimination. The ANN model may be used to determine the best deletion situation.

Numerous According to research, the application of artificial intelligence in WWTPs may greatly increase WWT productivity and administrative, reduce impact on environmental, and increase performance. This is due to the ability of AI algorithms to optimize a wide range of processes and operating variables, including the pH level, temperature, COD, chemical dose, flow management, and energy utilization [70]. This improves system performance overall, results in more effective resource utilization, lower operational costs, and uses less energy [71].

Additionally, real-time monitoring and control may be achieved by combining AI technology with sensors, which reduces reaction times, the danger of system failure, and the need for additional human resources. Overall, compared to WWTPs without AI, those that use AI technology perform better. Increased productivity, the use of AI enables continuous tracking, proactive upkeep, dataset-driven decision-making, and reduced impact on environment. The employment of AI in wastewater treatment plants can prime to improved productivity and long-term water resource management [72].

AI models may be utilized in wastewater pretreatment to improve pretreatment accurateness and system adaptive regulator accuracy, as well as applications in process performance prediction and process parameter optimization. To properly examine the water's quality of reservoirs, lakes, and rivers, several researchers have used different AI models and their variations.

5.1.11 Challenges and Future Perspectives

All of the aforementioned research have shown that AI models offer numerous benefits over conventional models and have had tremendous success in WWT-related sectors, but it is important to recognize that these models also have drawbacks and restrictions that prevent wide-scale deployment in WWT. The following summarizes the difficulties and prospects for AI applications in WWT:

1. Although ANN, an AI model, may represent complicated nonlinear interactions between it is basically a black-box and data-driven architecture with several inputs and outputs. That is, the machine learning algorithm can only give the relationship between inputs and outputs; it cannot provide any process specifics on the issue under investigation. AI models have shown to be effective tools with promising futures in engineering applications for WWT areas, but there is still more to be done in terms of scientific study. The fundamental reason is that many WWT-related problems in the current study still lack defined underlying processes. Although the AI models employed in the aforementioned research

perform well in solving certain issues and often demonstrate problem dependency, it is necessary to investigate if They may be used to various WWTP-related situations, and their relevance in WWTP is broad [73]. More crucially, white-box models, which describe the fundamental mechanics but struggle to articulate the subtle nonlinear connections between AI model inputs and outputs, are classic mathematical models that make use of well-known information, concepts, and equations. A classical mathematical model (knowledge-driven model) and an artificial intelligence (AI) model (data-driven model) combined can significantly minimize the amount of data needed and make relevant conclusions more quickly and easily accessible [74]. The combined black and white-box model is a valuable device for investigating the underlying mechanisms of WWT systems.

2. Since the characteristics of AI models, although neurons, layers that remain hidden, weights, and biases have no physical worth, they are frequently interpreted poorly. As a result of the random biases and weights, poor repeatability is another disadvantage of AI models. Furthermore, it is challenging to train NN, particularly DNN, making it challenging to determine the ideal network parameters. A NN that has been badly trained could reach a local minimum. Generally speaking, NN offers several solutions when faced with various network designs and parameter settings. The finest network design to date cannot be determined in a conventional manner since it depends on the specific problems at hand. The only option seems to be trial and error, although this is prone to over fitting or under fitting [45]. The correct choice of AI models, efforts, outcomes, model construction, and dataset for a specific real-world situation is critical for the results. Further study on how to develop AI models is required. More theoretic research on AI practices is necessary to solve the issues of NN exercise, variable optimization, reduced repeatability, and limited understanding to further stimulate the applications development for AI in WWT.
3. Big data are required for the development or developmental phases of artificial intelligence (AI) models in order to assure prediction or optimization accuracy. AI models are applied to different sectors related to WWT in the previously mentioned WWT applications, and a vast amount of data in diverse formats and types has been obtained, resulting in poor data management and challenges in reuse by other investigators. Furthermore, academic transparency is lacking since the information and source code used in the studies are rarely made public for a number of reasons. Another reason why it is challenging to duplicate the findings in the literature is because of this. Furthermore, in practically every publication evaluated in this journal, researchers have employed statistical techniques to assess the precision or accuracy of their AI models. It is challenging to objectively assess the effectiveness of various AI models because there is no open source code or data available [14]. Because there are no comparisons, standards, or equitable comparisons accessible, it is difficult for researchers to establish which AI model works better in a particular real-world circumstance. As a result, raw data and source code should be made accessible and shared, and benchmarking and standardization should be created to reduce trial costs, generate fair comparisons, and encourage wider use of AI models in WWT.

5.2 Summary

The technology-driven method known as Intelligent Data Acquisition and Analytics (IDA&A) collects and analyses data from wastewater treatment facilities (WWTPs) using sensors, data loggers, and sophisticated analytics. This data-driven strategy plays a key role in raising wastewater treatment's general effectiveness and sustainability. In conclusion, IDA&A uses sensors and data loggers to gather information on a number of crucial WWTP characteristics, such as temperature, turbidity, pH, flow rate, volatile suspended solids, nutrients, and chemical oxygen demand. The information gathered is used to track the WWTP's performance over time and spot possible problems before they become serious. When pH levels depart from the intended range or flow rates increase unexpectedly, IDA&A can notify operators of possible issues, such as spills or overflows. Early problem identification enables operators to stop treatment process interruptions and take prompt corrective action. By figuring out the best chemical dose and seeing ways to cut back on energy and water use, IDA&A helps to optimize the treatment process. The WWTP operations' sustainability and efficiency are improved by this data-driven optimization. In order to improve the efficiency of WWTPs, the Public Utilities Board uses IDA&A, concentrating on maximizing chemical doses and lowering energy use. IDA&A is used by large communities to keep an eye on recycled water quality and make sure it complies with drinking water regulations. Research organisations that use IDA&A to develop and optimize novel wastewater treatment methods for efficient pollutant removal include the National Environmental Engineering Research Institute. The area of IDA&A is fast developing, with new applications and ongoing advancements. It is anticipated that as IDA&A technology advances, it will contribute significantly to improving the sustainability and efficiency of wastewater treatment systems. In the wastewater treatment industry, intelligent data acquisition and analytics, or IDA&A, is a crucial technology. It enables operators to monitor, identify, and resolve problems early, optimize treatment procedures, and save operating costs by utilizing data from sensors and advanced analytics. The practical applications in a range of industries highlight its increasing significance. The efficiency and sustainability of wastewater treatment systems are expected to be significantly and permanently improved as IDA&A develops.

5.3 Conclusions

This paper provides an overview of the most often applied Artificial intelligence models and their applications in WWTP, including process design, laboratory-scale research, and water quality monitoring. A growing number of WWT-related sectors are using AI models due to their superior learning and prediction capabilities. Modelling WWT systems, process parameter optimization, process performance prediction, and contaminant identification and detection have all been done with success. Although AI models are highly beneficial for treating wastewater and have numerous benefits, it is important to be aware of their drawbacks and limits. The necessity for large amounts of data, bad data administration, Poor interpretability, model consistency, and an absence of physical importance, mechanistic clarification, academic clearness, standardization, and fair assessment are major hurdles to the adoption of AI in critical WWT disciplines.

To overwhelmed these challenges and effectively apply artificial intelligence models to WWT, scientists, environmentalists, scientists, and workstation scientists should collaborate to develop novel approaches or cutting-edge technology. More study at all scales, from laboratory to field, is also necessary to recognize the difficult behaviors of WWT system with a range of impact factors and to examine the physics of the primary challenges related with WWT. In addition, cross AI models, which integrate the benefits of multiple AI models, as well as recently created attention-based AI models, are being developed, may offer answers to challenging issues in the field of water treatment. In sectors relevant to WWT, there is growing interest in a novel and promising approach called the combination of data-driven and knowledge-based AI models.

Ethical Acceptance Not Applicable.

Participant Consent Not Applicable.

Contributions of Authors Each author provided ideas and contributed to the study's design. Prashantkumar B. Sathvara, J. Anuradha, Sandeep Tripathi, and R. Sanjeevi prepared the materials, collected data, and analyzed the results. J. Anuradha and R. Sanjeevi wrote the paper's first draught, while other writers offered comments on earlier draughts. The final version has been examined and accepted by all writers.

Funding No Research grants from funding agencies.

Competing Interests Declaration The authors declare that none of the work revealed in this study seems to have been impacted by any known conflicting financial concerns or personal relationships.

Data and Resources Are Available Not applicable.

References

1. Ray SS, Verma RK, Singh A, Ganesapillai M, Kwon YN (2023) A holistic review on how artificial intelligence has redefined water treatment and seawater desalination processes. Desalination 15(546):116221
2. Boretti A, Rosa L (2019) Reassessing the projections of the World Water Development Report. npj Clean Water 31;2(1):1–6
3. Zaibel I, Arnon S, Zilberg D (2022) Treated municipal wastewater as a water source for sustainable aquaculture: a review. Rev Aquac 14(1):362–377
4. Malviya A, Jaspal D (2021) Artificial intelligence as an upcoming technology in wastewater treatment: a comprehensive review. Environ Technol Rev 10(1):177–187

5. Ma Z, Cheah WY, Ng IS, Chang JS, Zhao M, Show PL (2022) Microalgae-based biotechnological sequestration of carbon dioxide for net zero emissions. Trends Biotechnol 40(12):1439–1453
6. Impact of societal development and infrastructure on biodiversity decline methodology and approaches for biodiversity conservation. IGI Global, pp. 271–299
7. Safeer S, Pandey RP, Rehman B, Safdar T, Ahmad I, Hasan SW et al (2022) A review of artificial intelligence in water purification and wastewater treatment: recent advancements. J Water Process Eng 1(49):102974
8. Rathod Ab, Sathvara Pb, Tripathi A, Anuradha J, Tripathi S, Sanjeevi R (2022) Vegetational cartography analysis utilizing multi-temporal ndvi data series: a case study From Rajkot District (Gujarat), India. J Tianjin Univ Sci Technol 55(4):490–497
9. Rathod AB, Anuradha J, Sathvara PB, Tripathi S, Sanjeevi R (2023) Vegetational change detection using machine learning in GIS technique: a case study from Jamnagar (Gujarat). J Data Acquisit Process 38(1):1046–1061
10. Shirkoohi MG, Tyagi RD, Vanrolleghem PA, Drogui P (2022) Artificial intelligence techniques in electrochemical processes for water and wastewater treatment: a review. J Environ Health Sci Engineer. 20(2):1089–1109
11. Bhardwaj A, Kishore S, Pandey DK (2022) Artificial intelligence in biological sciences. Life 12(9):1430
12. Impact of societal development and infrastructure on biodiversity decline current scenario of biodiversity loss due to developmental activities. IGI Global, pp. 50–69
13. Basheer IA, Hajmeer M (2000) Artificial neural networks: fundamentals, computing, design, and application. J Microbiol Methods 43(1):3–31
14. Ding S, Li H, Su C, Yu J, Jin F (2013) Evolutionary artificial neural networks: a review. Artif Intell Rev 39(3):251–260
15. Sherstinsky A (2020) Fundamentals of recurrent neural network (RNN) and long short-term memory (LSTM) network. Physica D 1(404):132306
16. Samuel AL (1959) Some studies in machine learning using the game of checkers. IBM J Res Dev 3(3):210–229
17. Mamandipoor B, Majd M, Sheikhalishahi S, Modena C, Osmani V (2020) Monitoring and detecting faults in wastewater treatment plants using deep learning. Environ Monit Assess 192(2):148
18. Sensors for environmental monitoring identification and assessment water pollutants sensor types and their advantages and challenges. IGI Global, pp. 78–101
19. Han J, Pei J, Kamber M (2006) Data mining. Elsevier, Southeast Asia Edition
20. Sensors for environmental monitoring identification and assessment air sensors and their capabilities. IGI Global, pp. 10–24
21. Cura T (2012) A particle swarm optimization approach to clustering. Expert Syst Appl 39(1):1582–1588
22. Biodiversity loss assessment for ecosystem protection biodiversity and its importance an introduction. IGI Global, pp. 1–17
23. Ding W, Abdel-Basset M, Hawash H, Ali AM (2022) Explainability of artificial intelligence methods, applications and challenges: a comprehensive survey. Inf Sci 1(615):238–292
24. Impact of climate change and its importance on human performance. In: Insights on impact of climate change and adaptation of biodiversity, 1st edn. KD Publication, pp. 1–9
25. Scientific American [Internet]. [cited 2023 Oct 10]. Genetic algorithms. Available from https://www.scientificamerican.com/article/genetic-algorithms/
26. Espejo PG, Ventura S, Herrera F (2010) A survey on the application of genetic programming to classification. IEEE Trans Syst Man Cybernet Part C (Applications and Reviews) 40(2):121–144
27. Singh NK, Yadav M, Singh V, Padhiyar H, Kumar V, Bhatia SK et al (2023) Artificial intelligence and machine learning-based monitoring and design of biological wastewater treatment systems. Biores Technol 1(369):128486

28. Lee MW, Hong SH, Choi H, Kim JH, Lee DS, Park JM (2008) Real-time remote monitoring of small-scaled biological wastewater treatment plants by a multivariate statistical process control and neural network-based software sensors. Process Biochem 43(10):1107–1113
29. Zhu J, Jiang Z, Feng L (2022) Improved neural network with least square support vector machine for wastewater treatment process. Chemosphere 308(Pt 1):136116
30. Qi J, Hou Y, Hu J, Ruan W, Xiang Y, Wei X (2020) Decontamination of methylene Blue from simulated wastewater by the mesoporous rGO/Fe/Co nanohybrids: artificial intelligence modeling and optimization. Mater Today Commun 1(24):100709
31. Alam G, Ihsanullah I, Naushad M, Sillanpää M (2022) Applications of artificial intelligence in water treatment for optimization and automation of adsorption processes: recent advances and prospects. Chem Eng J 1;427:130011
32. Altowayti WAH, Shahir S, Othman N, Eisa TAE, Yafooz WMS, Al-Dhaqm A et al (2022) The role of conventional methods and artificial intelligence in the wastewater treatment: a comprehensive review. Processes 10(9):1832
33. Abidli A, Huang Y, Ben Rejeb Z, Zaoui A, Park CB (2022) Sustainable and efficient technologies for removal and recovery of toxic and valuable metals from wastewater: recent progress, challenges, and future perspectives. Chemosphere 1(292):133102
34. Fan C, Lv C, Wang Z, Wu S, Jin Z, Bei K et al (2022) Influence of regular addition of ore on treatment efficiency and aquatic organisms in living machine system for black water treatment. J Clean Prod 20(341):130928
35. Zhang W, Tooker NB, Mueller AV (2020) Enabling wastewater treatment process automation: leveraging innovations in real-time sensing, data analysis, and online controls. Environ Sci Water Res Technol 6(11):2973–2992
36. Carreres-Prieto D, García JT, Carrillo JM, Vigueras-Rodríguez A (2023) Towards highly economical and accurate wastewater sensors by reduced parts of the LED-visible spectrum. Sci Total Environ 1(871):162082
37. Prashantkumar Sathvara et al. (2021) Chlorination treatment in common effluent treatment plant: evaluation of effectiveness and acute toxicity effects. Mendeley Data. https://doi.org/10.17632/f7jctrd8jj.3
38. Emaminejad SA, Morgan VL, Kumar K, Kavathekar A, Ragush C, Shuai W et al (2022) Statistical and microbial analysis of bio-electrochemical sensors used for carbon monitoring at water resource recovery facilities. Environ Sci: Water Res Technol 8(10):2052–2064
39. Zhang W, Wei SP, Winkler MKH, Mueller AV (2022) Design of a soft sensor for monitoring phosphorous uptake in an EBPR process. ACS EST Eng 2(10):1847–1856
40. Haimi H, Mulas M, Corona F, Vahala R (2013) Data-derived soft-sensors for biological wastewater treatment plants: an overview. Environ Model Softw 1(47):88–107
41. Ching PML, So RHY, Morck T (2021) Advances in soft sensors for wastewater treatment plants: a systematic review. J Water Process Eng 1(44):102367
42. Schneider MY, Carbajal JP, Furrer V, Sterkele B, Maurer M, Villez K (2019) Beyond signal quality: the value of unmaintained pH, dissolved oxygen, and oxidation-reduction potential sensors for remote performance monitoring of on-site sequencing batch reactors. Water Res 15(161):639–651
43. Kamali M, Appels L, Yu X, Aminabhavi TM, Dewil R (2021) Artificial intelligence as a sustainable tool in wastewater treatment using membrane bioreactors. Chem Eng J 1(417):128070
44. Sathvara PB, Anuradha J, Sanjeevi R, Tripathi S, Rathod AB (2023) Spatial analysis of carbon sequestration mapping using remote sensing and satellite image processing. In: Multimodal biometric and machine learning technologies [Internet], pp 71–83. Available from https://doi.org/10.1002/9781119785491.ch4
45. Jawad J, Hawari AH, Javaid ZS (2021) Artificial neural network modeling of wastewater treatment and desalination using membrane processes: a review. Chem Eng J 1(419):129540
46. Narayanan CM, Narayan V (2019) Biological wastewater treatment and bioreactor design: a review. Sustain Environ Res 29(1):33

47. Iorhemen OT, Hamza RA, Tay JH (2016) Membrane bioreactor (MBR) technology for wastewater treatment and reclamation: membrane fouling. Membranes 6(2):33
48. Ramakrishnan S, Jayaraman A (2019) Global warming and pesticides in water bodies. In: Handbook of research on the adverse effects of pesticide pollution in aquatic ecosystems [Internet]. IGI Global, pp 421–436. Available from https://www.igi-global.com/chapter/global-warming-and-pesticides-in-water-bodies/213519
49. Ramakrishnan S, Jayaraman A (2019) Pesticide contaminated drinking water and health effects on pregnant women and children. In: Handbook of research on the adverse effects of pesticide pollution in aquatic ecosystems [Internet]. IGI Global, pp 123–36. Available from https://www.igi-global.com/chapter/pesticide-contaminated-drinking-water-and-health-effects-on-pregnant-women-and-children/213500
50. Rahman TU, Roy H, Islam MR, Tahmid M, Fariha A, Mazumder A et al (2023) The advancement in membrane bioreactor (MBR) technology toward sustainable industrial wastewater management. Membranes 13(2):181
51. Sanjeevi R (2011) Studies on the treatment of low-strength wastewaters with upflow anaerobic sludge blanket (UASB) reactor: with emphasis on granulation studies. Centre for Pollution Control and Environmental Engineering, Pondicherry
52. Tomczak W, Gryta M (2022) Energy-efficient AnMBRs technology for treatment of wastewaters: a review. Energies 15(14):4981
53. Kamali M, Gameiro T, Costa MEV, Capela I (2016) Anaerobic digestion of pulp and paper mill wastes—an overview of the developments and improvement opportunities. Chem Eng J 15(298):162–182
54. Tomczak W, Grubecki I, Gryta M (2021) The use of NaOH solutions for fouling control in a membrane bioreactor: a feasibility study. Membranes 11(11):887
55. Aslam M, Charfi A, Lesage G, Heran M, Kim J (2017) Membrane bioreactors for wastewater treatment: a review of mechanical cleaning by scouring agents to control membrane fouling. Chem Eng J 1(307):897–913
56. Guo W, Ngo HH, Li J (2012) A mini-review on membrane fouling. Biores Technol 1(122):27–34
57. Abbasi T, Sanjeevi R, Makhija M, Abbasi SA (2012) Role of vitamins B-3 and C in the fashioning of granules in UASB reactor sludge. Appl Biochem Biotechnol 167(2):348–357
58. Jin Z, Lv C, Zhao M, Zhang Y, Huang X, Bei K et al (2018) Black water collected from the septic tank treated with a living machine system: HRT effect and microbial community structure. Chemosphere 1(210):745–752
59. Zaghloul MS, Iorhemen OT, Hamza RA, Tay JH, Achari G (2021) Development of an ensemble of machine learning algorithms to model aerobic granular sludge reactors. Water Res 1(189):116657
60. Ren N, Yan X, Chen Z, Hu D, Gong M, Guo W (2007) Feasibility and simulation model of a pilot scale membrane bioreactor for wastewater treatment and reuse from Chinese traditional medicine. J Environ Sci (China) 19(2):129–134
61. Cai Y, Ben T, Zaidi AA, Shi Y, Zhang K, Lin A et al (2019) Effect of pH on pollutants removal of ship sewage treatment in an innovative aerobic-anaerobic micro-sludge MBR system. Water Air Soil Pollut 230(7):163
62. Nayak M, Dhanarajan G, Dineshkumar R, Sen R (2018) Artificial intelligence driven process optimization for cleaner production of biomass with co-valorization of wastewater and flue gas in an algal biorefinery. J Clean Prod 10(201):1092–1100
63. Martín de la Vega PT, Jaramillo-Morán MA (2018) Obtaining key parameters and working conditions of wastewater biological nutrient removal by means of artificial intelligence tools. Water 10(6):685
64. Abbasi T, Sanjeevi R, Anuradha J, Abbasi SA (2013) Impact of Al^{3+} on sludge granulation in UASB reactor [Internet]. Available from https://nopr.niscpr.res.in/handle/123456789/19284
65. Nourani V, Elkiran G, Abba SI (2018) Wastewater treatment plant performance analysis using artificial intelligence—an ensemble approach. Water Sci Technol 78(10):2064–2076
66. Xie Y, Chen Y, Lian Q, Yin H, Peng J, Sheng M et al (2022) Enhancing real-time prediction of effluent water quality of wastewater treatment plant based on improved feedforward neural network coupled with optimization algorithm. Water 14(7):1053

67. El-Rawy M, Abd-Ellah MK, Fathi H, Ahmed AKA (2021) Forecasting effluent and performance of wastewater treatment plant using different machine learning techniques. J Water Process Eng 1(44):102380
68. Yang Y, Kim KR, Kou R, Li Y, Fu J, Zhao L et al (2022) Prediction of effluent quality in a wastewater treatment plant by dynamic neural network modeling. Process Saf Environ Prot 1(158):515–524
69. Nnaji PC, Anadebe VC, Onukwuli OD, Okoye CC, Ude CJ (2022) Multifactor optimization for treatment of textile wastewater using complex salt–Luffa cylindrica seed extract (CS-LCSE) as coagulant: response surface methodology (RSM) and artificial intelligence algorithm (ANN–ANFIS). Chem Pap 76(4):2125–2144
70. Chan H, Nai-He Y (2020) A pretreatment method of wastewater based on artificial intelligence and fuzzy neural network system. J Intell Fuzzy Syst 39(2):1711–1720
71. Şenol H (2021) Methane yield prediction of ultrasonic pretreated sewage sludge by means of an artificial neural network. Energy 15(215):119173
72. Song C, Yao L, Hua C, Ni Q (2021) Comprehensive water quality evaluation based on kernel extreme learning machine optimized with the sparrow search algorithm in Luoyang River Basin, China. Environ Earth Sci 80(16):521
73. Guo Q, Wang Y, Dai C, Wang L, Liu H, Li J et al (2023) Dynamics of a stochastic nutrient–plankton model with regime switching. Ecol Model 1(477):110249
74. Li K, Duan H, Liu L, Qiu R, van den Akker B, Ni BJ et al (2022) An integrated first principal and deep learning approach for modeling nitrous oxide emissions from wastewater treatment plants. Environ Sci Technol 56(4):2816–2826

Chapter 6
AI-Enabled Process Optimization for Sustainable Wastewater Treatment Solutions

Payal Gupta, Garima Bhardwaj, Shivani Dubey, Tarun Tayal, Abhishek Sengupta, and Priyanka Narad

Abstract Wastewater treatment stands as a critical pillar in environmental management, safeguarding public health and ensuring sustainable water use amid global population growth and urbanization. This manuscript explores the complexities and advancements in wastewater treatment methodologies, emphasizing the integration of artificial intelligence (AI) and machine learning (ML) to enhance efficiency, reduce costs, and minimize environmental impact. Figure 6.1 illustrates the diverse components involved in wastewater treatment, encompassing physical, chemical, and biological processes essential for pollutant removal and resource recovery. Traditional wastewater treatment methods, while effective, face challenges such as resource intensity and limited adaptability to varying wastewater compositions. The introduction of AI and ML models offers promising solutions by optimizing processes like predictive modeling, real-time monitoring, and resource allocation. These technologies not only improve treatment efficacy but also pave the way for sustainable practices in wastewater management. AI's ability to mimic human intelligence in learning, reasoning, and decision-making is pivotal in optimizing wastewater treatment operations. By leveraging AI-enabled methods, treatment facilities can achieve higher efficiency levels while reducing energy consumption and operational costs. The integration of data sources and sensor technologies further enhances AI applications, enabling precise monitoring and control of treatment processes. The manuscript reviews AI applications in wastewater treatment, highlighting the versatility of neural networks, fuzzy logic, and genetic algorithms in optimizing treatment efficiency and tackling multi-objective optimization challenges. It assesses the performance and limitations of these models, providing insights into their real-world implementations and benefits. Moreover, the chapter discusses current trends and future directions for AI in wastewater treatment, addressing challenges such as data infrastructure requirements and ethical considerations. Despite the transformative

P. Gupta · G. Bhardwaj · S. Dubey · T. Tayal · A. Sengupta
Amity Institute of Biotechnology, Amity University, Noida, Uttar Pradesh, India

P. Narad (✉)
Indian Council of Medical Research, Ansari Nagar, New Delhi, India
e-mail: priyanka.narad@gmail.com

M. C. Garg (ed.), *The AI Cleanse: Transforming Wastewater Treatment Through Artificial Intelligence*, Springer Water, https://doi.org/10.1007/978-3-031-67237-8_6

potential of AI, significant investments and ethical frameworks are necessary to maximize its benefits responsibly. This manuscript underscores the pivotal role of AI in advancing wastewater treatment towards sustainability and efficiency. It navigates through the intricate landscape of wastewater management, emphasizing the transformative impact of AI on conventional approaches. By embracing AI technologies, the wastewater treatment sector can evolve towards cleaner, more effective practices that align with ecological preservation and resource conservation in the era of technological advancement.

Keywords Artificial intelligence (AI) · Wastewater treatment · Machine learning (ML) · Environmental management · Resource recovery · Sustainability

6.1 Introduction

Wastewater treatment is a key pillar of environmental management and an essential protection for public health. As the world's population grows and urbanization rises, so does the requirement for effective wastewater treatment solutions. This dilemma, however, is not isolated; it intersects with broader imperatives such as sustainability, resource conservation, and ecological preservation. The different components involved in wastewater treatment in represented in Fig. 6.1. The proper management of our natural resources and the preservation of environmental integrity depend heavily on wastewater treatment. The treatment processes and methods majorly depend on the physical and chemical properties of the effluent. Various treatment options are available, exploring the combination of physical, chemical, and biological processes that form the basis of traditional treatment facilities [1]. Despite the fact that these techniques have achieved a great deal of success, their drawbacks and resource-intensive nature have opened the door for the incorporation of AI and machine learning (ML) models, leading to a period of increased productivity and adaptability. With the increasing water demands and limited water resources, it is imperative to explore AI incorporated wastewater treatment methods that may reduce the pollution to a safety level and are energy efficient.

Artificial intelligence is widely used in all aspects of society. Artificial intelligence is the capacity of a computer program to mimic human intelligence in terms of autonomous learning, reasoning, judgment, and decision-making. By utilizing AI enabled methods in wastewater treatment, we can explore the diverse range of applications, from process optimization to real-time monitoring and resource allocation, highlighting their potential to increase efficiency, lower costs, and have a smaller environmental impact. The introduction of AI into wastewater treatment is a transformative endeavor. It's important to recognize the obstacles and constraints that come with integrating AI, such as the necessity for substantial investments in data infrastructure, data quality, and ethical considerations. Nonetheless, AI appears as a ray of hope for a future in wastewater treatment that is cleaner, more effective, and sustainable in this dynamic age of technological progress.

Fig. 6.1 Representation of various wastewater treatment components

This chapter provides an outline of wastewater treatment methods, highlighting their immense significance and complex problems. It discusses the data sources and sensor technologies in AI optimized wastewater treatment. We have explored the AL-ML models for predictive modeling, real-time monitoring and control, and resource allocation [2]. Current trends and future directions for AI applications in wastewater treatment are also outlined.

6.2 Wastewater Treatment Processes

The wastewater treatment is crucial for protecting human health, preserving the environment, and assuring the sustainable use of water resources. The field of wastewater treatment is dynamic and complex. It is a process of effectively removing, separating and disposing of pollutants present in the wastewater. The treatment of wastewater handles various problems that wastewater presents, such as eliminating toxins and safeguarding both the environment and public health [2]. The overall classification of treatment methods includes physical, chemical, and biological treatment as shown in Fig. 6.2. It also comprises techniques like sludge treatment and advanced techniques like membrane filtration, ozonation, and UV Disinfection.

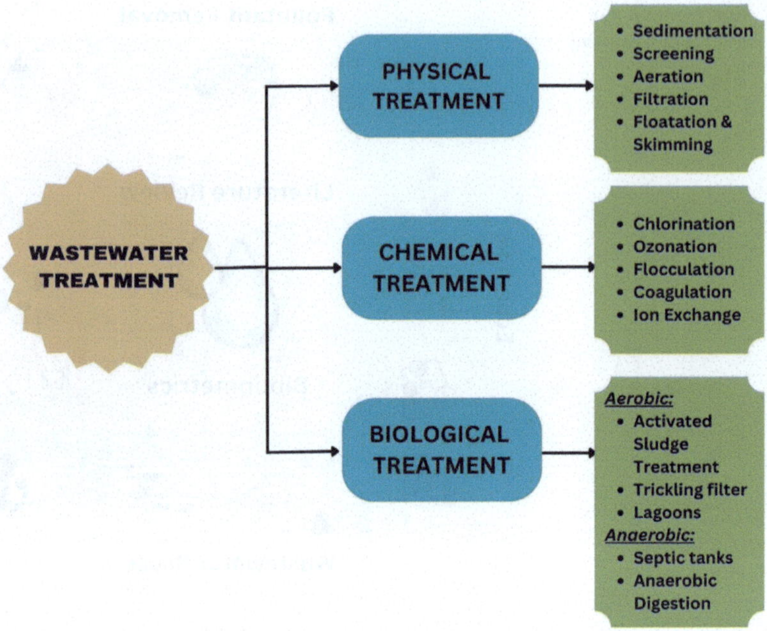

Fig. 6.2 Classification of the wastewater treatment methods

6.2.1 Physical Treatment

Large particles and solids are removed from wastewater using physical treatment methods. By reducing the organic matter and solid load in wastewater, it becomes more receptive to further chemical and biological treatment. The steps involved in the physical treatment are screening, sedimentation, and filtration of the wastewater using simple types of equipment [3].

The first and foremost line of defense in the treatment of wastewater is screening. Preventing big solids like sticks, leaves, plastic, and other waste from entering the treatment process, entails the use of screens or filters with varied mesh sizes. In order to safeguard downstream machinery and avoid jams, screening is essential. After the screening process, the effluent goes through sedimentation. Gravity-based sedimentation enables particles to separate from the effluent. The flow velocity is decreased in sedimentation tanks, enabling heavier materials to sink to the bottom. As a result, the solids and liquid phases are separated, making it simpler to remove and process the solids later in the process. Sedimentation is followed by the process of filtration. Filtration is the process of removing suspended matter and small particles from wastewater by using porous materials like gravel, sand, or fabric. It is particularly effective in getting rid of tiny particles that might have gotten past earlier treatment steps. Both municipal as well as industrial wastewater treatment frequently employ filtration.

6.2.2 Chemical Treatment

In chemical treatment methods, chemicals like ferric chloride, aluminium sulphate, and sodium carbonate are added to wastewater to help remove contaminants, particularly those that are difficult to eliminate through physical means. Chemical processing can improve the effectiveness of later phases of treatment [4].

Fluctuation and Coagulation

Waterwaste is coagulated by adding coagulants like ferric chloride or aluminium sulphate (alum). These substances make the electrical charges on the suspended particles disappear, which makes them stick together. Following coagulation, flocculation involves moderate mixing to create larger flocs. The removal of these flocs through sedimentation or filtration is then made simpler.

Chemical Precipitation

Precipitation by chemicals is used for eliminating particular pollutants, especially phosphorus and heavy metals. Wastewater is treated with precipitating agents like lime or sodium carbonate, which cause the target pollutants to condense into insoluble forms of precipitates that can be filtered or sedimented out of the water.

6.2.3 Biological Treatment

Utilizing microorganisms' ability to break down organic material and remove chemicals from sewage is the basis of biological treatment technologies. When compared to chemical approaches, these technologies produce less residual sludge while being just as successful at decreasing the organic and nutritional load in sewage. Biological treatment includes Processes for activated sludge, fluctuating filters, and lagoon systems [5].

Process for Activated Sludge

In the activated sludge process, wastewater is aerated to promote the development of aerobic bacteria. These bacteria break down organic matter into carbon dioxide, water, and biomass after consuming it. The residual biomass can then be recycled back into the treatment process after being separated from the treated water in a clarifier.

Fluctuating Filters

Trickling filters are media beds, frequently constructed with plastic or pebbles, that encourage the development of adherent microbial biofilms. Microorganisms in the biofilm metabolize and eliminate organic materials and nutrients as wastewater is trickling over the media. At modest wastewater treatment facilities, trickling filters are frequently employed.

Lagoon Systems

Lagoon systems employ natural processes in small, artificial basins. Lagoons hold wastewater for a set amount of time, allowing organic matter to break down through microbial action. Due to its ease of use and affordable operation, lagoon systems are frequently employed in rural or distant places.

6.2.4 Sludge Treatment

Sludge treatment, associated with the solid waste produced during the treatment process, is an important part of wastewater treatment. Effective byproduct utilization and environmental impact reduction depend on proper sludge management. The systematic representation of sludge treatment is shown in Fig. 6.3. The major steps involved in sludge treatment are dewatering, anaerobic digestion, and thermal treatment [6]. The dewatering process involves dewatering the sludge waste to reduce its volume, reduce its weight, and make it simpler to handle and transport. Centrifugation, belt pressing, and filter pressing are often used in dewatering processes. Dewatering is followed by the process of anaerobic digestion. Anaerobic digestion is an oxygen-free biological process that breaks down organic sludge, generating biogas (methane and carbon dioxide) as a byproduct. The digested sludge can subsequently undergo additional processing or be applied to the soil. Additional processing can include thermal treatment. In thermal treatment, Sludge is heated to high temperatures using processes like incineration and drying in order to reduce its volume and eliminate microorganisms. In addition to producing electricity, incineration lessens the demand for landfill disposal.

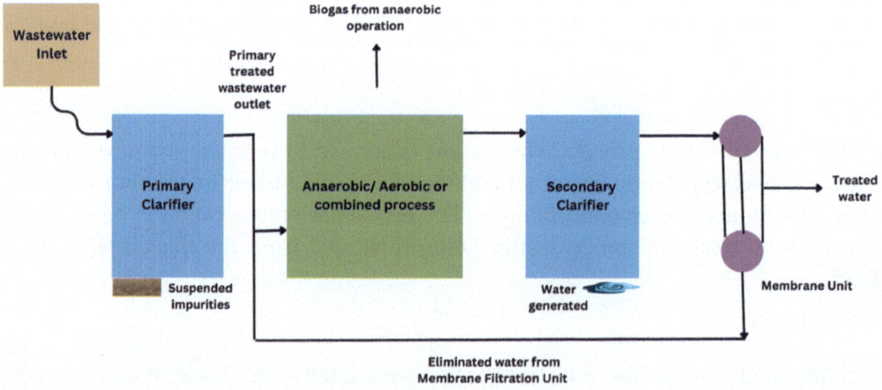

Fig. 6.3 Systematics flowchart of Sludge treatment for wastewater

6.2.5 *Advanced Treatment*

When stricter water quality requirements are needed or when a specific set of contaminants has to be eliminated, advanced treatment techniques are used [7]. To attain higher effluent quality, these techniques make use of cutting-edge technologies. The three major advance treatment methods are depicted in Fig. 6.4.

Membrane Filtration

To physically stop particles, bacteria, and dissolved chemicals from going through, membrane filtration uses membranes with tiny pores. High-level filtration is provided by methods like microfiltration and ultrafiltration, which result in water of extraordinary purity.

UV (Ultraviolet) Disinfection

Ultraviolet disinfection exposes wastewater to ultraviolet light, which breaks germs' genetic material and prevents them from reproducing or spreading disease. UV light disinfection is a common last stage in the treatment of wastewater and is an efficient way to get rid of pathogens.

Ozonation

Injecting ozone gas into wastewater is known as ozonation. A potent oxidizing agent, ozone can degrade organic molecules and eliminate microorganisms. It is efficient at removing emerging pollutants as well as taste and odor chemicals.

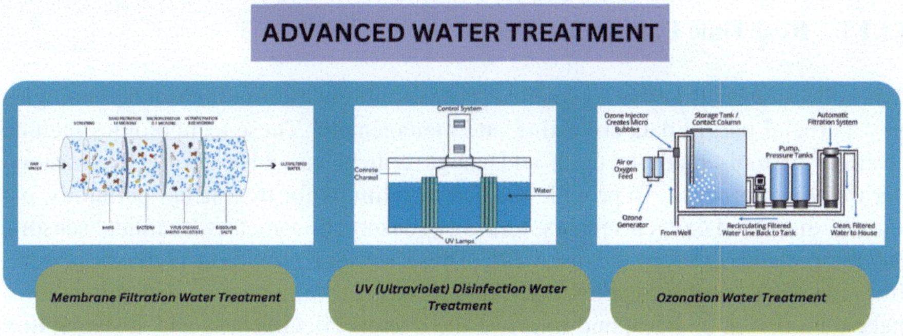

Fig. 6.4 Advanced wastewater treatment including membrane filtration water treatment; UV disinfection water treatment; and ozonation water treatment

6.3 AI in Wastewater Treatment

The complex composition of wastewater affects how it is treated. The traditional methods have been successful in treating the wastewater, but they are resource and energy intensive. To improve the efficiency of treatment plants and reduce the burden of ever-growing water demands, it is imperative to implement innovative technologies. The incorporation of artificial intelligence (AI) into wastewater treatment has emerged as a powerful tool. AI has shown significant promise in optimizing wastewater treatment processes, improving efficiency, and enhancing decision-making by analyzing complex data patterns and facilitating real-time monitoring and control. It represents the nexus of technology and environmental awareness, ensuring a future with clean water and sustainability [8].

When it comes to optimising various processes of wastewater treatment, AI models are crucial.

These models increase productivity, cut costs, and provide insights by using algorithms and data-driven insights. The AI-ML models process and analyse the data and provide results for process optimization. For optimization and prediction, these models require a large amount of data that can be collected through the different types of sensors in real time. From the viewpoint of research, the data can be collected from publicly available repositories as well.

6.3.1 Data Collection and Quality

6.3.1.1 Real Time Data Collection

Data collection is the key to AI enabled process optimization and sensor technologies play a crucial part in data collecting and management. These technologies include a variety of tools made to track and gauge critical variables during the wastewater treatment process. To provide the best conditions for treatment operations, pH sensors, dissolved oxygen sensors, turbidity sensors, flow meters, chemical sensors, and temperature sensors are used [9]. These sensors provide a plethora of real-time data when combined, allowing AI algorithms to optimize treatment procedures, make data-driven choices, and improve the overall effectiveness and sustainability of wastewater treatment plants. Figure 6.5 depicts the usage of water level sensors for reducing water wastage and improving usage factors.

With real-time data that enables precise control and ongoing monitoring of treatment processes, sensor technologies are essential for wastewater treatment. Wastewater treatment plants use a variety of sensor technology, including:

- **pH Sensors**: For monitoring and regulating the acidity or alkalinity of wastewater, pH sensors are essential. Chemical treatment operations must be optimized by maintaining the proper pH level, and these sensors make sure that modifications are done in real-time to guarantee effective treatment.

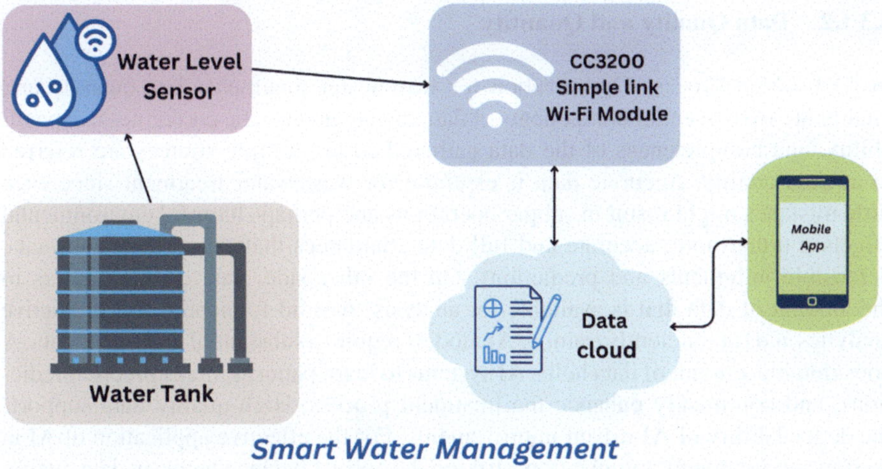

Fig. 6.5 Smart water management using AI

- **Dissolved Oxygen Sensors**: These devices gauge wastewater's level of dissolved oxygen, a crucial element in biological treatment procedures. In order to sustain the microbial activity that breaks down organic matter, appropriate oxygen levels must be maintained.
- **Turbidity Sensors**: Measure the degree of clarity or cloudiness in wastewater. High levels of turbidity might obstruct treatment procedures like UV disinfection. Operators can maintain ideal conditions by adjusting treatment settings by monitoring turbidity.
- **Flow Meters**: Monitor how quickly wastewater is moving through the treatment facility. It is possible to avoid overloading or underutilizing treatment resources by ensuring that treatment procedures are calibrated to manage the volume of wastewater using accurate flow data.
- **Chemical Sensors**: By accurately dosing treatment chemicals, these sensors may measure the concentration of certain chemicals or pollutants in wastewater. As a result, treatment procedures are customized to the unique makeup of the influent wastewater.
- **Temperature Sensors**: Temperature sensors give information on the temperature of the water, which can affect how well different treatment methods work. Keeping an eye on the temperature aids in maintaining the ideal conditions for chemical and biological processes.

These sensor technologies offer real-time data analysis, predictive analytics, and dynamic process optimization when combined with AI systems [10]. They enable wastewater treatment plants to operate more efficiently, use less energy, and guarantee environmental requirements are followed, eventually resulting in more efficient and sustainable wastewater treatment solutions.

6.3.1.2 Data Quality and Quantity

In AI-Enabled Process Optimization for wastewater treatment, data quantity and quality are two essential components of data management. The correctness, dependability, and completeness of the data gathered from multiple sources are referred to as data quality. Accurate data is essential for wastewater treatment since even little mistakes might result in subpar operations and perhaps harmful environmental effects. Furthermore, accurate and full data guarantees that AI systems can make defensible judgments and predictions. On the other side, data quantity refers to the amount of data that is available for analysis. A solid foundation for predictive analytics and for efficiently training AI models require a substantial amount of data. A considerable amount of data helps AI systems to learn patterns, make precise predictions, and continually enhance the treatment process. High-quality data supports the dependability of AI-driven improvements. For the effective application of AI in wastewater treatment optimization, striking the correct balance between data quality and quantity is crucial [11].

Different AI models including neural network like Recurrent Neural Network (RNN), Convolutional Neural Network (CNN), Fuzzy Neural Network (FNN), Deep Neural Network (DNN) and machine learning algorithms such as Decision Tree (DT), Random Forest (RF) and Adaptive-Network-Based Fuzzy Inference System (ANBFIS) can be utilized for the analysis and prediction of the water treatment [12].

6.3.2 Artificial Neural Network (ANN) Models

A mathematical model known as an Artificial Neural Network (ANN) is recognised to simulate how organic neural networks process information. In ANN models, each unit node simulates a neuron, and data processing is accomplished by varying the weights of connections between a large number of unit nodes (neurons) in the network [13]. An input layer, an output layer, and multiple hidden layers positioned in between the input and output layers are the typical components of ANNs (shown in Fig. 6.6). To carry out complex nonlinear computations, ANNs use different weight values between neurons and activation functions, such as sigmoid, tanh, and ReLU functions. The ability of ANNs to represent relationships between inputs and outputs increases as hidden layers are added, allowing them to build more complicated nonlinear models.

As a result, by creating and improving loss functions, ANNs can be trained to learn and optimise complex nonlinear relationships. Recurrent neural networks (RNNs), convolutional neural networks (CNNs), fuzzy neural networks (FNNs), and deep neural networks (DNNs) are the most popular ANN models.

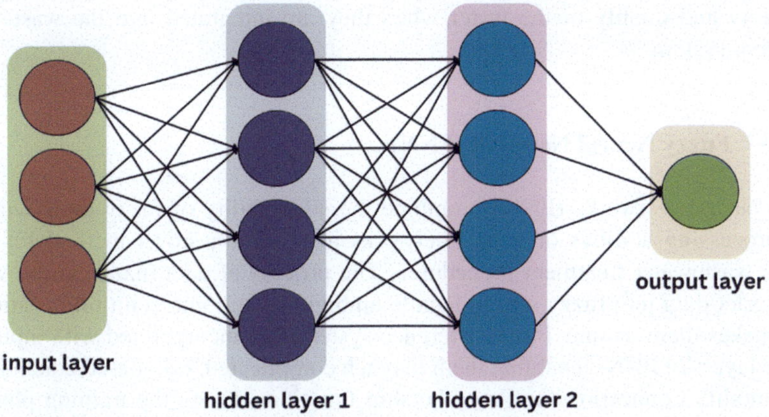

Fig. 6.6 Overview of artificial neural network structure

6.3.2.1 Recurrent Neural Network (RNN)

Recurrent Neural Networks (RNNs) are dynamic wastewater treatment systems that use feedback connections to efficiently analyse sequential data. RNNs start by gathering and preparing data relevant to different treatment procedures, which frequently include influent characteristics, water quality metrics, and equipment sensor data. RNNs group this data into sequences and have input and hidden layers that make modelling and enhancing these sequences easier. RNNs modify the weights of the connections, as well as the learning patterns and relationships in the data, during the training phase. They are excellent in predictive modelling, which predicts future water quality parameters based on historical and present-day data, allowing treatment procedures to be adjusted in real-time [14]. Additionally, RNNs have strong anomaly detection abilities that enable them to quickly spot deviations or unexpected changes in influent quality and procedure.

6.3.2.2 Convolutional Neural Network (CNN)

By analysing visual data from treatment procedures, Convolutional Neural Networks (CNNs), which are generally employed for image analysis, find utility in wastewater treatment. Image data gathered from crucial treatment plant locations, including equipment and sedimentation tanks, is supplied to these networks. CNNs are taught to classify images after preprocessing, identifying different states or circumstances, such as identifying equipment faults or detecting pollutants [15]. By initiating actions based on picture classifications after being trained, CNNs can support real-time process optimisation. They are excellent at spotting anomalies and highlighting odd circumstances. CNNs improve performance, lower maintenance costs, and guarantee the safety and quality of the water when they are integrated into the wastewater treatment system.

6.3.2.3 Fuzzy Neural Network (FNN)

Fuzzy Neural Networks (FNNs) combine the adaptability of fuzzy logic with the data processing abilities of neural networks to provide a novel method for optimising wastewater treatment procedures. The process of data fuzzification, which turns exact data into fuzzy sets to handle ambiguous and uncertain information, is what makes them unique. Fuzzy inference systems are incorporated with input and hidden layers of FNNs, enabling them to reach complicated, rule-based conclusions. They modify connections and membership functions during the training phase to uncover patterns in the fuzzily collected data. FNNs are crucial tools for increasing the efficacy and efficiency of treatment facilities because they are particularly useful for predictive modelling, anomaly detection, and real-time process optimisation in wastewater treatment [16].

6.3.2.4 Deep Neural Network (DNN)

Deep Neural Networks (DNNs) are proving to be effective tools for jobs involving data processing and optimisation in wastewater treatment. They start with data collecting and then preprocess the data to ensure consistency and quality. Multiple hidden layers in DNNs enable sophisticated data processing and pattern recognition. They become proficient at predictive modelling through a training phase where they learn to reduce errors by changing the links in the network. DNNs, for instance, can forecast water quality metrics, enabling process modifications in real time [17]. In addition, DNNs are excellent at detecting anomalies in data patterns. Their optimisation ability enables the improvement of treatment processes while lowering operating expenses and increasing productivity. DNNs have enormous promise for data-intensive tasks in wastewater treatment, helping to create efficient and long-lasting treatment methods.

6.3.3 Machine Learning Models

The core of machine learning, a branch of artificial intelligence, is the creation of statistical models and techniques that enable computers to learn on their own from data without explicit programming. It can be categorized into three types: supervised, unsupervised, and reinforcement learning (Fig. 6.7). Machine learning's main objective is to build predictive models that are capable of making precise predictions or judgements based on the information learned from data. Decision Tree (DT), Random Forest (RF), and the Adaptive-Network-Based Fuzzy Inference

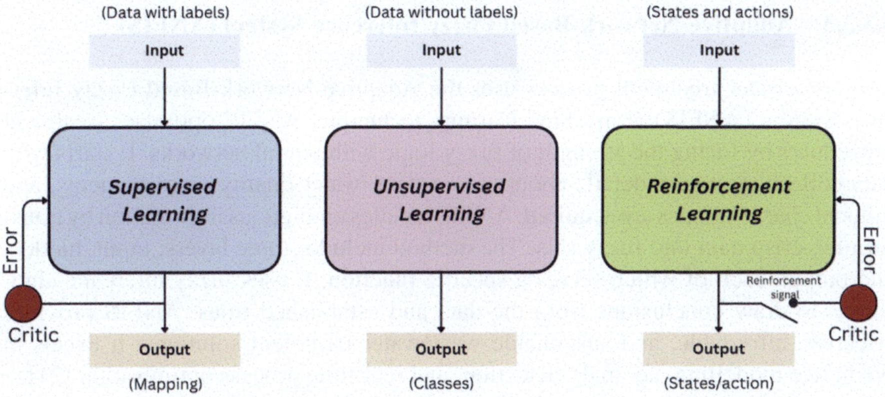

Fig. 6.7 Types of learning models

System (ANFIS) are some of the machine learning models that are most frequently used [18].

6.3.3.1 Decision Tree (DT)

Machine learning techniques called decision trees (DT) are used in wastewater treatment to increase effectiveness and provide predictive control. They start with data collection, which entails gathering and quality-checking information. The goal of the DT is to build a tree that predicts various treatment outcomes by choosing qualities for decision nodes. This prediction model is used to identify deviations from predicted patterns and control processes in real time. Dynamic modifications based on data are possible thanks to integration with Supervisory Control and Data Acquisition (SCADA) systems [19]. Transparency and interpretability provided by DTs support successful, economical, and sustainable wastewater treatment procedures.

6.3.3.2 Random Forest (RF)

An effective machine learning technique used in wastewater treatment is called Random Forest (RF). Data such as influent characteristics, water quality indicators, and equipment sensor readings are first gathered and pre-processed. RF is a collection of decision trees. Predictive modelling for treatment effectiveness and water quality is made possible by the algorithm's learning of the correlations between input data and output labels during training. RF excels in managing intricate, connected variables. It is capable of making future predictions, real-time anomaly detection, and treatment process optimisation [20]. The ensemble approach used by the RF algorithm improves forecast accuracy and helps to create wastewater treatment methods that are more effective and long-lasting.

6.3.3.3 Adaptive-Network-Based Fuzzy Inference System (ANFIS)

The wastewater treatment process uses the Adaptive-Network-Based Fuzzy Inference System (ANFIS), a machine learning technique. ANFIS optimises treatment procedures by fusing the strength of fuzzy logic with neural networks. It starts with data collection, where details about sensor data, water quality measurements, and influent characteristics are acquired. ANFIS handles ambiguous information by transforming crisp data into fuzzy sets. The method includes three layers: input, hidden, and output, each of which serves a specific function. It uses fuzzy inference algorithms to draw conclusions from the data and established rules. ANFIS provides effective, affordable, and sustainable wastewater treatment solutions. It excels in predictive modelling, anomaly detection, and real-time process optimisation [21].

6.4 Process Optimization with AI

AI-enabled process optimization provides a substantial improvement in terms of efficacy, cost-effectiveness, and minimising environmental impact in the context of sustainable wastewater treatment systems. Wastewater treatment facilities may optimise their procedures with accuracy by utilising AI. Massive datasets are analysed in real-time by machine learning algorithms, which find patterns, anomalies, and prospective improvement areas. In order to obtain the best outcomes, operators may use this data-driven approach to optimise a variety of factors, including chemical doses, treatment periods, and energy usage [22]. In addition, AI systems may recommend preventative maintenance and forecast equipment breakdowns, saving downtime and optimising resource efficiency. The overall workflow for the optimization of wastewater treatment utilizing AI methods is depicted in Fig. 6.8.

6.4.1 Predictive Modelling

Predictive modelling is the key to long-term wastewater treatment systems through AI-enabled process optimization. Being able to foresee and react to changes is essential in the intricate and dynamic field of wastewater treatment. Based on historical data and real-time inputs, predictive modelling uses the ability of artificial intelligence to foresee system behaviour and consequences. Several AI algorithms can be used in predictive modeling for wastewater treatment to forecast important characteristics and optimize the treatment procedure. These algorithms include time series forecasting strategies like ARIMA (AutoRegressive Integrated Moving Average) or exponential smoothing, which may forecast the properties of wastewater and the results of its treatment based on past data [23]. These models can offer priceless insights into how numerous operational elements, like influent characteristics, meteorological conditions, and equipment performance, will affect the treatment process

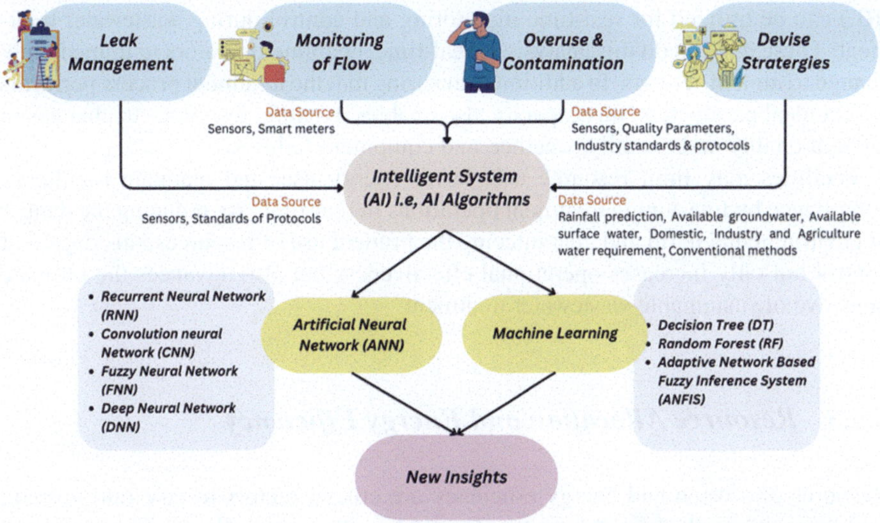

Fig. 6.8 Workflow for the optimization of wastewater treatment in different ways that shows the utility of AI

by utilising sophisticated algorithms. With this foresight, wastewater treatment plants may proactively modify their procedures, assuring peak performance, lowering the chance of compliance violations, and avoiding resource waste.

Predictive modelling also enables scenario analysis and "what-if" simulations, enabling operators to examine several techniques and choose the most environmentally friendly and economically practical methods for wastewater treatment. Predictive modelling is an essential tool that not only improves the effectiveness of wastewater treatment but also helps to achieve sustainability goals by assisting in the development of informed decisions that maximise resource use and reduce environmental impact.

6.4.2 Real-Time Monitoring and Control

Real-time monitoring and control is crucial for optimization for environmentally friendly wastewater treatment solutions. Conditions may quickly change in the complex world of wastewater treatment, and deviations from planned parameters can have a substantial impact on both operational effectiveness and the environment. Data is continually collected on many elements of the treatment process, such as water quality, flow rates, and equipment performance, via AI systems that are coupled with real-time monitoring sensors [24]. Artificial intelligence (AI) methods including Proportional-Integral-Derivative (PID) Control Model, Predictive Control (MPC), Fuzzy Logic Control, Supervised Learning methods, and Reinforcement Learning

(RL) can be utilized for real-time monitoring and control during wastewater treatment. These devices do this analysis in real-time, enabling operators to immediately change treatment settings. In addition to ensuring that the treatment process performs within ideal parameters, this capacity also enables dynamic responses to changes in influent quality and quantity, weather, and equipment failures.

Facilities may limit resource use, lower energy use, and maintain regulatory compliance by fine-tuning treatment operations in real-time. By reducing the danger of environmental harm and guaranteeing the prudent use of resources, this degree of control not only increases operational effectiveness but also advances the ultimate objective of sustainable wastewater treatment.

6.4.3 Resource Allocation and Energy Efficiency

Resource allocation and energy efficiency are crucial factors to take into account for environmentally friendly wastewater treatment systems. Wastewater treatment plants frequently produce considerable energy demands while using huge amounts of resources, such as chemicals, power, and water. AI is essential for maximising how these resources are distributed [10]. In order to reduce unnecessary use and related expenses, machine learning algorithms examine enormous datasets to calculate the precise quantities of chemicals needed for effluent treatment. Additionally, AI-driven predictive algorithms improve equipment operating schedules to reduce energy usage during peak hours by anticipating energy demands. Linear Programming, Reinforcement Learning, and Energy Efficiency Models are some of the AI techniques for Resource Allocation and Energy Efficiency in wastewater treatment. AI decreases operating costs and diminishes the environmental impact of wastewater treatment by optimising resource allocation and improving energy efficiency.

This strategy completely satisfies sustainability objectives by assuring responsible resource use, assisting in the reduction of greenhouse gas emissions, and improving the economic and environmental viability of wastewater treatment plants.

6.5 AI Enabled Sustainability in Wastewater Treatment

Sustainable wastewater treatment emphasizes water conservation, lowering greenhouse gas emissions, and minimizing resource usage [25]. Through data-driven insights and predictive analytics, AI technologies play a crucial role in attaining these objectives. The Al enabled sustainable wastewater treatments aim at Reducing Carbon Footprint, Minimising Chemical Usage, and Enhancing Ecosystem Health.

6.5.1 Reducing Carbon Footprint

The use of AI in the field of wastewater treatment is a ray of hope for dramatically lowering carbon footprints. The utilisation of energy-intensive technologies in traditional wastewater treatment techniques frequently results in significant greenhouse gas emissions. Nevertheless, real-time data analytics and predictive modelling made possible by AI technology are revolutionising this field. AI algorithms improve energy use by precisely calibrating procedures like aeration, pumping, and chemical dosing, greatly reducing the carbon emissions caused by excessive energy use [26]. Additionally, AI-driven predictive maintenance makes sure that machinery runs as efficiently as possible, preventing wasteful energy usage and lowering the need for frequent replacements, which reduces the requirement for energy-intensive industrial processes. Additionally, AI improves treatment effectiveness while lowering the carbon footprint associated with the production and delivery of chemicals by limiting chemical usage through data-driven insights. The combination of AI with wastewater treatment essentially guarantees better water while also making a big contribution to the global fight against climate change by dramatically reducing the carbon footprint of the entire wastewater treatment process.

6.5.2 Minimising Chemical Usage

The delicate balancing act between efficient filtration and chemical usage in wastewater treatment is one of the main issues. In this delicate balance, AI-enabled process optimization is crucial. Artificial intelligence (AI) systems examine sizable datasets containing the composition of incoming wastewater by utilising cutting-edge algorithms and machine learning. These studies allow for accurate chemical dosage forecasts and real-time adjustments, maintaining the treatment process' high level of efficacy [27]. AI technology greatly lowers the overall volume of chemicals needed for treatment by tailoring chemical usage to the precise needs of the wastewater at any given time. This decrease decreases the environmental effect caused by the production, distribution, and disposal of these chemicals while simultaneously yielding significant financial savings. Additionally, the intelligent use of chemicals results in cleaner treated water and a far lower release of dangerous contaminants into the environment. This decrease in chemical use is evidence of the environmental awareness made possible by AI-enabled technologies, which makes wastewater treatment procedures both more effective and naturally sustainable.

6.5.3 *Enhancing Ecosystem Health*

AI-driven technologies ensure that treated water released back into rivers, lakes, and oceans satisfies strict quality standards through significant accuracy and real-time analysis. AI reduces the presence of contaminants, preserving aquatic life and the delicate balance of underwater ecosystems by precisely regulating chemical dosages and optimising treatment procedures. Additionally, AI indirectly helps to sustain natural water sources, which are essential for the existence of innumerable species, by conserving water through effective operations. AI-optimized wastewater treatment results in better water quality, which promotes healthier aquatic environments and the resurgence of flora and fauna. In essence, AI-driven wastewater treatment improves ecosystem health in a way that not only safeguards wildlife but also strengthens the balance of the environment as a whole, highlighting the interdependence between human actions and the natural world [28].

6.6 Opportunities and Future Directions

Although integrating AI-enabled process optimization in wastewater treatment offers a promising path toward sustainability, widespread adoption faces a number of obstacles. The initial investment necessary for installing AI systems, which includes hardware, software, and personnel training, is a considerable barrier. The key issues are overcoming financial limitations and assuring affordable for smaller wastewater treatment facilities. Additionally, maintaining data security and privacy is of utmost importance, particularly when dealing with sensitive information of water treatment plants. Another issue is standardisation and interoperability across multiple wastewater treatment facilities because AI solutions need to be flexible enough to work with different operating scales and infrastructures. Additionally, ongoing research and development are required to improve AI algorithms and make them better able to handle difficult situations. The development of explainable AI models, which allows operators to comprehend and believe AI-driven judgments, is what the future of AI-enabled wastewater treatment holds. The development of legislative frameworks that promote the deployment of AI technology in wastewater treatment depends critically on collaborations between technologists, environmental scientists, and legislators. In order to promote a shared knowledge of the advantages and best practices associated with AI integration, it is essential to emphasize education and awareness among stakeholders and industry professionals. The future of AI-enabled process optimization in sustainable wastewater treatment appears promising, giving potential solutions to some of the most serious environmental problems of our day. By tackling these issues and setting a roadmap for ongoing research and cooperation.

6.6.1 Regulatory and Ethical Considerations

The legislative and ethical environment surrounding these technologies is a crucial consideration that cannot be ignored as the usage of AI/ML in wastewater treatment increases. To make sure that AI is used in wastewater treatment facilities responsibly, regulatory frameworks must be put in place [29]. To protect the community, these policies should cover data privacy, security, and environmental norms. Additionally, it is crucial to take ethical issues into account when using AI algorithms for decision-making. For AI algorithms to make decisions that are consistent with moral and ethical standards, transparency and accountability are crucial. To find a balance between technical innovation and ethical responsibility, collaboration is essential between technology innovators, legislators, and environmental organizations. The incorporation of AI in wastewater treatment can be ethically sound, socially responsible, and technologically sophisticated as well as promote a sustainable and inclusive future by deftly negotiating these regulatory and ethical issues.

6.6.2 Technology Problems and Their Solutions

Although implementing AI-enabled process optimization in wastewater treatment is unquestionably revolutionary, there are still certain difficulties involved. The complexity and variety of wastewater components represent one of the main obstacles. It is difficult to develop algorithms for universal AI since different sources produce different pollution. Deep learning and neural networks, two developments in machine learning approaches, allow systems to adapt to and learn from a variety of datasets, improving their ability to adapt to different effluent compositions [30]. The integration of AI systems into current infrastructures is another difficulty. Since many wastewater treatment facilities still use antiquated technology, smooth integration is a difficult undertaking. To overcome this obstacle, technology providers and plant operators must work together to provide specialized solutions that can integrate with old systems and guarantee a smooth transition to AI-driven processes. Privacy and data security issues are also quite important. Strong cybersecurity safeguards, encryption techniques, and stringent access controls are required when handling sensitive information about wastewater facilities [30]. Modern encryption techniques and strict data governance regulations must be adopted in order to address these challenges. Another difficulty is finding qualified individuals who can design, deploy, and maintain AI systems. A new generation of professionals capable of handling the complexities of AI-enabled wastewater treatment solutions can be fostered by investing in educational initiatives and professional development opportunities. The use of AI in wastewater treatment procedures can overcome these difficulties by coming up with creative answers.

6.6.3 The Path Forward: Emerging Trends

A number of new trends are emerging and influencing the direction of sustainable wastewater management as the relationship between AI technology and wastewater treatment continues to develop. The combination of Internet of Things (IoT) sensors and AI algorithms is one of the major developments, enabling real-time monitoring of numerous factors including water quality, flow rates, and equipment status. The precision of AI-driven optimizations is improved by this interconnected network, which also makes sure that treatment procedures are always in line with the demands of the moment. Furthermore, as machine learning models advance, predictive analytics may now be performed with greater precision. These cutting-edge algorithms maximize efficiency and save resource usage by anticipating changes in influent composition and making proactive treatment strategy adjustments. Decentralized wastewater treatment solutions are also becoming more popular when combined with AI. These systems profit from AI's capacity to quickly adjust to changing demands because they are frequently utilized in smaller communities or industrial settings. In line with environmental objectives, decentralization lowers the energy required to transport wastewater to centralized treatment facilities. Additionally, the integration of renewable energy sources like solar and wind power is expected to grow. By guaranteeing that energy-intensive processes are powered by clean sources and constantly adjusting operations based on the availability of renewable energy, AI algorithms can further reduce the environmental impact. Exploration of natural and nature-inspired solutions—often referred to as "green infrastructure"—in conjunction with AI is another key trend. These technologies, when enhanced by AI optimizations, replicate natural processes to increase treatment efficiency and decrease operational costs. By examining the intricate relationships between microbes, plants, and contaminants, for example, AI algorithms might enhance artificial wetlands or biological filters to ensure optimal pollution removal effectiveness [31].

6.7 Transforming Wastewater Treatment with Predictive Control: A Case Study

This case study examines the practical use of AI in wastewater treatment, showing the radical change it has brought about at a municipal wastewater treatment plant in a medium-sized city. It also depicts how the use of technology in conjunction with cutting-edge AI algorithms not only enhances plant performance but also paves the way for the operation of predictive wastewater treatment plants.

The wastewater treatment facility at Ulu Pandan Water Reclamation Plant, Singapore was under a lot of pressure to process an increasing amount of wastewater effectively while retaining high-quality effluent. Concerns over energy usage, maintenance expenses, and non-compliance issues arose as a result of manual operation and conventional control systems' inability to meet these objectives. In order to resolve these issues, the Aquasuite® software from Royal Haskoning DHV was introduced in March 2019. Advanced AI algorithms such as Deep Neural Network (DNN), Regression Models, and other Machine learning methods, were integrated with Aquasuite PURE to create a system that improved plant performance while offering predictive insights. The use of Aquasuite PURE software resulted in the following:

- **Real-Time Data Gathering**

Aquasuite PURE gathered data on the plant's flows as well as qualitative measurements of elements such as ammonia, nitrates, oxygen, phosphates, and dry solids. It created an extensive historical database that served as the basis for predictive modelling.

- **Advanced Analytics and Machine Learning**

To forecast many elements of plant operations, the software used advanced analytics and machine learning techniques. Forecasting wastewater flows, loads, oxygen needs, chemical dosing requirements, and other crucial characteristics was part of this.

- **Real-Time Process Control**

Based on its projections and the plant's previous performance, the system was built to control important treatment operations. The effectiveness and efficiency of the treatment were increased by dynamic optimisation.

- **Anomaly Detection**

Quantile regression techniques were included by Aquasuite PURE to find irregularities in the plant's operations. This function guaranteed the early detection and remediation of problems.

- **Integration with SCADA**

To collect data and manage crucial processes, the software smoothly integrated with the plant's Supervisory Control and Data Acquisition (SCADA) system. After that, data was transferred to the Aquasuite cloud platform for additional analysis.

- **Digital Twin and Enhanced Analytics**

To monitor the on-premise solution's real-time performance, a digital twin—a virtual duplicate of the physical system—was deployed. Operators were given access to advanced insights through the cloud, while on-premise components kept improving real-time performance.

- **Maximum Effectiveness Set-points**

Machine learning algorithms were used to determine each process's efficiency. The most effective set-points for treatment processes were found using the learned relationships in conjunction with predictions of influent load.

The Aquasuite PURE software was able to work as an autopilot for unattended operation and it learned and anticipated plant operations many days in advance. According to preliminary findings, the aeration flow was significantly reduced, which could save up to 15% in energy.

This case study serves as an example of the efficacy, cost-saving, and sustainable treatment procedures that AI-powered predictive control systems can offer in the wastewater treatment industry, all while ensuring regulatory compliance.

6.8 Conclusion

The wastewater treatment is necessary for clean environment and human health. With the increasing use of water as the population increases, the burden false on the wastewater treatment plants to recycle the wastewater. Even though the conventional methods have been proven successful in water treatment, but they are resource and energy intensive. The exploration of innovative techniques to solve these issues is imperative. Integration of AL/ML models with wastewater treatment machines and software can reduce the cost and increase efficiency. Artificial neural network models and machine learning algorithms can be used to develop integrated software which can analyse and predict the desired outcomes. These models can provide benefits like predictive modelling, real-time control, and resource allocation without human intervention. Despite the advantages of AL models, it is important to consider that these models are technologically costly and require expertise. The regulation and ethical norms need to be considered as well. Nevertheless, AI gives us hope for a cleaner, better, and more sustainable future in wastewater treatment as technology keeps moving forward.

Acknowledgements We would like to acknowledge Dr. Ashok K. Chauhan, Founder, and President, Amity University, Uttar Pradesh for providing us the opportunity to conduct research. We would also like to thank Centre for Computational Biology and Bioinformatics, Amity Institute of Biotechnology, Amity University for providing us with necessary resources.

Competing Interests
No potential conflict of interest was reported by the authors.

Contributions
GB, SD, TT, and PG conceived the context of the chapter; PN & AS designed the outline of the chapter; GB, SD, and TT performed the case study analysis; PG, AS and PN authored and reviewed drafts of the chapter.

References

1. Saravanan A, Senthil Kumar P, Jeevanantham S et al (2021) Effective water/wastewater treatment methodologies for toxic pollutants removal: processes and applications towards sustainable development. Chemosphere 280:130595. https://doi.org/10.1016/j.chemosphere.2021.130595
2. Safder U, Kim J, Pak G et al (2022) Investigating machine learning applications for effective real-time water quality parameter monitoring in full-scale wastewater treatment plants. Water 14:3147. https://doi.org/10.3390/w14193147
3. Ahmed SF, Mofijur M, Nuzhat S et al (2021) Recent developments in physical, biological, chemical, and hybrid treatment techniques for removing emerging contaminants from wastewater. J Hazard Mater 416:125912. https://doi.org/10.1016/j.jhazmat.2021.125912
4. Obaideen K, Shehata N, Sayed ET et al (2022) The role of wastewater treatment in achieving sustainable development goals (sdgs) and sustainability guideline. Energy Nexus 7:100112. https://doi.org/10.1016/j.nexus.2022.100112
5. Hamza RA, Iorhemen OT, Tay JH (2016) Advances in biological systems for the treatment of high-strength wastewater. J Water Process Eng 10:128–142. https://doi.org/10.1016/j.jwpe.2016.02.008
6. Di Capua F, Spasiano D, Giordano A et al (2020) High-solid anaerobic digestion of sewage sludge: challenges and opportunities. Appl Energy 278:115608. https://doi.org/10.1016/j.apenergy.2020.115608
7. Hiller CX, Hübner U, Fajnorova S et al (2019) Antibiotic microbial resistance (AMR) removal efficiencies by conventional and advanced wastewater treatment processes: a review. Sci Total Environ 685:596–608. https://doi.org/10.1016/j.scitotenv.2019.05.315
8. Wang Y, Cheng Y, Liu H et al (2023) A review on applications of artificial intelligence in wastewater treatment. Sustainability 15:13557. https://doi.org/10.3390/su151813557
9. Al-Dasoqi N, Mason A, Alkhaddar R, Al-Shamma'a A (2011) Use of sensors in wastewater quality monitoring—a review of available technologies. World Environ Water Resour Congress 2011. https://doi.org/10.1061/41173(414)354
10. Su B, Lin Y, Wang J et al (2022) Sewage treatment system for improving energy efficiency based on particle swarm optimization algorithm. Energy Rep 8:8701–8708. https://doi.org/10.1016/j.egyr.2022.06.053
11. Wang R, Pan Z, Chen Y et al (2021) Influent quality and quantity predictionin wastewater treatment plant: modelconstruction and evaluation. Pol J Environ Stud 30:4267–4276. https://doi.org/10.15244/pjoes/132821
12. Khataee AR, Kasiri MB (2011) Modeling of biological water and wastewater treatment processes using artificial neural networks. Clean Soil Air Water 39:742–749. https://doi.org/10.1002/clen.201000234
13. Chen Y, Song L, Liu Y et al (2020) A review of the artificial neural network models for water quality prediction. Appl Sci 10:5776. https://doi.org/10.3390/app10175776
14. Asghari M, Dashti A, Rezakazemi M et al (2018) Application of neural networks in membrane separation. Rev Chem Eng 36:265–310. https://doi.org/10.1515/revce-2018-0011
15. Zhu M, Wang J, Yang X et al (2022) A review of the application of machine learning in water quality evaluation. Eco Environ Health 1:107–116. https://doi.org/10.1016/j.eehl.2022.06.001
16. Sundui B, Ramirez Calderon OA, Abdeldayem OM et al (2021) Applications of machine learning algorithms for biological wastewater treatment: updates and perspectives. Clean Technol Environ Policy 23:127–143. https://doi.org/10.1007/s10098-020-01993-x
17. Radović S, Pap S, Turk Sekulić M (2022) Machine learning as a support tool in wastewater treatment systems—a short review. In: Proceedings—the eleventh international symposium GRID 2022. https://doi.org/10.24867/grid-2022-p89
18. Andrade Cruz I, Chuenchart W, Long F et al (2022) Application of machine learning in anaerobic digestion: perspectives and challenges. Biores Technol 345:126433. https://doi.org/10.1016/j.biortech.2021.126433

19. Dalmau M, Rodriguez-Roda I, Ayesa E et al (2013) Development of a decision tree for the integrated operation of nutrient removal MBRS based on simulation studies and expert knowledge. Chem Eng J 217:174–184. https://doi.org/10.1016/j.cej.2012.11.060
20. Cheng Q, Chunhong Z, Qianglin L (2023) Development and application of random forest regression soft sensor model for treating domestic wastewater in a sequencing batch reactor. Sci Rep. https://doi.org/10.1038/s41598-023-36333-8
21. Wu G-D, Lo S-L (2008) Predicting real-time coagulant dosage in water treatment by artificial neural networks and adaptive network-based Fuzzy Inference system. Eng Appl Artif Intell 21:1189–1195. https://doi.org/10.1016/j.engappai.2008.03.015
22. Alam G, Ihsanullah I, Naushad M, Sillanpää M (2022) Applications of artificial intelligence in water treatment for optimization and automation of adsorption processes: recent advances and prospects. Chem Eng J 427:130011. https://doi.org/10.1016/j.cej.2021.130011
23. Huo J, Seaver WL, Robinson RB, Cox CD (2005) Application of time series models to analyze and forecast the influent components of wastewater treatment plants (wwtps). Impacts Global Clim Change. https://doi.org/10.1061/40792(173)97
24. Saboe D, Ghasemi H, Gao MM et al (2021) Real-time monitoring and prediction of water quality parameters and algae concentrations using microbial potentiometric sensor signals and machine learning tools. Sci Total Environ 764:142876. https://doi.org/10.1016/j.scitotenv.2020.142876
25. Viet ND, Jang D, Yoon Y, Jang A (2021) Enhancement of membrane system performance using artificial intelligence technologies for sustainable water and wastewater treatment: a critical review. Crit Rev Environ Sci Technol 52:3689–3719. https://doi.org/10.1080/10643389.2021.1940031
26. Chen J, Dai L, Mataya D et al (2022) Enhanced sustainable integration of CO2 utilization and wastewater treatment using microalgae in circular economy concept. Biores Technol 366:128188. https://doi.org/10.1016/j.biortech.2022.128188
27. Icke O, van Es DM, de Koning MF et al (2020) Performance improvement of wastewater treatment processes by application of machine learning. Water Sci Technol 82:2671–2680. https://doi.org/10.2166/wst.2020.382
28. Fels EAE, A, Mandi L, Kammoun A, et al (2023) Artificial Intelligence and wastewater treatment: a global scientific perspective through text mining. Water 15:3487. https://doi.org/10.3390/w15193487
29. Doorn N (2021) Artificial Intelligence in the water domain: opportunities for responsible use. Sci Total Environ 755:142561. https://doi.org/10.1016/j.scitotenv.2020.142561
30. Richards CE, Tzachor A, Avin S, Fenner R (2023) Rewards, risks and responsible deployment of artificial intelligence in water systems. Nat Water 1:422–432. https://doi.org/10.1038/s44221-023-00069-6
31. Bahramian M, Dereli RK, Zhao W et al (2023) Data to intelligence: the role of data-driven models in wastewater treatment. Expert Syst Appl 217:119453. https://doi.org/10.1016/j.eswa.2022.119453

Chapter 7
Real-Time Monitoring and Fault Detection in AI-Enhanced Wastewater Treatment Systems

Anita Mohanty, Subrat Kumar Mohanty, and Ambarish G. Mohapatra

Abstract Wastewater treatment, a pivotal aspect of environmental engineering, faces escalating demands for efficiency as well as compliance with stringent regulations. This chapter elucidates the significance of real-time monitoring and fault detection in AI-enhanced wastewater treatment systems, leveraging technological advancements to meet contemporary challenges. Detailed exploration of wastewater treatment components sets the context, elucidating the criticality of real-time monitoring. A comparative analysis illuminates the limitations of conventional monitoring methodologies, highlighting the transformative potential of AI integration. Crucial parameters, including chemical oxygen demand (COD), biological oxygen demand (BOD), and pH levels, are scrutinized for enhanced real-time monitoring and predictive analytics. Advanced AI applications, encompassing machine learning algorithms and neural networks, are discussed, showcasing their role in predictive modeling and data analysis. Comprehensive insights into cutting-edge sensors, remote sensing technologies, and IoT devices establish a foundation for real-time data acquisition. The chapter delves into fault detection and diagnosis, employing AI-driven algorithms for the timely identification of deviations and anomalies. Through informative case studies, the efficacy of AI implementations is underscored, demonstrating tangible benefits in wastewater treatment optimization. However, challenges such as privacy, data security, and integration complexities necessitate strategic resolutions. Forward-looking trends forecast an evolution toward holistic, AI-centric wastewater treatment paradigms, reiterating the indispensability of real-time monitoring and fault detection.

Keywords Wastewater treatment · Fault detection · Environmental engineering · Compliance · Conventional monitoring · AI integration · Real-time monitoring · Chemical oxygen demand (COD) · pH levels · Predictive analytics · Machine

---

A. Mohanty (✉) · A. G. Mohapatra
Electronics Engineering, Silicon University, Bhubaneswar, Odisha, India
e-mail: anitamohanty776@gmail.com

S. K. Mohanty
Electronics and Communication Engineering, Einstein Academy of Technology and Management, Bhubaneswar, Odisha, India

M. C. Garg (ed.), *The AI Cleanse: Transforming Wastewater Treatment Through Artificial Intelligence*, Springer Water, https://doi.org/10.1007/978-3-031-67237-8_7

learning algorithms · Biological oxygen demand (BOD) · Neural networks · AI-enhanced systems · Data analysis · Sensors · Remote sensing technologies · IoT devices

7.1 Introduction

This chapter starts by providing an overview of wastewater treatment systems, emphasizing the importance of real-time monitoring and fault detection. Integration of AI in these systems is highlighted as a transformative approach.

Wastewater treatment stands as a cornerstone of environmental engineering, playing a pivotal role in safeguarding ecological sustainability and public health [1]. As populations grow and industrialization continues, the demands on wastewater treatment systems intensify. Meeting these escalating demands requires a shift from conventional practices to innovative, technologically advanced solutions [2, 3]. In this context, the integration of Artificial Intelligence (AI) into wastewater treatment systems emerges as a transformative paradigm. This chapter addresses the critical imperatives of real-time monitoring and fault detection within the context of AI-enhanced wastewater treatment systems. The need for real-time monitoring is underscored by the complexity of modern treatment facilities and the ever-stricter environmental regulations governing effluent quality. Simultaneously, the integration of AI technologies, including machine learning and neural networks, offers unprecedented opportunities for enhancing the efficiency, reliability, and sustainability of wastewater treatment processes.

7.1.1 Overview of Wastewater Treatment Systems

To provide a comprehensive understanding of the subject matter, this chapter is structured to guide the reader through a multifaceted exploration. It commences by elucidating the fundamental components of wastewater treatment, establishing the foundation for the subsequent discussions. The limitations of conventional monitoring methodologies are juxtaposed against the potential advantages of real-time monitoring and AI-driven approaches. Key parameters, such as chemical oxygen demand (COD), biological oxygen demand (BOD), and pH levels, are scrutinized in the context of real-time monitoring and predictive analytics [4, 5]. Advanced AI applications, ranging from machine learning algorithms to neural networks, are examined for their role in predictive modeling and data analysis, with a focus on their tangible benefits in wastewater treatment optimization [6].

7.1.2 Importance of Real-Time Monitoring and Fault Detection

This chapter also delves into the realm of real-time monitoring sensors and devices, showcasing contemporary technologies such as IoT devices, remote sensing technologies, and online sensors. These technological advancements are integral to the acquisition of real-time data, enabling timely and informed decision-making. Furthermore, the imperative of fault detection and diagnosis within wastewater treatment systems is addressed comprehensively. Common faults that plague wastewater treatment processes are identified, and AI-driven fault detection algorithms are explored as an effective means to mitigate operational disruptions and ensure process stability [7].

7.1.3 Integration of AI in Wastewater Treatment Systems

The efficacy of AI implementations in wastewater treatment is exemplified through case studies, providing insights into successful real-world applications and the tangible benefits realized in terms of operational efficiency and regulatory compliance. Nonetheless, these advancements are not devoid of challenges. Privacy concerns, data security, and the complexities of integration remain significant hurdles that demand strategic resolutions.

Finally, the chapter concludes by examining the future trends and prospects in the realm of AI-enhanced wastewater treatment, reaffirming the central role of real-time monitoring and fault detection in driving the evolution of more holistic and efficient wastewater treatment paradigms.

7.2 Components of Wastewater Treatment Systems

Components of wastewater treatment systems are essential in ensuring the effective removal of contaminants and pollutants from wastewater before its safe discharge into the environment [8, 9]. These systems typically consist of several key components working in tandem as shown in Fig. 7.1. Primary treatment involves physical processes such as screening and sedimentation, where large solids are removed, and the wastewater is allowed to settle. Secondary treatment employs biological processes, using microorganisms to break down organic matter. Tertiary treatment, if necessary, further refines the water quality through advanced processes like filtration and chemical disinfection. Throughout this process, monitoring and control systems play a crucial role in regulating the treatment parameters, ensuring that the effluent meets environmental standards. Additionally, sludge management and disposal are essential elements to handle the solid by-products generated during

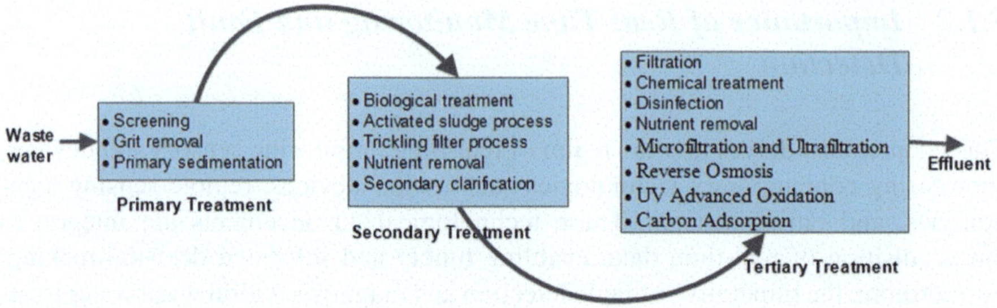

Fig. 7.1 Componemts of wastewater treatment system

treatment. Together, these components form a comprehensive wastewater treatment system that safeguards public health and preserves the ecological balance of natural water bodies.

7.2.1 *Primary Treatment*

It is the initial stage of wastewater treatment and plays a critical role in the removal of physical contaminants and some organic matter from raw wastewater. This process is essential for protecting downstream treatment units and ensuring the effectiveness of subsequent treatment steps. The steps of primary treatment in wastewater treatment systems are:

- **Screening**: In primary treatment, the first step is removing large debris from incoming wastewater. This debris, like sticks, leaves, and plastics, can clog pumps and harm equipment downstream. Screening passes wastewater through coarse screens or bars with openings for water to pass, trapping larger items. Collected debris is compacted and sent to a landfill.
- **Grit Removal**: After screening, the wastewater may still contain small, dense particles like sand, gravel, and coffee grounds, collectively known as "grit." Grit can wear down equipment and interfere with subsequent treatment processes. To remove grit, the wastewater is typically directed into a grit chamber, where the velocity is reduced to allow the grit to settle to the bottom. The settled grit is then removed and disposed of.
- **Primary Sedimentation**: After screening and grit removal, wastewater enters primary sedimentation tanks. These tanks slow down the flow, causing suspended solids to settle. Heavier solids, including organic matter and some inorganic particles, form sludge at the tank bottom, while clear water rises to the top. Periodically, the sludge is removed for further treatment and disposal.

- **Effluent Discharge**: The clarified water, cleaner than raw wastewater, exits the primary sedimentation tanks. However, it still contains dissolved and fine particulate contaminants, requiring additional treatment in secondary and, if needed, tertiary processes.

So, Primary treatment is a vital first step in wastewater treatment, focused on removing physical contaminants and reducing organic load. While it improves wastewater quality, secondary and, sometimes, tertiary treatments follow to further enhance water quality and meet environmental regulations before discharge.

7.2.2 Secondary Treatment

Secondary treatment is a critical stage in wastewater treatment that follows primary treatment and focuses on the removal of dissolved and colloidal organic matter, as well as nutrients like nitrogen and phosphorus. This step further purifies the wastewater, making it safer for discharge into the environment or for reuse. The steps of secondary treatment in wastewater treatment systems are:

- **Biological Treatment**: Secondary treatment relies on biological processes to degrade and eliminate organic pollutants. These processes employ microorganisms, mainly bacteria (and occasionally fungi), to consume and transform organic material into biomass, carbon dioxide, and water. The two primary biological treatment methods are the activated sludge process and the trickling filter process.
- **Activated Sludge Process**: At the core of secondary treatment lies the utilization of biological processes to degrade and eliminate organic pollutants. These processes employ microorganisms, mainly bacteria and occasionally fungi, to digest and transform organic substances into biomass, carbon dioxide, and water. The activated sludge process and the trickling filter process are the two prevalent biological treatment methods.
- **Trickling Filter Process**: In the trickling filter process, wastewater flows over a bed of solid media, like rocks, gravel, or plastic, where microorganisms attached to the media surface consume organic matter. Introducing air supplies oxygen for microbial activity. After treatment, the wastewater goes through a clarifier to separate solids from the liquid before release.
- **Nutrient Removal**: Besides organic matter, secondary treatment can also tackle nutrient removal. Certain treatment plants employ processes like biological nutrient removal (BNR) to lower nitrogen and phosphorus levels in the effluent. This is crucial in regions where nutrient pollution can cause water quality problems.
- **Secondary Clarification**: After biological treatment, the wastewater undergoes a secondary clarification step to separate the treated water from the remaining biomass (activated sludge). This clarifier allows the settled biomass to be returned to the biological reactor (if using the activated sludge process) or to be processed further for sludge management.

Hence secondary treatment is a crucial step in wastewater treatment that relies on biological processes to remove dissolved and colloidal organic pollutants and, in some cases, nutrients. It plays a pivotal role in improving water quality, protecting the environment, and ensuring compliance with water quality regulations.

7.2.3 Tertiary Treatment

Tertiary treatment is the last phase in wastewater treatment, following primary and secondary steps. Its main aim is to further enhance treated wastewater quality, meeting specific water quality standards or making it suitable for various uses like irrigation, industrial processes, or even drinking water supply, depending on the required treatment level. Tertiary treatment involves advanced and specialized processes to achieve these objectives. Here's a detailed explanation of tertiary treatment in wastewater systems:

- **Filtration**: Filtration plays a vital role in tertiary treatment, where treated wastewater is passed through fine materials like sand, anthracite coal, or membranes. This process effectively removes any remaining suspended particles, colloids, and microorganisms, resulting in clear, high-quality water.
- **Chemical Treatment**: Chemical treatment is also frequently employed during tertiary treatment to eliminate contaminants that are resistant to biological processes. Common chemical treatments include coagulation, flocculation, and the addition of chemicals like ferric chloride or alum to precipitate and remove impurities.
- **Disinfection**: Disinfection is vital in tertiary treatment to rid wastewater of harmful microorganisms like bacteria, viruses, and pathogens. Common methods are chlorine, UV radiation, or ozone, chosen based on water quality needs. Each method has its advantages and is chosen based on specific water quality requirements.
- **Nutrient Removal (Advanced)**: In some cases, especially when wastewater discharge can cause nutrient pollution in receiving waters, advanced nutrient removal may be part of tertiary treatment. This includes extra steps to reduce nitrogen and phosphorus levels in the effluent, often using biological methods.
- **Microfiltration and Ultrafiltration**: These membrane-based processes remove even smaller particles and microorganisms from treated wastewater, producing very high-quality effluent.
- **Reverse Osmosis (RO)**: An advanced membrane process, RO removes nearly all contaminants from treated wastewater, including salts, organic matter, and microorganisms. It's used when extremely pure water is needed, like for drinking water or industrial processes.
- **UV Advanced Oxidation**: UV advanced oxidation combines UV radiation with advanced oxidants like hydrogen peroxide to effectively break down and eliminate persistent organic pollutants and trace contaminants.

- **Carbon Adsorption**: This method passes treated wastewater through activated carbon beds to adsorb and remove trace organic compounds, as well as taste and odor compounds. It is commonly used to enhance water quality for drinking water production.

So tertiary treatment is the final step in wastewater treatment, designed to further improve water quality, remove remaining contaminants, and meet specific water quality standards or reuse requirements. It employs advanced processes such as filtration, chemical treatment, disinfection, and membrane technologies to produce high-quality effluent suitable for various purposes, from industrial processes to potable water supply.

7.3 Need for Real-Time Monitoring as Well as Fault Detection Methodology

Continuous real-time monitoring and fault detection are essential in wastewater treatment. They promptly spot and address anomalies, enhancing treatment efficiency and minimizing environmental risks. Real-time data acquisition allows quick intervention, ensuring compliance with environmental regulations by monitoring key parameters like pollutant levels, pH, and turbidity [10]. Adjustments can maintain legal limits. This data-driven approach optimizes processes, conserves resources, and improves sustainability while preventing equipment failures and environmental risks. Real-time monitoring's crucial role is in protecting public health, resource allocation, and wastewater treatment's environmental integrity.

7.3.1 Environmental Compliance and Regulations

Environmental compliance and regulations are important aspects of real-time monitoring and fault detection in wastewater treatment. To ensure the protection of natural resources, ecosystems, and public health, wastewater treatment facilities must stick to strict environmental standards. Real-time monitoring systems always collect data on parameters like pollutant levels, pH, turbidity, and flow rates [11]. These systems help wastewater treatment plants conform with regulatory limits by giving instant alerts when parameters exceed allowable thresholds. In the occurrence of non-compliance, immediate corrective actions can be taken to avoid environmental harm and legal consequences.

Also, environmental regulations permit the reporting of data to regulatory authorities and real-time monitoring systems to ease this process by automatically producing and transmitting compliance reports. This ensures secure transparency and accountability, which are vital in demonstrating adherence to environmental laws.

Table 7.1 Environmental compliance and regulations in real-time monitoring and fault detection in wastewater treatment

Regulation/aspect	Description
Regulatory authorities	Government agencies responsible for overseeing and enforcing environmental regulations (e.g., EPA)
Key parameters for monitoring	Parameters include pollutant levels, pH, turbidity, temperature, flow rates, and dissolved oxygen
Real-time monitoring systems	Continuously monitor parameters and provide immediate alerts for deviations from regulatory limits
Compliance reporting	Regulations require facilities to report real-time monitoring data to regulatory authorities regularly
Fault detection methodologies	Continuous monitoring of equipment performance to detect malfunctions or failures in real-time
Preventative maintenance	Timely fault detection allows for proactive maintenance, preventing equipment failures and environmental hazards
Legal consequences	Non-compliance with regulations can result in fines, shutdown orders, or legal liability for environmental harm

Fault detection methodologies play a complementary role in environmental compliance. By continuously monitoring equipment performance, these methodologies can identify malfunctions or failures in real-time. Timely detection of faults is essential to prevent environmental hazards, as equipment failures can lead to wastewater spills or suboptimal treatment, violating environmental regulations. Rapid fault detection enables swift maintenance or shutdown procedures to rectify issues before they escalate, thus maintaining compliance with regulations.

Overall, the integration of real-time monitoring and fault detection methodologies in wastewater treatment is vital not only for optimizing processes and resource utilization but also for ensuring strict compliance with environmental regulations, ultimately protecting the environment and public health. Environmental Regulations in Real-Time Monitoring and Fault Detection in Wastewater Treatment are mentioned in Table 7.1.

7.3.2 *Operational Efficiency and Cost Reduction*

Operational efficiency and cost reduction are significant benefits of implementing real-time monitoring and fault detection systems in wastewater treatment facilities [12]. Here in Table 7.2, the key points related to these advantages are summarised.

Table 7.2 Operational efficiency and cost reduction in real-time monitoring and fault detection in wastewater treatment

Aspect	Description
Real-time process optimization	Continuous data collection and analysis optimize treatment processes, reducing resource wastage and energy consumption
Improved resource allocation	Data-driven insights enable better allocation of chemicals, energy, and labor, minimizing operational costs
Energy efficiency	Real-time monitoring identifies energy-intensive operations and suggests energy-saving measures, reducing electricity costs
Proactive maintenance	Fault detection allows for timely equipment maintenance, preventing costly breakdowns and minimizing downtime
Reduced repair and replacement costs	Early detection of equipment issues lowers repair and replacement costs, contributing to long-term cost savings
Preventing environmental penalties	Compliance with regulations through real-time monitoring avoids fines and penalties for non-compliance
Lower risk of accidental spills	Timely detection of faults reduces the risk of accidental spills, saving cleanup costs and potential legal liabilities
Data-driven decision-making	Data analytics support informed decision-making, optimizing processes, and reducing inefficiencies
Predictive maintenance	Fault detection systems can predict equipment failures, allowing for proactive maintenance and reducing unexpected costs
Extended asset lifespan	Early fault detection and maintenance extend the lifespan of treatment plant assets, reducing capital expenditures

7.3.3 Public Health and Safety

Public health and safety are paramount considerations in the implementation of real-time monitoring and fault detection systems in wastewater treatment facilities [13]. Table 7.3 summarizes the key points related to public health and safety.

Table 7.3 Public health and safety in real-time monitoring and fault detection systems in wastewater treatment facilities

Aspect	Description
Early detection of contamination	Real-time monitoring detects changes in water quality, ensuring early identification of contamination events that could harm public health
Protection of drinking water sources	Continuous monitoring safeguards the quality of source waters, ensuring the safety of drinking water supplies
Mitigation of disease spread	Prompt detection of wastewater treatment issues prevents the release of harmful pathogens into the environment, reducing the risk of waterborne diseases
Emergency response	Real-time systems trigger immediate responses to abnormal conditions, allowing for rapid containment and mitigation of hazards
Compliance with health regulations	Adherence to environmental and health regulations through real-time monitoring ensures the protection of public health
Minimizing exposure to harmful substances	Fault detection prevents accidental spills or releases of hazardous materials, minimizing public exposure to harmful substances
Preventing contamination events	Early fault detection prevents equipment failures that could lead to contamination events, safeguarding public health
Transparency and accountability	Public access to real-time data fosters transparency and ensures that facilities are held accountable for their impact on public health and safety
Disaster preparedness	Real-time monitoring systems enhance disaster preparedness by providing data for emergency response planning and management
Verification of treatment effectiveness	Continuous monitoring verifies that wastewater treatment processes effectively remove contaminants, ensuring public health protection

7.4 Traditional Monitoring Versus Real-Time Monitoring

Traditional monitoring involves periodic data collection with significant delays and potentially missing short-term changes. In contrast, real-time monitoring offers immediate, continuous data, crucial for applications like environmental monitoring, industrial control, and infrastructure management [14, 15]. Though costlier to implement, it provides early warnings and accuracy, enabling proactive optimization and risk mitigation.

7.4.1 *Limitations of Traditional Monitoring*

Traditional monitoring, while valuable in many contexts, comes with several limitations that can affect its effectiveness in various applications. Some of the key limitations of traditional monitoring are shown in Fig. 7.2.

- **Data Lag**: Conventional monitoring collects data at set intervals, leading to delays before analysis. This can impede prompt decision-making, particularly in situations demanding rapid responses.
- **Sampling Bias**: Traditional monitoring relies on sample-based data collection, which can introduce sampling bias. The selected samples may not fully represent the entire population or the complete range of conditions, leading to potentially skewed results.
- **Limited Temporal Resolution**: Traditional monitoring may miss short-term fluctuations or events that occur between sampling intervals. This limitation can be critical in applications where real-time or near-real-time insights are crucial.
- **Resource-Intensive**: Traditional monitoring, involving sample collection and lab analysis, demands considerable time, labor, and equipment resources. This complexity can result in high costs and logistical difficulties, particularly in long-term monitoring endeavors.
- **Reduced Sensitivity**: Traditional monitoring may not be sensitive enough to detect subtle changes or low-level variations in data. This limitation can lead to underestimating the significance of certain trends or issues.
- **Limited Spatial Coverage**: Traditional monitoring often involves collecting data from specific sampling locations. This can limit the spatial coverage and may not capture variations in different areas or regions.

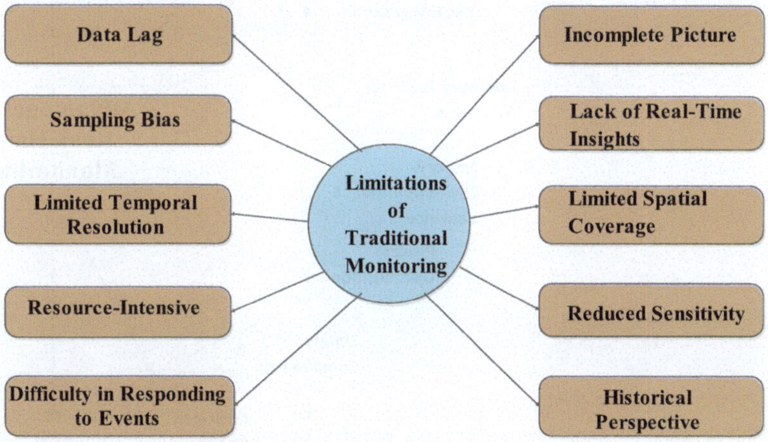

Fig. 7.2 Factors affecting the effectiveness of traditional monitoring

- **Lack of Real-Time Insights**: In situations requiring real-time insights (e.g., environmental monitoring, healthcare), traditional methods fall short by lacking immediate data access and early event warnings.
- **Incomplete Picture**: Traditional monitoring offers a glimpse of data at specific moments, potentially missing the intricate nature of dynamic systems. This constraint can lead to an incomplete comprehension of the monitored phenomena.
- **Difficulty in Responding to Events**: Traditional monitoring may not enable quick responses to rapidly changing conditions or emergencies. This limitation can pose challenges in situations where immediate action is necessary.
- **Historical Perspective**: While traditional monitoring is valuable for historical data analysis and trend identification, it may not support real-time decision-making, making it less suitable for applications requiring agile responses.

7.4.2 Advantages of Real-Time Monitoring

Real-time monitoring offers numerous advantages across various industries and applications due to its ability to continuously collect and analyze data at high frequencies. The key advantages of real-time monitoring are shown in Fig. 7.3.

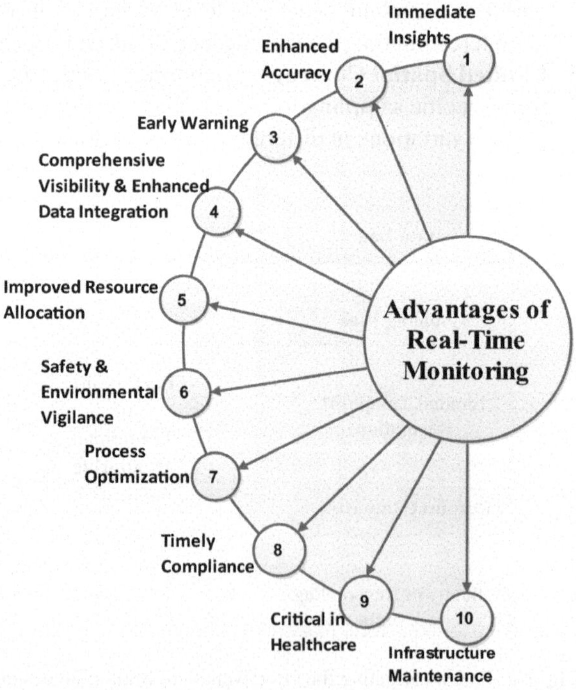

Fig. 7.3 Key advantages of real-time monitoring

- **Immediate Insights**: Real-time monitoring provides instant data access for quick decision-making in emergency situations or process control.
- **Enhanced Accuracy**: Continuous data collection decreases information gaps, ensuring a more specific understanding of the system's current status.
- **Early Warning**: Real-time systems identify anomalies as they occur, allowing proactive interventions to avoid problems and decrease risks.
- **Comprehensive Visibility and Enhanced Data Integration**: Continuous data streams put forward a holistic understanding of processes, revealing secret trends. Real-time data mix seamlessly with control systems and predictive tools, increasing system efficiency and performance.
- **Improved Resource Allocation**: Real-time data can assist organizations optimize resource allocation by finding areas of inefficiency, overuse, or underuse. This can lead to cost savings and better resource utilization.
- **Safety and Environmental Vigilance**: Real-time monitoring provides safety in crucial applications, offering immediate alerts to avoid accidents and protect assets. It also aids environmental stewardship by swiftly detecting pollution events and supporting sustainability efforts.
- **Process Optimization**: Real-time monitoring aids in process optimization by continuously tracking variables, making adjustments in real-time, and maximizing efficiency. This is particularly beneficial in industrial processes, energy management, and quality control.
- **Timely Compliance**: Industries subject to regulatory requirements can benefit from real-time monitoring to ensure compliance with environmental, safety, and quality standards. Immediate data availability facilitates timely reporting and regulatory adherence.
- **Critical in Healthcare**: In healthcare, real-time patient monitoring is essential for tracking vital signs, detecting medical emergencies, and providing timely interventions. It improves patient outcomes and reduces the risk of adverse events.
- **Infrastructure Maintenance**: Real-time monitoring of critical infrastructure, such as bridges, dams, and pipelines, can detect structural issues or anomalies in real-time, allowing for preventive maintenance and minimizing the risk of catastrophic failures.

7.5 AI Applications in Wastewater Treatment

Artificial Intelligence (AI) is revolutionizing wastewater treatment, boosting efficiency, sustainability, and effectiveness. Key AI applications include predictive maintenance to reduce downtime, process optimization to enhance treatment efficiency and reduce energy use, and energy management for cost and environmental impact reduction [16, 17]. AI excels in anomaly detection, ensuring water quality and quick issue response, and continuous water quality monitoring for regulatory compliance and pollution detection. These AI applications, along with asset management, plant design, resource recovery, and data analytics, promote sustainable, efficient wastewater treatment, and reduce chemical usage. As AI technology evolves, it promises to further transform treatment, improve water quality, and conserve resources in an environmentally conscious world.

7.5.1 Machine Learning Algorithms

Machine learning algorithms are finding diverse applications in wastewater treatment, offering data-driven solutions to optimize processes, enhance efficiency, and improve water quality [18]. Here are some key applications of machine learning algorithms in wastewater treatment:

- Predictive maintenance for reducing equipment downtime and extending infrastructure life.
- Real-time process optimization to improve treatment efficiency.
- Energy management for minimizing consumption and costs.
- Anomaly detection to identify issues promptly.
- Water quality prediction to ensure regulatory compliance.
- Asset monitoring for effective maintenance planning.
- Treatment process design for efficiency.
- Resource recovery for sustainability.
- Data analytics for informed decision-making.
- Remote control for improved responsiveness.
- Compliance assurance for regulatory adherence.
- Reduced chemical usage for cost and environmental benefits.

7.5.2 Neural Networks

Neural networks, a subset of artificial intelligence, are increasingly being applied to wastewater treatment processes to improve efficiency, optimize operations, and enhance water quality [19]. Here are several key applications of neural networks in wastewater treatment:

- They adjust treatment parameters using real-time sensor data, improving efficiency and cutting costs.
- They forecast parameters, aiding decision-making and compliance.
- They spot irregularities, enabling swift responses for water quality maintenance.
- Neural networks control energy use for sustainability and cost reduction.
- They predict water quality for regulatory adherence.
- Neural networks detect potential equipment issues, reducing downtime.
- They analyze data for informed decisions and regulatory reporting.
- They enable remote control, reducing on-site staff needs.

- Neural networks identify resource opportunities for sustainability.
- They ensure regulatory adherence through continuous monitoring.

7.5.3 Data Analysis and Predictive Modeling

Data analysis and predictive modeling are powerful tools applied in wastewater treatment to improve operational efficiency, optimize processes, and ensure compliance with environmental regulations [20]. Here are key applications of data analysis and predictive modeling in wastewater treatment:

- Ongoing data analysis tracks parameters, aiding pollution detection and regulatory compliance.
- Data analysis and predictive modeling enhance treatment efficiency, lowering costs.
- Predictive models schedule proactive maintenance, reducing downtime.
- Data analysis optimizes energy usage, cutting costs and promoting sustainability.
- Data analysis and modeling automate reporting, ensuring compliance.
- Data analysis identifies resource opportunities for sustainability.
- Predictive modeling detects equipment issues for proactive maintenance.
- Data analysis optimizes water flow for efficient treatment.
- Predictive modeling provides real-time insights for quick decisions.
- Data analysis reduces inefficiencies, lowering operational costs.
- Predictive modeling creates early warning systems for timely corrective actions.
- Predictive models integrate into control systems for consistent treatment quality.

7.6 Key Parameters for Monitoring

Monitoring key parameters in wastewater treatment is crucial to ensuring the effective and environmentally responsible treatment of sewage and industrial wastewater. These parameters help operators maintain compliance with regulatory standards, optimize treatment processes, and safeguard the environment [21]. Here are some of the key parameters that are typically monitored in wastewater treatment:

- **Flow Rate**: Monitoring the flow of wastewater helps in controlling and optimizing treatment processes.
- **pH Level**: pH indicates the acidity or alkalinity of wastewater, which is crucial for the efficiency of chemical treatments and biological processes.
- **Dissolved Oxygen (DO)**: DO levels are essential for aerobic biological treatment processes, and low DO can indicate insufficient aeration.
- **Temperature**: Temperature affects the rate of biological reactions and can affect treatment efficiency.
- **Turbidity**: Turbidity measures the vagueness or clarity of water and is an indicator of suspended solids in wastewater.

- **Chemical Oxygen Demand (COD)**: COD estimates the amount of oxygen needed to chemically oxidize organic and inorganic matter in wastewater, providing insights into its pollutant load.
- **Biological Oxygen Demand (BOD)**: BOD calculates the oxygen demand of microorganisms decomposing organic matter in wastewater, reflecting its organic content and biodegradability.
- **Total Suspended Solids (TSS)**: TSS estimates the concentration of suspended solids in wastewater, which can influence treatment processes and water quality.
- **Total Dissolved Solids (TDS)**: TDS measures the concentration of dissolved solids in wastewater, which can contain salts, minerals, as well as other substances.
- **Nutrient Levels (Nitrogen and Phosphorus)**: Monitoring nutrient levels is vital for preventing nutrient pollution as well as eutrophication in receiving waters.
- **Heavy Metals**: Monitoring heavy metal concentrations is crucial because of their toxicity as well as their potential for environmental harm.
- **Pathogen Levels**: Monitoring the existence of pathogens like bacteria and viruses is crucial for public health and regulatory compliance.
- **Specific Pollutants**: Depending on the industrial or municipal source of wastewater, distinguished pollutants such as oil and grease, industrial chemicals, or pharmaceuticals may require monitoring.
- **Residual Chlorine**: In cases where chlorine is used for disinfection, monitoring residual chlorine levels ensures effective disinfection without exceeding regulatory limits.
- **Sulphides and Ammonia**: These parameters can indicate anaerobic conditions or nutrient imbalances in wastewater.
- **Soluble Organic Compounds**: Monitoring specific organic compounds may be necessary to assess the efficiency of treatment for certain industrial wastewater streams.
- **Odor and Colour**: Although not specific parameters, monitoring odor and colour can provide valuable information about the quality of wastewater and treatment effectiveness.
- **Sedimentation and Settling Characteristics**: Monitoring the settling characteristics of wastewater can help assess the performance of sedimentation processes.

7.6.1 Chemical Oxygen Demand (COD)

Chemical Oxygen Demand (COD) is a critical water quality parameter used to measure the amount of organic and inorganic pollutants in a water sample. It quantifies the oxygen required to chemically oxidize the organic and oxidizable inorganic compounds present in the water [22]. COD is typically expressed in milligrams per liter (mg/L) or parts per million (ppm) of oxygen.

Key points about Chemical Oxygen Demand (COD) include:

1. **Purpose**: COD testing is accomplished to assess the pollution level of water bodies, especially wastewater. It gives information about the number of pollutants that can be oxidized and helps to find out the treatment requirements for water and wastewater.
2. **Measurement Principle**: COD is found by oxidizing the organic and inorganic substances in a water sample with a strong chemical oxidizing agent, usually potassium dichromate (K2Cr2O7), in the presence of a strong acid. The amount of oxygen absorbed at the time of this chemical reaction is proportional to the COD.
3. **Units**: COD is typically expressed in milligrams of oxygen absorbed per liter (mg/L) of the sample. It can also be written as parts per million (ppm).
4. **Applications**:

   (a) **Wastewater Treatment**: COD is generally used in wastewater treatment plants to determine the organic load in influent and effluent streams.
   (b) **Environmental Monitoring**: It is also used to assess the impact of industrial discharges and evaluate the water quality of natural water bodies.
   (c) **Quality Control**: COD is used in industries to check the water quality used in processes and to establish compliance with environmental regulations.
   (d) **Research**: Scientists use COD measurements to analyze water chemistry, determine the effectiveness of water treatment techniques, and conduct environmental research.

5. **Importance**:

   (a) COD gives a more comprehensive assessment of water pollution than Biological Oxygen Demand (BOD) testing, which can take several days to complete. COD results are available much more quickly.
   (b) It helps analyze both biodegradable and non-biodegradable pollutants, containing organic chemicals, ammonia, and oxidizable inorganic compounds.
   (c) COD measurements can help wastewater treatment plants in optimizing chemical doses, aeration, and other treatment processes to take out pollutants.

6. **Limitations**:

   (a) COD does not provide information about specific pollutants or their toxicity.
   (b) Chloride ions and certain reducing agents can affect COD results.
   (c) The COD test involves the use of hazardous chemicals.

7. **Environmental Impact**: High COD levels in wastewater can deplete dissolved oxygen in receiving water bodies when the treated wastewater is discharged. This oxygen depletion can harm aquatic life and impact the ecological balance.

7.6.2 *Biological Oxygen Demand (BOD)*

Biological Oxygen Demand (BOD) is a crucial water quality parameter used to assess the level of organic pollution in natural water bodies and wastewater. It measures the amount of dissolved oxygen (DO) consumed by microorganisms as they biologically decompose organic matter present in the water [23]. BOD is a fundamental test in environmental science and wastewater treatment.

Key points about Biological Oxygen Demand (BOD) include:

1. **Purpose**: BOD testing is accomplished to find the amount of biodegradable organic matter in water. It helps to assess the pollution level, measure the potential effect on aquatic ecosystems, and calculate the effectiveness of wastewater treatment.
2. **Measurement Principle**: BOD is calculated by incubating a water sample in a sealed container for a specific period (commonly 5 days at 20 °C) under controlled environmental conditions. During incubation, microorganisms exist in the sample metabolize the organic matter, and consume oxygen. The dissimilarity in dissolved oxygen concentration before and after incubation is utilized to evaluate BOD.
3. **Units**: BOD is typically conveyed in milligrams of oxygen consumed per liter (mg/L) of the sample. It can also be known as parts per million (ppm).
4. **Applications**:

   (a) **Environmental Monitoring**: It is used to find the impact of wastewater discharges on natural water bodies, such as rivers, lakes, and streams.
   (b) **Wastewater Treatment**: BOD is a vital parameter in wastewater treatment plants to calculate the efficiency of biological treatment processes.
   (c) **Compliance and Regulation**: Regulatory agencies regularly set limits on BOD levels in wastewater discharge permits to take care of aquatic ecosystems.
   (d) **Research**: Scientists utilize BOD measurements to analyze water quality, evaluate the performance of treatment processes, and carry on ecological research.

5. **Importance**:

   (a) BOD gives information about the biodegradable organic content in water, which is crucial for understanding the potential for oxygen depletion and harm to aquatic life.
   (b) It helps wastewater treatment plants optimize aeration and biological treatment processes so as to efficiently remove organic pollutants.
   (c) BOD testing is mostly used for determining the self-purification capacity of natural water bodies.

6. **Limitations**:

   (a) BOD testing is time-consuming, as it typically requires a 5-day incubation period.

(b) The Correctness of BOD results can be affected by components such as temperature, nutrient availability, and the types of microorganisms present in the sample.
(c) BOD measures only biodegradable organic matter and does not report for non-biodegradable pollutants or inorganic substances.

7. **Environmental Impact**: Increased BOD levels in natural water bodies can point to oxygen depletion (hypoxia) when microbial decomposition of organic matter absorbs large amounts of dissolved oxygen. This oxygen depletion can spoil aquatic organisms, potentially killing fish and creating ecological imbalances.

7.6.3 Suspended Solids (SS)

Suspended Solids (SS), also known as Total Suspended Solids (TSS), are a fundamental water quality factor that evaluates the concentration of solid particles that are suspended in water. These particles can have a wide range of materials, such as silt, clay, organic matter, microorganisms, and other fine particulate matter. Monitoring suspended solids is crucial in a variety of environmental, industrial, and water treatment contexts [24].

Basic points about Suspended Solids (SS) include:

1. **Purpose**: SS measurement is applied to verify the turbidity, clarity, and overall quality of water. It is a crucial parameter for checking the effect of sediment and particulate matter on aquatic ecosystems and water treatment processes.
2. **Measurement Principle**: The intensity of suspended solids is generally estimated by allowing a known volume of water through a filter with a specific pore size (usually 0.45 μm). The solid particles in the water are caught on the filter and the increase in mass of the filter after filtration is applied to estimate the concentration of suspended solids in the original water sample.
3. **Units**: SS is generally determined in milligrams per liter (mg/L) or parts per million (ppm) of the sample.
4. **Applications**:

   (a) **Environmental Monitoring**: SS measurements are utilized to detect the impact of land runoff, erosion, and sedimentation in natural water bodies like rivers, lakes, and coastal areas.
   (b) **Wastewater Treatment**: SS levels are supervised in wastewater treatment plants to detect the efficiency of solids removal processes, such as sedimentation and filtration.
   (c) **Industrial Processes**: Various industries, such as mining and construction, monitor SS levels to prevent the release of suspended solids into the environment.
   (d) **Water Quality Assessment**: SS is an essential parameter in water quality assessments and compliance with regulatory standards.

5. **Importance**:
   (a) Elevated levels of suspended solids can reduce water clarity, hinder light penetration, and disrupt aquatic ecosystems by affecting the habitat and food sources of aquatic organisms.
   (b) In wastewater treatment, effective removal of suspended solids is crucial to prevent clogging of pipes, pumps, and equipment, and to ensure treated water meets quality standards.
6. **Limitations**:
   (a) The accuracy of SS measurements can be influenced by factors like the filter pore size, filtration time, and sample handling procedures.
   (b) SS measurements do not differentiate between types of particles; they provide a total concentration of all suspended materials.
   (c) Very small particles or colloids may not be effectively captured by standard filtration methods.
7. **Environmental Impact**: High levels of suspended solids in natural water bodies can lead to reduced light penetration, decreased photosynthesis, and sediment deposition, impacting aquatic flora and fauna. It can also degrade water quality and reduce the aesthetic value of water bodies.

7.6.4 pH Levels

pH levels, often simply referred to as pH, are a measure of the acidity or alkalinity of a solution. The pH scale ranges from 0 to 14, with 7 being considered neutral. Values below 7 indicate acidity, while values above 7 indicate alkalinity. The pH scale is logarithmic, meaning that each whole number change on the scale represents a tenfold difference in acidity or alkalinity [25]. Here are some key points about pH levels:

1. **Definition**: pH stands for "power of hydrogen" and shows the concentration of hydrogen ions (H^+) in a solution. The larger the concentration of H^+ ions, the more acidic the solution, and the lower the pH. Conversely, the lower the concentration of H^+ ions, the more alkaline (or basic) the solution, and the higher the pH.
2. **Measurement**: pH is typically estimated using a pH meter or pH indicator strips (pH paper). The pH meter gives a numerical pH value, while pH paper changes colour to indicate the approximate pH of a solution based on a colour chart.
3. **Scale**: The pH scale ranges from 0 to 14:
   (a) pH 0–6: Acidic (e.g., lemon juice, vinegar)
   (b) pH 7: Neutral (e.g., pure water)
   (c) pH 8–14: Alkaline (e.g., baking soda, bleach)
4. **Importance**: pH is a crucial parameter in various fields and applications, including:

(a) **Environmental Monitoring**: Determining the acidity or alkalinity of natural waters, soils, and ecosystems.
(b) **Water Treatment**: Making sure that drinking water is within an acceptable pH range.
(c) **Chemical and Biological Processes**: pH affects the behaviour of chemicals and enzymes in chemical reactions and biological processes.
(d) **Industrial Processes**: Monitoring and controlling pH in various manufacturing processes, like food production, pharmaceuticals, and chemical industries.
(e) **Aquariums and Pools**: Preserving pH levels in aquatic environments to help the health of aquatic organisms and prevent corrosion or scaling.

5. **Buffer Solutions**: pH levels can be affected by substances called buffers, which oppose changes in pH and help to maintain stability in solutions. Buffer solutions are commonly utilized in laboratory work as well as some industrial processes.
6. **Health**: In the human body, pH plays a vital role in continuing the proper functioning of biological systems. Blood, for example, has a pH range of approximately 7.35–7.45.
7. **Environmental Impact**: Extreme deviations from natural pH levels in ecosystems can harm aquatic life. Acid rain, for instance, can lower the pH of bodies of water, negatively impacting aquatic organisms.
8. **Adjusting pH**: pH can be adjusted using acids or bases to bring a solution into a desired pH range. For example, adding sodium hydroxide (a base) can increase the pH of a solution, while adding hydrochloric acid (an acid) can decrease it.
9. **pH in Foods**: pH is a critical factor in food preservation and flavour. It influences the texture and taste of various foods and is utilized in processes like fermentation and canning.

7.7 Real-Time Monitoring Sensors and Devices

In wastewater treatment, real-time sensors are vital for precise process management. They continuously gather data on parameters such as flow rate, pH, DO, turbidity, and pollutant concentrations, allowing operators to make immediate adjustments, optimize resources, and ensure compliance with regulations [26]. pH sensors keep optimal pH levels for effective biological treatment, while DO sensors provide oxygen to microorganisms. Sensors for turbidity, TSS, COD, and nutrients offer insights for quick corrective actions. Overall, real-time monitoring improves precision, responsiveness, and environmental protection in wastewater treatment.

7.7.1 *Online Sensors*

Online sensors for real-time monitoring are advanced instruments designed to provide continuous, instantaneous, and remote data collection and analysis across various industries and applications. These sensors are vital for maintaining process efficiency, ensuring safety, and making informed decisions in real time. Some common types of online sensors for real-time monitoring are:

1. **Online Water Quality Sensors**: These sensors are used in environmental monitoring and water treatment applications. They include:
   (a) **Online pH Sensors**: Continuously monitor the acidity or alkalinity of water.
   (b) **Online Dissolved Oxygen (DO) Sensors**: Measure the concentration of oxygen dissolved in water, crucial for aquatic ecosystems and wastewater treatment.
   (c) **Online Turbidity Sensors**: Assess water clarity by measuring the scattering of light caused by suspended particles.
   (d) **Online Conductivity Sensors**: Monitor the electrical conductivity of water, providing insights into ion concentrations and salinity.
2. **Online Gas Sensors**: These sensors detect and quantify the presence of specific gases in real-time. Applications include air quality monitoring, industrial safety, and emissions control.
3. **Online Chemical Sensors**: Used in industries, these sensors measure chemical concentrations, ensuring product quality and safety, e.g., monitoring chlorine in water or chemicals in manufacturing.
4. **Online Particle Counters:** Count and size particles in fluids to detect contamination, ensuring quality in pharmaceuticals, semiconductors, and hydraulics.
5. **Online Temperature and Pressure Sensors:** Monitor variations in processes, HVAC systems, and equipment for optimal conditions.
6. **Online Flow Sensors**: Continuously measure fluid flow rates in pipelines, allowing for real-time control of processes and efficient resource management.
7. **Online Vibration and Condition Monitoring Sensors**: Used in equipment, sensors detect vibrations and assess equipment health.
8. **Online Surveillance Cameras**: Surveillance cameras offer real-time video monitoring for security, traffic, and facility management.
9. **Online Energy and Power Monitoring Sensors**: Energy monitoring sensors track consumption, enabling efficient energy management.
10. **Online Weather Sensors**: Gather live data on meteorological factors like temperature, humidity, wind speed, direction, and precipitation. Valuable in agriculture, weather prediction, and environmental surveillance.
11. **Online GPS and Location Sensors**: Provide real-time location data for applications like fleet management, logistics, and geospatial mapping.

12. **Online Wireless Sensor Networks**: These networks comprise interconnected wireless sensors, facilitating extensive and distributed monitoring across diverse environments.

7.7.2 Remote Sensing Technologies

Remote sensing technologies have limited direct applications within wastewater treatment facilities themselves due to the enclosed nature of these facilities and the need for sensors to be in close proximity to the wastewater [27]. However, remote sensing can still be valuable in various indirect ways in the context of wastewater treatment and management:

1. **Monitoring Wastewater Discharge**: Utilize remote sensing to oversee the release of treated wastewater into natural water bodies like rivers, lakes, or coastal areas. Satellite or aerial imagery tracks changes in water quality, turbidity, and pollutant spread.
2. **Infrastructure and Pipeline Inspection**: Employ drones with remote sensing instruments to inspect wastewater infrastructure, including pipes, treatment plants, and sewage systems. LiDAR and thermal imaging identify leaks, blockages, and structural issues.
3. **Environmental Impact Assessment**: Assess the environmental impact of wastewater discharges through remote sensing. Monitor changes in land cover, vegetation health, and water quality near treatment plants and discharge points to evaluate effects on ecosystems.
4. **Land Use Planning and Urban Development**: Utilize remote sensing data for urban planning and land use management, identifying suitable locations for wastewater treatment plants based on terrain, proximity to water bodies, and population density. Monitor urban expansion's impact on water resources.
5. **Flood Monitoring and Early Warning**: Remote sensing, including radar and satellite imagery, aids flood monitoring and early warning systems. Track rainfall, river water levels, and soil moisture to anticipate flooding events and protect wastewater infrastructure.
6. **Assessment of Receiving Water Bodies**: Evaluate natural water bodies receiving treated effluent using remote sensing. Monitor water quality, aquatic vegetation, and impacts on aquatic ecosystems. Autonomous underwater vehicle (AUV) sensors provide in situ data.
7. **Climate Resilience and Adaptation**: Assess wastewater treatment facilities' vulnerability to climate change using remote sensing data. Consider factors like sea-level rise, extreme weather, and precipitation changes to inform adaptation strategies.

7.7.3 IoT Devices

IoT (Internet of Things) devices are increasingly being integrated into wastewater treatment systems to enhance efficiency, improve monitoring, and optimize resource management. These devices facilitate real-time data collection, remote control, and data analytics, leading to more efficient wastewater treatment processes [7]. Here are some ways IoT devices are used in wastewater treatment:

1. **Flow Monitoring Sensors**: IoT-enabled flow meters as well as sensors continuously monitor the flow of wastewater incoming and leaving treatment facilities. This data helps operators manage the flow rates, detect leaks or blockages, and optimize treatment processes for varying flow settings.
2. **Water Quality Sensors**: IoT water quality sensors estimate parameters like pH, dissolved oxygen, turbidity, and pollutant concentrations in real-time. This data permits for quick detection of anomalies and the adjustment of treatment processes to control water quality standards.
3. **Level Sensors**: IoT-level sensors mark the depth of wastewater in treatment tanks, clarifiers, as well as other components. This data aids in optimizing sludge removal, chemical dosing, and in general treatment efficiency.
4. **Pump and Equipment Monitoring**: IoT sensors attached to pumps and other equipment give information on performance and operational status. Predictive maintenance algorithms can be used to minimize downtime and prevent equipment failures.
5. **Chemical Dosing Control**: IoT devices can mechanize the dosing of chemicals in wastewater treatment, ensuring precise and timely additions of coagulants, flocculants, disinfectants, as well as pH adjusters based on real-time conditions.
6. **Energy Management**: IoT-based energy monitoring systems help in tracking energy consumption within water treatment plants. This data helps in getting opportunities for energy conservation, optimizing equipment operation, and managing peak demand.
7. **Remote Control and Automation**: IoT devices allow remote monitoring as well as control of treatment processes. Operators can make adjustments and respond to alarms from anywhere, enhancing operational flexibility and minimizing the need for on-site presence.
8. **Data Analytics and Reporting**: IoT platforms collect and analyze sensor data to create informative reports and dashboards, aiding decision-making, process optimization, and compliance.
9. **Predictive Analytics and AI**: IoT data can be used in conjunction with artificial intelligence (AI) and machine learning algorithms to predict equipment failures, optimize treatment processes, and improve overall system performance.
10. **Compliance and Reporting**: IoT devices assist in meeting regulatory requirements by continuously monitoring and recording critical parameters. This data is essential for reporting and demonstrating compliance with environmental standards.

11. **Security and Alarm Systems**: IoT devices can enhance security through surveillance cameras, access control systems, and intrusion detection sensors. Alarms can be triggered in the event of unauthorized access or security breaches.
12. **Environmental Impact Assessment**: IoT devices can collect data on the environmental impact of wastewater discharges, helping facilities comply with regulations and minimize their ecological footprint.

7.8 Fault Detection and Diagnosis

Fault detection and diagnosis (FDD) in wastewater treatment involves continuous sensor data monitoring, predefined threshold alarms, pattern recognition, statistical and machine learning analysis, model-based and data-driven FDD methods, and causal analysis [28]. When deviations are detected, FDD systems prioritize alarms by severity, employ expert systems for root cause analysis, and offer user-friendly interfaces with data visualization. Integration with control strategies enhances efficiency, reducing downtime, optimizing energy use, and ensuring regulatory compliance for reliable and efficient wastewater treatment.

7.8.1 Common Faults in Wastewater Treatment

Common faults and issues in wastewater treatment plants can disrupt operations, lead to inefficiencies, and impact the quality of treated water [29]. Table 7.4 summarizes some common faults.

Effective monitoring, regular maintenance, best practices, and advanced technologies like fault detection systems can mitigate and prevent common faults in wastewater treatment, ensuring reliable and consistent wastewater treatment processes.

7.8.2 AI-Based Fault Detection Algorithms

AI-based fault detection algorithms are increasingly used in wastewater treatment plants to boost efficiency, reliability, and sustainability. These algorithms employ artificial intelligence and machine learning to identify and diagnose faults in real or near-real-time [30]. They analyze data from sensor networks monitoring parameters like flow rates, pH levels, and chemical concentrations, and detecting deviations that may indicate anomalies. AI models use pattern recognition to identify abnormal trends, aiding in early fault detection. Additionally, they predict equipment maintenance needs based on historical data, minimizing downtime and breakdowns. These systems classify faults, conduct root cause analysis, and optimize treatment

Table 7.4 Some common faults in wastewater treatment

Fault	Description
Flow variations	Irregular flow rates are caused by factors such as stormwater inflows or equipment failures
pH imbalance	Fluctuations in pH levels can disrupt biological treatment processes and reduce chemical treatment efficiency
Dissolved oxygen (DO) deficiency	Inadequate DO levels can deplete oxygen-dependent microorganisms, leading to process upsets
Excessive foaming	Foam formation in aeration tanks or clarifiers can hinder settling and affect solids-liquids separation
Bulking sludge	Microbial imbalances or nutrient deficiencies can cause sludge bulking, leading to poor settling
Sludge dewatering issues	Problems with sludge dewatering equipment can result in high moisture content and increased disposal costs
Equipment failures	Mechanical failures in pumps, blowers, mixers, and other equipment can disrupt treatment processes
Chemical overdosing or underdosing	Incorrect dosing of chemicals can lead to treatment inefficiencies or inadequacies
Bypassing	Bypassing occurs when wastewater isn't properly directed through treatment processes due to malfunctions
Clogged or fouled filters	Filters and screens can become clogged, reducing their effectiveness in removing solids
Odor issues	Inadequate odor control systems can result in malodorous emissions, leading to complaints and concerns
Safety hazards	Chemical spills or gas leaks can pose risks to personnel and the environment
Regulatory non-compliance	Failure to meet regulatory standards can result in fines and legal consequences
Corrosion and wear	Corrosion and wear of infrastructure components can lead to leaks and structural failures
Energy inefficiencies	Inefficient operation of equipment can increase energy consumption and operational costs
Foam and scum accumulation	Accumulation of foam and scum on treatment tank surfaces can hinder processes
Microbial upsets	Imbalances in microbial populations can result from shock loads or toxic influents, affecting treatment
Inadequate nutrient removal	Failing to remove nutrients like nitrogen and phosphorus can lead to environmental pollution
Aging infrastructure	Aging treatment plants may require costly repairs or upgrades due to deteriorating infrastructure

processes, all while improving alarm management and energy efficiency. They also ensure compliance with environmental regulations, fuse data from various sources for comprehensive insights, adapt to changing conditions, and integrate seamlessly with SCADA systems for holistic monitoring and control. By leveraging AI-based fault detection algorithms, wastewater treatment facilities can proactively address issues, reduce operational costs, optimize resource usage, ensure regulatory compliance, and enhance the overall performance and sustainability of the treatment process [17]. These advanced systems contribute to more efficient and reliable wastewater treatment operations, ultimately benefiting both the environment and the communities served by these facilities.

7.8.3 *Early Warning Systems*

Early warning systems (EWS) in wastewater treatment are crucial for identifying potential issues or imminent failures in treatment processes at an early stage. By providing timely alerts and actionable information, these systems help operators and managers take preventive measures, reduce the risk of disruptions, and ensure the efficient and effective treatment of wastewater [31]. Here are the key components and benefits of early warning systems in wastewater treatment:

Components of Early Warning Systems:

(a) **Sensor Networks**: EWS relies on strategically placed sensors continuously monitoring parameters like flow rates, pH levels, dissolved oxygen, turbidity, and chemical concentrations.
(b) **Data Acquisition and Monitoring**: Sensor data is collected in real-time and transmitted to a central monitoring system or SCADA for continuous analysis.
(c) **Algorithms and Analytics**: Advanced algorithms process sensor data, identifying trends, patterns, or anomalies indicating potential issues or faults.
(d) **Thresholds and Trigger Values**: Predefined threshold values for parameters trigger alarms and alerts when crossed, notifying operators and decision-makers.
(e) **Visualization and Reporting**: User-friendly interfaces and dashboards provide real-time data, alarms, and diagnostic information, aiding quick decision-making.

Benefits of Early Warning Systems:

(a) **Preventive Maintenance**: EWS detects equipment issues early, preventing breakdowns and failures.
(b) **Process Optimization**: Timely alerts enable adjustments to treatment processes, ensuring optimal performance and regulatory compliance.
(c) **Cost Reduction**: Early detection minimizes downtime and reduces costly emergency repairs, lowering operational and maintenance expenses.
(d) **Environmental Protection**: EWS maintains effluent quality, preventing pollutant release into the environment, safeguarding ecosystems, and public health.

(e) **Regulatory Compliance**: Continuous monitoring of critical parameters helps meet regulatory requirements and avoid fines or penalties.
(f) **Resource Conservation**: Proactive process management and optimization promote efficient resource use, including chemicals and energy.
(g) **Safety Enhancement**: EWS provides alerts for potential safety hazards, enhancing facility safety by detecting issues like gas leaks or chemical spills.
(h) **Data-Driven Decision-Making**: Data collected supports informed decision-making, enabling operators to respond to changing conditions and optimize processes.
(i) **Resilience to External Factors**: EWS helps treatment plants prepare for and respond to external factors such as extreme weather events or influent quality fluctuations.
(j) **Public Health Protection**: Reliable wastewater treatment safeguards public health by preventing the release of harmful contaminants into the environment.

7.9 Case Studies

Wastewater treatment case studies provide valuable lessons about addressing real-world challenges in water resource management and environmental sustainability. They often feature innovative technologies, best practices, and successful strategies used by wastewater treatment facilities worldwide. By reviewing these cases, decision-makers, engineers, and environmentalists can gain insights into optimizing treatment processes, meeting regulations, minimizing environmental impact, and safely releasing treated effluent into natural water bodies. These case studies underscore the importance of ongoing development, resource preservation, and technology's role in meeting evolving wastewater treatment requirements.

7.9.1 Successful Implementation of AI for Monitoring and Fault Detection

Background: A leading industrial provision specializing in chemical production faced continuing challenges connected to energy consumption and equipment downtime. With a complicated and interconnected system of pumps, compressors, heat exchangers, and reactors, the service sought a result to optimize energy usage, minimize operational costs, and minimize unplanned shutdowns.

Objective: The primary goal was to start AI-based monitoring and fault detection to proactively recognize operational anomalies and inefficiencies, enabling the facility to get higher energy efficiency, minimize maintenance costs, and enhance overall reliability.

Implementation: The facility worked together with an AI and data analytics company to develop and set up a comprehensive AI-based monitoring and fault detection system. The key components of the solution consisted:

1. **Data Integration**: Information from various sensors and operational systems, containing flow meters, temperature sensors, pressure gauges, and historical maintenance logs, were joined into a centralized data platform.
2. **Machine Learning Models**: Advanced machine learning algorithms, like anomaly recognition and predictive maintenance models, were trained on historical data to find out normal operating patterns and identify deviations.
3. **Real-time Monitoring**: The system always monitored real-time data streams and found the health and performance of critical equipment and processes.
4. **Alerting Mechanism**: When anomalies or faults were found, the system generated immediate alerts for operators and maintenance teams, classifying problems by severity and providing diagnostic insights.

7.9.2 *Results and Benefits Obtained*

Results: The implementation of AI-based monitoring and fault detection gave significant improvements:

1. **Energy Efficiency**: The facility achieved a 15% drop in energy consumption, primarily by optimizing equipment function and addressing energy-wasting anomalies promptly.
2. **Reduced Downtime**: Unplanned shutdowns due to equipment failures were minimized by 30%, resulting in raised production output and higher revenue.
3. **Maintenance Cost Reduction**: Predictive maintenance recommendations approved for planned equipment servicing, minimizing maintenance costs by 25%.
4. **Safety Improvement**: The system's early warning capabilities improved safety by identifying potential safety hazards and starting safety protocols when required.
5. **Operational Insights**: In-depth data analytics gave valuable insights into process developments, enabling the facility to fine-tune the operational parameters and increase product quality.

Benefits: The successful implementation of AI for monitoring and fault detection gives a large range of associated benefits across various industries and applications:

1. **Improved Operational Efficiency**: AI systems can constantly analyze large datasets and find anomalies in real-time, enabling organizations to optimize operations, streamline processes, and decrease resource wastage.
2. **Cost Reduction**: By proactively finding faults and addressing issues early, organizations can significantly minimize maintenance costs, reduce unplanned downtime, and lengthen the lifespan of equipment and assets.

3. **Enhanced Productivity**: AI-based monitoring permits for more efficient resource allocation, which advances to increased production capacity and higher output.
4. **Energy Efficiency**: Organizations can accomplish substantial energy savings by identifying and rectifying energy-wasting abnormalities and inabilities in real-time.
5. **Increased Reliability**: AI systems increase the reliability of critical systems and processes, ensuring smoother operations and lowering the risk of equipment failures.
6. **Predictive Maintenance**: AI-powered predictive maintenance models can recognize when equipment is likely to fail, allowing organizations to carry out maintenance tasks before a breakdown occurs, reducing downtime and associated costs.
7. **Safety Improvement**: Early detection of faults and safety hazards increases workplace safety, protecting employees and minimizing the risk of accidents.
8. **Data-Driven Decision-Making**: AI systems give data-driven insights, empowering organizations to build informed decisions, optimize strategies, and get used to changing conditions effectively.
9. **Compliance Assurance**: Organizations can make sure compliance with regulatory standards and environmental needs by continuously monitoring critical parameters and processes.
10. **Environmental Impact Reduction**: AI can help organizations reduce their environmental footprint by minimizing emissions, pollutants, and waste.
11. **Enhanced Customer Satisfaction**: Enhanced operational efficiency and product quality can proceed to increased customer satisfaction and loyalty.
12. **Competitive Advantage**: Organizations that select AI-based monitoring and fault detection, gain a competitive edge by optimizing their operations, minimizing costs, and delivering higher-quality products and services.
13. **Resource Conservation**: AI-driven development leads to more efficient resource usage, decreasing the consumption of water, energy, and raw materials.
14. **Process Innovation**: The insights generated by AI systems can ride innovation by getting opportunities for process improvements and product development.
15. **Scalability**: AI solutions are scalable, making them appropriate for organizations of all sizes and adaptable to changing business requirements.
16. **Operational Resilience**: AI can increase the resilience of operations by giving early warning of potential disruptions and enabling quick response.

Conclusion: The successful implementation of AI-based monitoring and fault detection not only significantly enhanced energy efficiency and operational reliability but also improved safety and decreased maintenance costs. This case study demonstrates how AI-powered solutions can give tangible advantages to industrial facilities by leveraging data-driven insights to optimize operations and drive substantial improvements in key performance metrics.

7.10 Challenges and Future Directions

The challenges in wastewater treatment, driven by factors like population growth, aging infrastructure, emerging contaminants, and climate change, require innovative solutions [32]. Future wastewater treatment relies on advanced tech, decentralized systems, and resource recovery for sustainable growth. Smart wastewater management with IoT and AI boosts efficiency and reliability, while a circular economy approach minimizes waste and maximizes resources. Nature-based solutions and climate-resilient infrastructure protect against climate change impacts for continuous operation [33]. Public education and global collaboration are vital for responsible water use, ensuring access to clean water and environmental protection in a world with water scarcity and environmental concerns.

7.10.1 Privacy and Data Security

Privacy and data security are increasingly important considerations in the context of challenges and future directions in wastewater treatment. As the industry adopts more digital technologies and data-driven solutions, ensuring the confidentiality, integrity, and availability of sensitive data becomes crucial. Here's a brief discussion of privacy and data security in wastewater treatment:

Privacy Concerns:

- **Data Collection**: Wastewater treatment facilities gather extensive data, encompassing operational details, water quality, and even industrial process information. Safeguarding this data's privacy is crucial, particularly when it contains sensitive industrial or personal data.
- **Regulatory Compliance**: Meeting data privacy regulations like GDPR (General Data Protection Regulation) in Europe necessitates strong data protection measures and transparency for individuals about data use in wastewater treatment facilities.

Data Security Challenges:

- **Cyber security Threats**: As wastewater treatment systems become more interconnected and reliant on digital technologies, they become susceptible to cyber attacks. Protecting against data breaches, ransom ware, and other cyber security threats is paramount.
- **Secure Data Transmission**: Transmitting data between sensors, control systems, and central databases must be done securely to prevent interception or tampering by unauthorized parties.

Future Directions for Privacy and Data Security:

- **Encryption**: Implementing strong encryption protocols for data at rest and in transit can protect sensitive information from unauthorized access or tampering.

- **Access Control**: Implementing strict access control mechanisms ensures that only authorized personnel have access to sensitive data, and data is only shared with those who have a legitimate need.
- **Anonymization and De-Identification**: Implementing techniques like data anonymization and de-identification can help protect privacy while still allowing for the use of data in research and analysis.
- **Continuous Monitoring**: Employing continuous monitoring and intrusion detection systems can help identify and respond to potential security threats in real-time.
- **Training and Awareness**: Educating staff about the importance of data security and privacy and providing training on best practices can reduce the risk of security breaches caused by human error.
- **Vendor Due Diligence**: When implementing third-party solutions or outsourcing data management, conducting thorough vendor due diligence to ensure their commitment to data security and privacy is essential.

As the wastewater treatment industry evolves, it must tackle privacy and data security issues to safeguard sensitive information, defend against cyber threats, and uphold stakeholder trust while harnessing the advantages of data-driven technologies for enhanced efficiency and sustainability.

7.10.2 Integration Challenges

Integration challenges in wastewater treatment revolve around the need to harmonize legacy systems with modern technologies, break down data silos, ensure interoperability among diverse devices, and tackle cybersecurity concerns. Overcoming these challenges requires standardization, middleware solutions, cloud-based platforms, and open interfaces to facilitate seamless integration. The goal is to create a cohesive ecosystem that enables efficient data management and supports both small-scale decentralized and large centralized wastewater treatment facilities. Emphasizing cybersecurity from the outset is crucial, given the increasing risk of cyberattacks. By addressing these integration challenges and embracing future-oriented strategies, the wastewater treatment industry can leverage digital technologies to optimize operations, meet regulatory compliance, and advance sustainability goals effectively.

7.10.3 Future Trends and Prospects

The future of wastewater treatment is shaped by societal needs, environmental concerns, technology advancements, and sustainability. Here are key upcoming trends:

1. **Advanced Treatment Tech**: Continued development and use of advanced methods like membrane bioreactors and photo catalysis to improve efficiency and remove contaminants.
2. **Decentralized Systems**: On-site and small-scale treatment systems will grow, especially in areas lacking centralized plants, offering cost-effective solutions.
3. **Resource Recovery**: Plants will focus on extracting resources like phosphorus and energy from wastewater, aligning with the circular economy.
4. **Smart Management**: IoT, sensors, and AI for real-time monitoring and analytics will become standard, enhancing efficiency.
5. **Water Reuse**: Treated wastewater will be used for non-potable purposes, addressing water scarcity.
6. **Energy Efficiency**: Plants will prioritize energy recovery and optimization to reduce costs and environmental impact.
7. **Climate-Resilient Infrastructure**: Facilities will be designed to withstand climate change impacts.
8. **Circular Economy**: A shift towards a circular economy approach will minimize waste and promote recycling.
9. **Data-Driven Decisions**: Data will drive decision-making for process optimization and compliance.
10. **Public Awareness**: Education campaigns will promote responsible water use and pollution prevention.
11. **International Collaboration**: Global cooperation will accelerate research and knowledge exchange.
12. **Environmental Stewardship**: Sustainability will guide practices, ensuring wastewater treatment protects ecosystems and public health.

These future trends and prospects in wastewater treatment reflect a growing emphasis on sustainability, innovation, and resilience in addressing the challenges posed by population growth, urbanization, pollution, and climate change. Wastewater treatment is assured to play an important role in ensuring clean water resources and environmental protection in an increasingly water-scarce and environmentally conscious world.

7.11 Conclusion

The real-time monitoring and fault detection in AI-enhanced wastewater treatment systems depict a transformative advancement in the field of environmental management. By utilizing the power of artificial intelligence, these systems enable continuous, data-driven oversight of treatment processes, leading to enhanced efficiency, reliability, and sustainability. The ability to find anomalies, predict equipment failures, and optimize operations in real-time not only decreases operational costs and energy consumption but also improves environmental protection and regulatory compliance. Furthermore, these systems cover the way for proactive decision-making and resource recovery, such as energy and nutrient capture, contributing to a more circular and resource-efficient technique in wastewater treatment. As wastewater treatment facilities hold AI-driven solutions, they position themselves as pioneers in the journey toward a cleaner, extra sustainable, and water-secure future.

References

1. Wang Y, Cheng Y, Liu H (2023) A review on applications of artificial intelligence in wastewater treatment. Sustainability 15(18)
2. Zhu M, Wang J, Yang X, Zhang Y, Zhang L, Ren H, Wu B, Ye L (2022) A review of the application of machine learning in water quality evaluation. Eco-Environ Health 1(2):107–116
3. Alzahrani AIA, Chauhdary SH, Alshdadi AA (2023) Internet of Things (IoT)-based wastewater management in smart cities. Electronics 12(12)
4. Zhou M, Zhang Y, Wang J, Xue T, Dong Z, Zhai W (2023) Fault detection of wastewater treatment plants based on an improved kernel extreme learning machine method. Water 15(11)
5. Hamed MM, Khalafallah MG, Hassanien EA (2004) Prediction of wastewater treatment plant performance using artificial neural networks. Environ Modell Softw 19(10)
6. Zhang Z, Chang X, Chen A (2016) Determination of chemical oxygen demand based on photoelectrocatalysis of nanoporous TiO_2 electrodes. Sens Actuators B Chem 223:664–670
7. Kumar PM, Hong CS (2022) Internet of things for secure surveillance for sewage wastewater treatment systems. Environ Res 203
8. Aghalari Z, Dahms H, Sillanpää M, Sosa-Hernandez JE, Parra-Saldívar R (2020) Effectiveness of wastewater treatment systems in removing microbial agents: a systematic review. Glob Health 16(13):1–11
9. Mojiri A, Bashir MJK (2022) Wastewater treatment: current and future techniques. Water 14(3). https://doi.org/10.3390/w14030448
10. Peng C, Ying X, FanChao M (2023) Efficient fault monitoring in wastewater treatment processes with time stacked broad learning network. Expert Syst Appl 233:120958
11. Obaideen K, Shehata N, Sayed ET, Abdelkareem MA, Mahamoud MS, Olabi AG (2022) The role of wastewater treatment in achieving sustainable development goals (SDGs) and sustainability guideline. Energy Nexus 7
12. Cassidy J, Silva T, Semião N, Ramalho P, Santos A, Feliciano J (2020) Improving wastewater treatment plants operational efficiency and effectiveness through an integrated performance assessment system. H_2Open J 3(1):276–287
13. Simon-Varhelyi M, Cristea VM, Luca AV (2020) Reducing energy costs of the wastewater treatment plant by improved scheduling of the periodic influent load. J Environ Manag 262
14. Wade MJ, Sánchez A, Katebi MR (2005) On real-time control and process monitoring of wastewater treatment plants: real-time process monitoring. Trans Inst Meas Control 27(3):173–193. https://doi.org/10.1191/0142331205tm140oa
15. Sharma R, Verma N, Lugani Y, Kumar S, Asadnia M (2021) Chapter 1—conventional and advanced techniques of wastewater monitoring and treatment. Green Sustain Process Chem Environ Eng Sci 1–48
16. Wang Y, Cheng Y, Liu H, Guo Q, Dai C, Zhao M, Liu D (2023) A review on applications of artificial intelligence in wastewater treatment. Sustainability 15(18). https://doi.org/10.3390/su151813557
17. Zhang S, Jin Y, Chen W, Wang J, Wang Y, Ren H (2023) Artificial intelligence in wastewater treatment: a data-driven analysis of status and trends. Chemosphere 336
18. Sundui B, Calderon OAR, Abdeldayem OM, Lavaro-Gil J, Rene ER, Sambuu U (2021) Applications of machine learning algorithms for biological wastewater treatment: updates and perspectives. Clean Technol Environ Policy 23:127–143

19. Jawad J, Hawari AH, Zaidi SJ (2021) Artificial neural network modeling of wastewater treatment and desalination using membrane processes: a review. Chem Eng J 419
20. Bahramian M, Dereli RK, Zhao W, Giberti M, Casey E (2023) Data to intelligence: the role of data-driven models in wastewater treatment. Expert Syst Appl 217
21. Aghdam E, Mohandes SR, Manu P, Cheung C, Yunusa-Kaltungo A, Zayed T (2023) Predicting quality parameters of wastewater treatment plants using artificial intelligence techniques. J Cleaner Prod 405
22. Abba SI, Elkiran G (2017) Effluent prediction of chemical oxygen demand from the wastewater treatment plant using artificial neural network application. Proced Comput Sci 120:156–163
23. Baki OT, Aras E, Akdemir UO, Yilmaz B (2019) Biochemical oxygen demand prediction in wastewater treatment plant by using different regression analysis models. Desalin Water Treat 157:79–89
24. Liang G, Zhao Q, Liu B, Du Z, Xia X (2021) Treatment and reuse of process water with high suspended solids in low-grade iron ore dressing. J Cleaner Prod 278
25. Sanchez M, Gonzalo OG, Yanez S, Ruiz I, Soto M (2021) Influence of nutrients and pH on the efficiency of vertical flow constructed wetlands treating winery wastewater. J Water Process Eng 42
26. Chafa AT, Chirinda GP, Matope S (2022) Design of a real–time water quality monitoring and control system using Internet of Things (IoT). Cogent Eng 9(1)
27. Li H, Zech J, Hong D, Ghamisi P, Schultz M, Zipf A (2022) Leveraging openstreetmap and multimodal remote sensing data with joint deep learning for wastewater treatment plants detection. Int J Appl Earth Observ Geoinform 110
28. Chi B, Guo L (2019) Wastewater treatment sensor fault detection using RBF neural network with set membership estimation. 2019 Chinese Control And Decision Conference (CCDC), China, pp 2685–2690
29. Chen Z, Chu H (2021) Analysis of common problems in industrial wastewater treatment and countermeasures. IOP Conf Series Earth Environ Sci 937
30. Zhao L, Dai T, Qiao Z, Sun P, Hao J, Yang Y (2019) Application of artificial intelligence to wastewater treatment: a bibliometric analysis and systematic review of technology, economy, management, and wastewater reuse. Process Safety Environ Prot
31. Trigo-Tasende N et al (2023) Wastewater early warning system for SARS-CoV-2 outbreaks and variants in a Coruña, Spain. Environ Sci Pollut Res 30:79315–79334. https://doi.org/10.1007/s11356-023-27877-3
32. Soares A (2020) Wastewater treatment in 2050: challenges ahead and future vision. Environ Sci Ecotechnol 2(1)
33. Mojiri A, Bashir MJK (2022) Wastewater treatment: current and future techniques. Water 14(3)

Chapter 8
Innovations in Nutrient Removal and Resource Recovery Through Artificial Intelligence Techniques

AI and Water: A Sustainable Approach to Environmental Restoration

Jyoti Chowdhry, Sheetal Kumari, and Meenakshi Nandal

Abstract Wastewater treatment is pivotal for environmental sustainability, particularly in managing nutrients like nitrogen and phosphorus to prevent ecological degradation. This manuscript explores the integration of artificial intelligence (AI) in advancing nutrient removal and resource recovery in wastewater treatment. Beginning with an overview of traditional methods and their limitations, the chapter emphasizes the transformative potential of AI technologies in optimizing processes and enhancing environmental outcomes. AI's predictive capabilities enable real-time monitoring and adaptive management, crucial for efficient nutrient removal and reducing environmental impact. Key applications include AI-driven dosing strategies in chemical treatment and adaptive control of biological processes. While AI offers significant benefits in improving treatment efficiency and resource recovery, challenges such as data quality and ethical considerations must be addressed to realize its full potential.

Keywords Artificial intelligence (AI) · Wastewater treatment · Nutrient removal · Resource recovery · Environmental management

---

J. Chowdhry · M. Nandal (✉)
Department of Environmental Science, Maharishi Dayanand University, Rohtak, India
e-mail: meenakshi.env.sc@mdurohtak.ac.in

J. Chowdhry
e-mail: Jyotichoudhry94@gmail.com

S. Kumari
Amity Institute of Environmental Science (AIES), Amity University Uttar Pradesh, Sector-125, Gautam Budh Nagar, Noida 201313, India
e-mail: Sheetalluhania99@gmail.com

M. C. Garg (ed.), *The AI Cleanse: Transforming Wastewater Treatment Through Artificial Intelligence*, Springer Water, https://doi.org/10.1007/978-3-031-67237-8_8

8.1 Introduction

Wastewater treatment is a critical aspect of environmental stewardship, aiming to purify water before its release into natural bodies. With the growing concerns over water scarcity and environmental pollution, there is an increasing need for innovative approaches in wastewater treatment [32]. This chapter delves into the intersection of cutting-edge technologies and wastewater management, specifically focusing on the role of artificial intelligence (AI) in advancing nutrient removal and resource recovery processes.

8.1.1 Background

Traditional wastewater treatment methods have long been effective but face challenges, particularly nutrient removal [48] Nutrients such as nitrogen and phosphorus, while essential for life, can lead to adverse environmental effects when present in excess in water bodies. Eutrophication, harmful algal blooms, and the degradation of aquatic ecosystems are among the consequences [13]. The need for more efficient and precise nutrient removal methods prompted the exploration of technological interventions, with a notable focus on AI [49].

The background section will explore the historical evolution of wastewater treatment, the emergence of challenges in nutrient removal, and the shifting paradigm towards embracing technological solutions. It will provide a context for the subsequent discussions on the specific objectives and the incorporation of AI in addressing these challenges.

8.1.2 Objectives of Nutrient Removal and Resource Recovery

The objectives of this chapter revolve around two primary facets: nutrient removal and resource recovery. Nutrient removal is imperative to mitigate the negative impacts of excess nutrients in water bodies. By exploring novel methods and technologies, the aim is to enhance the efficiency of nutrient removal processes and reduce their environmental footprint. Simultaneously, the chapter focuses on resource recovery, seeking to transform wastewater from a disposal challenge into a valuable resource. Through the intelligent use of technology, the goal is to recover and reuse valuable components present in wastewater, such as energy and nutrients, fostering a sustainable and circular approach to wastewater management.

The section detailing objectives will provide a roadmap for readers, highlighting the specific goals and areas of focus within the broader context of nutrient removal and resource recovery. It sets the stage for the subsequent sections that delve into the application of AI in achieving these objectives.

8.1.3 Significance of Artificial Intelligence in Wastewater Treatment

The significance of AI in wastewater treatment is multifaceted. AI technologies, including machine learning and data analytics, offer a paradigm shift in how we approach the complex challenges associated with nutrient removal and resource recovery. By harnessing the power of AI, wastewater treatment processes can become more adaptive, efficient, and tailored to dynamic environmental conditions [53]. AI brings predictive capabilities to the forefront, enabling the development of models that can forecast nutrient levels and optimize treatment processes in real-time. This predictive power enhances the precision of nutrient removal, reducing the likelihood of environmental harm and optimizing resource utilization [47]. Moreover, AI contributes to the intelligent design of monitoring systems, facilitating early detection of anomalies and enabling swift corrective actions [6, 46].

The section on the significance of AI will emphasize the transformative potential of these technologies in revolutionizing wastewater treatment practices. It will discuss how AI can address existing limitations, improve overall system performance, and contribute to the sustainable management of water resources. The introduction sets the stage for an in-depth exploration of the integration of AI in nutrient removal and resource recovery in wastewater treatment. It provides the necessary background, outlines specific objectives, and emphasizes the profound significance of AI in addressing the complex challenges of contemporary wastewater management.

8.2 Foundations of Nutrient Removal

Nutrient removal is a cornerstone in wastewater treatment, playing a pivotal role in safeguarding water ecosystems from the detrimental effects of excessive nitrogen and phosphorus [36]. Traditional nutrient removal methods have formed the bedrock of wastewater treatment practices, evolving over time to meet the increasing demands for water quality [20]. This section delves into the foundational aspects of nutrient removal, exploring traditional methods, the challenges they pose, and the overarching importance of effective nutrient management in wastewater treatment.

8.2.1 Traditional Nutrient Removal Methods

Historically, conventional wastewater treatment plants have employed physical, chemical, and biological processes for nutrient removal (Fig. 8.1). Biological nutrient removal (BNR) is one of the most common methods, leveraging the metabolic activities of microorganisms to convert nitrogen and phosphorus compounds into harmless gases [1]. Activated sludge processes and sequencing batch reactors are examples

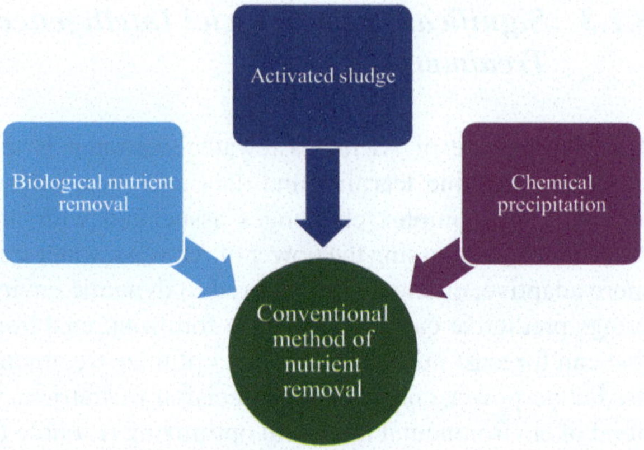

Fig. 8.1 Conventional methods for nutrient removal

of BNR methods widely used for nutrient removal. Chemical precipitation, another traditional method, involves adding chemicals that react with nutrients to form solid precipitates, which can be easily separated from the water [20]. This approach is effective but can be resource-intensive and may result in the production of additional waste streams [22].

While these methods have been successful to a certain extent, they often fall short in addressing the dynamic nature of nutrient levels in wastewater. Conventional approaches lack the adaptability needed to respond to fluctuating nutrient concentrations, seasonal variations, and emerging contaminants. This limitation has underscored the necessity for innovative approaches, leading to the exploration of advanced technologies such as artificial intelligence [2].

8.2.2 Challenges in Conventional Approaches

Conventional nutrient removal methods face several challenges that hinder their efficiency and environmental sustainability. One major challenge is the variability in influent wastewater composition [40]. Fluctuations in nutrient concentrations, as well as the presence of emerging contaminants, can strain the effectiveness of traditional treatment processes [16]. The inherent limitations of relying solely on fixed, rule-based systems become apparent in these dynamic conditions.

Additionally, traditional methods may not fully recover the valuable resources in wastewater, missing an opportunity for sustainable resource management. The discharge of nutrient-rich effluent into water bodies poses risks of eutrophication, impacting aquatic ecosystems and human health [29]. As environmental regulations become more stringent, there is a growing need for wastewater treatment solutions that meet compliance standards and contribute to resource recovery and conservation [18].

8.2.3 Importance of Nutrient Management in Wastewater Treatment

Effective nutrient management is crucial for several reasons, encompassing environmental preservation, public health, and sustainable resource utilization [10]. Nutrient-rich effluents can lead to eutrophication in receiving water bodies, resulting in the overgrowth of algae and depletion of oxygen levels. This, in turn, harms aquatic life and disrupts the ecological balance [24]. Proper nutrient management in wastewater treatment aims to prevent these adverse impacts, preserving the health and biodiversity of water ecosystems. Moreover, nutrient management is intricately linked to public health concerns. Excessive nutrients in drinking water sources can pose risks to human health, contributing to conditions such as methemoglobinemia and other waterborne diseases [17]. Addressing nutrient levels in wastewater is not only an environmental imperative but also a fundamental aspect of ensuring the safety and well-being of communities that rely on water resources.

So, understanding the foundations of nutrient removal is essential for appreciating the evolution of wastewater treatment practices. Traditional methods, while effective to a certain extent, face challenges that necessitate the exploration of advanced technologies [5]. The importance of nutrient management goes beyond compliance; it is a critical element in safeguarding ecosystems, public health, and the sustainable use of water resources [15]. This sets the stage for the subsequent exploration of how artificial intelligence can enhance nutrient removal processes [49].

8.3 Role of Artificial Intelligence in Nutrient Removal

As the complexities of wastewater treatment evolve, the integration of artificial intelligence (AI) emerges as a transformative force in optimizing nutrient removal processes. This section delves into the role of AI in nutrient removal, providing an overview of its applications in environmental management, specific use cases in nutrient removal processes, and an analysis of the advantages and limitations associated with the application of AI in nutrient management [32].

8.3.1 Overview of AI in Environmental Management

AI encompasses a suite of technologies, including machine learning, data analytics, and advanced algorithms, which collectively contribute to more intelligent decision-making processes. In the realm of environmental management, AI has found applications in diverse areas, ranging from air quality monitoring to waste management. In the context of wastewater treatment, AI is leveraged to enhance the efficiency and adaptability of nutrient removal processes.

One key aspect of AI in environmental management is its ability to process vast amounts of data rapidly. By analyzing real-time data from sensors and monitoring systems, AI algorithms can identify patterns, trends, and anomalies, enabling proactive decision-making [44]. This predictive capability is particularly valuable in the context of nutrient removal, where the dynamic nature of influent wastewater requires swift and adaptive responses.

8.3.2 Applications of AI in Nutrient Removal Processes

The application of AI in nutrient removal processes is multifaceted, offering innovative solutions to longstanding challenges. Machine learning models play a crucial role in predicting nutrient concentrations, optimizing dosing strategies, and enhancing the overall efficiency of biological and chemical treatment methods [3]. In biological nutrient removal, AI algorithms can be trained to recognize microbial behaviour patterns, allowing for more precise control of activated sludge processes. This adaptive approach ensures that the microbial community responds effectively to variations in nutrient levels, optimizing nutrient removal without overtreating or underperforming [49].

Chemical precipitation processes benefit from AI-driven dosing strategies, where algorithms analyze real-time data to determine the optimal dosage of chemicals based on influent nutrient concentrations (Fig. 8.2) [9, 37]. This not only improves the efficacy of nutrient removal but also minimizes the use of chemicals, reducing operational costs and environmental impact. Furthermore, AI contributes to the intelligent design of monitoring systems. Smart sensors, integrated with AI algorithms, enable real-time tracking of nutrient levels and the early detection of deviations from optimal conditions [19]. This proactive monitoring facilitates timely interventions, preventing potential environmental harm and ensuring compliance with regulatory standards.

8.3.3 Advantages and Limitations of AI in Nutrient Management

The integration of AI in nutrient management brings forth a range of advantages and disadvantages (Table 8.1). AI's predictive capabilities enhance wastewater treatment systems' responsiveness, allowing for real-time adjustments to changing conditions [14]. This adaptability is particularly valuable in addressing the variability in influent wastewater composition and mitigating the challenges associated with conventional, rule-based approaches.

AI-driven nutrient removal processes also contribute to resource efficiency. By optimizing dosing strategies, minimizing chemical usage, and enhancing the overall

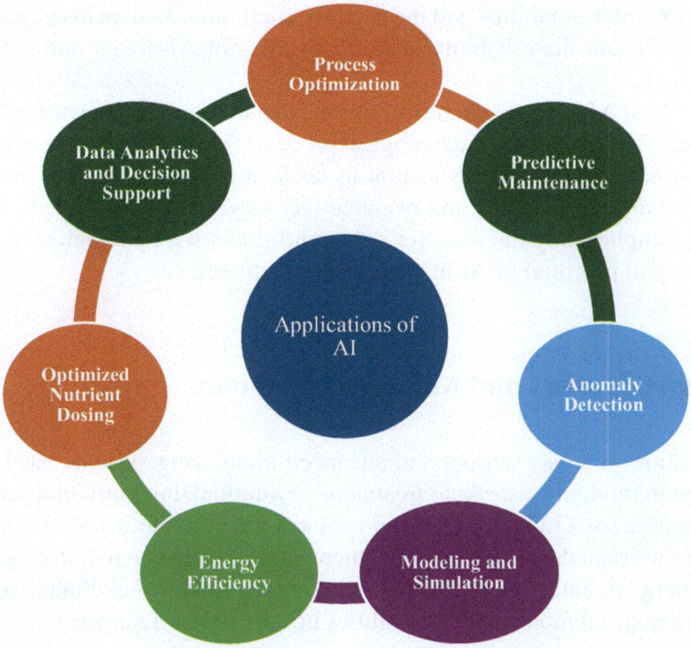

Fig. 8.2 Applications of AI in nutrient removal process

Table 8.1 Advantages and limitations of AI in nutrient management

Advantages	Limitations
Optimizes nutrient dosing in real-time	Relies on accurate data for effective predictions
Provides real-time monitoring and adjustment	Potential for algorithmic biases or errors
Enhanced process efficiency and accuracy	Requires significant initial investment in AI infrastructure
Reduced operational costs	Requires significant initial investment in AI infrastructure
Enhances environmental sustainability	Limited by the complexity of biological systems
Enables predictive maintenance and troubleshooting	Dependent on continuous updates and maintenance

performance of treatment systems, AI helps reduce operational costs and environmental impact [41]. Additionally, the ability to recover valuable resources from wastewater is augmented through the intelligent application of AI, aligning with the principles of circular economy and sustainable resource management [8, 41].

However, it's crucial to acknowledge the limitations of AI in nutrient management. The successful implementation of AI requires robust data infrastructure, accurate models, and ongoing maintenance and calibration. Challenges such as data quality, model interpretability, and the need for continuous human oversight must be addressed to ensure the reliability and effectiveness of AI-driven nutrient removal processes [21].

So, the role of AI in nutrient removal signifies a paradigm shift in wastewater treatment practices. From optimizing biological processes to intelligent dosing strategies, AI brings a new level of sophistication to nutrient management. The advantages, including enhanced efficiency and resource recovery, must be weighed against the limitations, emphasizing the need for a thoughtful and well-integrated approach to leverage the full potential of AI in wastewater treatment.

8.4 Smart Sensors and Monitoring Systems

The integration of smart sensors and advanced monitoring systems has become a cornerstone in modern wastewater treatment, revolutionizing nutrient detection and monitoring processes [33]. This section explores the intricate world of sensor technologies for nutrient detection, the implementation of real-time monitoring systems, and the synergistic integration of these technologies with artificial intelligence (AI) to achieve enhanced monitoring capabilities in wastewater treatment.

8.4.1 Sensor Technologies for Nutrient Detection

Traditional wastewater monitoring relied on periodic sampling and laboratory analysis, presenting challenges in providing real-time insights into nutrient concentrations (Table 8.2). Smart sensor technologies have emerged as a transformative solution, enabling continuous and instantaneous monitoring of nutrient levels [27]. These sensors utilize various technologies, including optical, electrochemical, and spectroscopic methods, to detect and quantify nutrients in wastewater [23, 27].

Optical sensors, for instance, leverage the interaction of light with specific molecules to measure nutrient concentrations [34]. Fluorescence-based sensors can target specific compounds like phosphates or nitrates, providing high sensitivity and selectivity [25]. Electrochemical sensors, on the other hand, measure electrical changes resulting from chemical reactions, offering a rapid and cost-effective means of nutrient detection [38]. Spectroscopic techniques, such as infrared or UV–ViS spectroscopy, provide detailed molecular information for accurate nutrient analysis [43].

The deployment of smart sensors facilitates real-time data acquisition, enabling a dynamic understanding of nutrient levels in wastewater. This real-time information is crucial for optimizing nutrient removal processes, responding promptly to fluctuations, and ensuring compliance with regulatory standards [55].

Table 8.2 Summarizing some sensor technologies commonly used for nutrient detection

Sensor technology	Description	Application
Electrochemical sensors	Measure the concentration of ions or molecules by generating an electrical signal	Measure the concentration of ions or molecules by generating an electrical signal
Optical sensors	Use light absorption, reflection, or fluorescence to detect target molecules	Detection of organic compounds, proteins, and pollutants
Piezoelectric sensors	Convert changes in pressure, acceleration, or force into an electrical charge	Detection of mass changes in biochemical reactions
Microfluidic sensors	Employ microfluidic channels to manipulate small volumes of fluids for analysis	Point-of-care diagnostics, environmental monitoring
Fluorescence sensors	Measure the emission of fluorescent light from molecules when excited by a specific wavelength	Detection of biomolecules, environmental pollutants
Biosensors	Integrate a biological sensing element with a physicochemical transducer for the detection	Monitoring of biomolecules, pathogens, and toxins

8.4.2 Real-Time Monitoring Systems

Real-time monitoring systems represent a significant leap forward in the evolution of wastewater treatment [30]. These systems continuously collect data from various sensors strategically placed within treatment facilities. The data encompass not only nutrient concentrations but also key operational parameters such as flow rates, pH levels, and temperature [52]. The real-time nature of these monitoring systems allows for immediate insights into the performance of treatment processes.

Continuous monitoring enables the identification of trends, patterns, and anomalies in nutrient levels, facilitating proactive decision-making. Operators can swiftly respond to changing conditions, adjusting treatment parameters to optimize nutrient removal without compromising system efficiency. Moreover, real-time monitoring enhances the ability to detect and address issues promptly, minimizing the risk of environmental harm and ensuring the overall effectiveness of the wastewater treatment plant [30, 52].

The data generated by real-time monitoring systems also serve as a valuable resource for trend analysis, system optimization, and compliance reporting. Regulatory authorities increasingly recognize the importance of real-time data in assessing and managing the environmental impact of wastewater discharges, underscoring the significance of these monitoring systems in contemporary wastewater treatment practices [28].

8.4.3 Integration with AI for Enhanced Monitoring

The true power of smart sensors and real-time monitoring systems is realized when integrated with artificial intelligence (AI) technologies [45]. AI algorithms can process vast amounts of data, identify complex patterns, and make predictions, thereby enhancing the monitoring capabilities of wastewater treatment processes [45, 51]. The integration of AI enables predictive modelling for nutrient levels, allowing for proactive adjustments to treatment parameters based on anticipated changes. Machine learning algorithms can learn from historical data, adapting to variations in influent wastewater composition and optimizing nutrient removal processes in real time [39]. This adaptive approach contributes to the efficiency and resilience of nutrient removal systems.

Furthermore, AI-driven anomaly detection becomes a crucial component of enhanced monitoring. By continuously analyzing sensor data, AI algorithms can identify deviations from expected patterns, signalling potential issues or abnormalities in the treatment process. This early warning system allows operators to take preventive measures, minimizing the risk of system failures or environmental noncompliance [45]. The synergy between smart sensors, real-time monitoring systems, and AI not only optimizes nutrient removal but also contributes to the overall sustainability of wastewater treatment [7]. The enhanced monitoring capabilities foster resource efficiency, reduce operational costs, and align with the principles of responsible environmental stewardship [45].

Therefore, the integration of smart sensors and real-time monitoring systems represents a transformative leap in the field of wastewater treatment. These technologies provide a dynamic understanding of nutrient levels, enabling operators to make informed decisions in real time. When coupled with AI, these monitoring systems evolve into intelligent, adaptive tools that not only optimize nutrient removal processes but also contribute to the broader goals of sustainable and efficient wastewater treatment.

8.5 Machine Learning Models for Nutrient Optimization

In the pursuit of more efficient and adaptive wastewater treatment processes, machine learning (ML) models have emerged as powerful tools for nutrient optimization [12]. This section explores the applications of predictive modeling for nutrient removal, the implementation of adaptive control strategies enabled by ML, and presents case studies showcasing successful ML-driven nutrient optimization in real-world wastewater treatment scenarios.

8.5.1 Predictive Modeling for Nutrient Removal

Predictive modeling using machine learning techniques has revolutionized the field of nutrient removal in wastewater treatment [41]. Traditional methods often struggle to adapt to the dynamic nature of influent wastewater, leading to challenges in maintaining optimal nutrient levels [14]. ML models, however, leverage historical data, process parameters, and real-time sensor information to predict future nutrient concentrations and optimize treatment processes accordingly.

Supervised learning algorithms, such as regression and neural networks, can be trained on historical data to establish relationships between influent characteristics and nutrient levels at the effluent [21]. This enables the creation of predictive models capable of forecasting nutrient concentrations based on varying influent conditions. The models learn from patterns in the data, improving their accuracy over time and adapting to changing environmental factors. Predictive modelling facilitates proactive decision-making by providing operators with insights into potential variations in nutrient levels. By anticipating these changes, treatment plants can adjust dosing strategies, modify operational parameters, and optimize biological or chemical processes to ensure effective nutrient removal [32]. This dynamic, data-driven approach enhances the overall efficiency of nutrient management in wastewater treatment.

8.5.2 Adaptive Control Strategies

Machine learning extends its impact beyond predictive modeling to the implementation of adaptive control strategies. Adaptive control leverages real-time data from sensors and predictive models to dynamically adjust treatment parameters, ensuring continuous optimization of nutrient removal processes [35].

Reinforcement learning, a subset of machine learning, plays a pivotal role in adaptive control. In a wastewater treatment context, reinforcement learning algorithms learn by interacting with the treatment system, receiving feedback on their actions, and adjusting strategies to maximize a predefined objective—in this case, optimal nutrient removal [42]. These algorithms continuously adapt to changing conditions, making them well-suited for the inherent variability in influent wastewater composition.

Adaptive control strategies enable treatment plants to respond in real-time to fluctuations in nutrient concentrations [4, 35]. For example, if a predictive model anticipates an increase in nitrogen levels, an adaptive control system can autonomously adjust aeration rates or dosing of chemicals to optimize nutrient removal without human intervention. This level of automation enhances the responsiveness of nutrient management systems, reducing the risk of over-treatment or under-treatment and ensuring compliance with environmental regulations [4, 35].

8.6 Resource Recovery Strategies

Wastewater, once viewed solely as a challenge for disposal, is now being recognized as a valuable resource [31]. Resource recovery strategies aim to extract value from wastewater by reclaiming and reusing its components, such as nutrients, energy, and water [50]. This section explores the paradigm shift toward extracting value from wastewater, the role of artificial intelligence (AI) in driving resource recovery techniques, and the economic and environmental impacts of these innovative approaches.

8.6.1 Extracting Value from Wastewater

Traditionally, wastewater treatment primarily focused on mitigating environmental impacts by removing pollutants [48]. However, the evolution of resource recovery strategies acknowledges that wastewater contains valuable components that can be harnessed for various purposes [51].

One key aspect of resource recovery is nutrient reuse. Nitrogen and phosphorus, essential nutrients for plant growth, are present in wastewater. By recovering these nutrients, treated wastewater can be transformed into a valuable fertilizer, closing the nutrient loop and reducing the dependence on traditional chemical fertilizers [48]. Additionally, energy recovery from organic matter in wastewater, such as through anaerobic digestion or microbial fuel cells, presents opportunities for harnessing renewable energy [50].

The concept of extracting value from wastewater aligns with the principles of a circular economy, where waste is minimized, and resources are utilized efficiently [20]. This shift towards resource recovery not only contributes to sustainable wastewater management but also supports broader environmental and economic goals [8].

8.6.2 AI-Driven Resource Recovery Techniques

Artificial intelligence plays a crucial role in optimizing resource recovery from wastewater. AI algorithms can analyze complex data sets, predict resource concentrations in wastewater, and optimize treatment processes for enhanced recovery [14, 41]. Machine learning models can adapt to variations in influent characteristics and make real-time decisions to maximize the extraction of valuable resources [51].

In nutrient recovery, AI-driven predictive modeling can optimize dosing strategies for precipitating agents, ensuring efficient removal of nutrients while minimizing chemical usage. Smart sensors, integrated with AI, enable real-time monitoring of nutrient concentrations, allowing for precise control over resource recovery processes

[27]. Moreover, machine learning algorithms can identify patterns in organic matter content, optimizing conditions for energy recovery processes like anaerobic digestion [50].

The intelligent application of AI in resource recovery not only enhances the efficiency of these processes but also opens up new possibilities for innovation. For example, AI can contribute to the selective extraction of specific nutrients or the identification of novel resource recovery pathways, further maximizing the value extracted from wastewater [4, 50].

8.6.3 Economic and Environmental Impacts

The implementation of resource recovery strategies driven by AI has significant economic and environmental implications [54]. From an economic perspective, extracting value from wastewater can transform treatment plants from cost centers to revenue generators. The recovered resources, whether in the form of fertilizers, energy, or purified water, can be sold or reused, contributing to the financial sustainability of wastewater treatment facilities [50].

AI-driven resource recovery also offers cost savings through optimized processes. By precisely controlling dosing, energy usage, and other operational parameters, treatment plants can reduce overall operational costs [26]. The predictive capabilities of AI enable preventive maintenance, minimizing downtime and extending the lifespan of equipment, further contributing to economic efficiency [51].

On the environmental front, resource recovery mitigates the environmental impact of wastewater discharges. Nutrient-rich effluents, when properly recovered and reused, reduce the need for synthetic fertilizers, mitigating the environmental consequences associated with their production and use. Energy recovery from organic matter in wastewater contributes to the development of renewable energy sources, aligning with global efforts to transition towards sustainable and carbon–neutral energy systems [50].

Furthermore, the reduction of waste and the transformation of wastewater into a resource align with broader sustainability goals [8]. The circular economy principles embedded in resource recovery strategies contribute to minimizing environmental pollution, conserving resources, and fostering a more sustainable and resilient water management system [54].

So, resource recovery strategies, facilitated by artificial intelligence, represent a transformative approach to wastewater management. By extracting value from wastewater, treatment plants can contribute to economic sustainability, reduce environmental impact, and align with the principles of a circular economy [50, 54]. As technology continues to advance, the integration of AI-driven resource recovery is poised to play a pivotal role in shaping the future of wastewater treatment, driving towards a more sustainable and resource-efficient paradigm [41].

8.7 Challenges and Future Directions

As the field of wastewater treatment undergoes a transformative phase with the integration of artificial intelligence (AI) in nutrient removal and resource recovery, it is crucial to address the current challenges, navigate ethical considerations, and envision future trends and potential innovations [50]. This section explores the multifaceted landscape of challenges and future directions in the realm of AI-integrated wastewater treatment.

8.7.1 Current Challenges in Nutrient Removal

Despite the advancements in AI-driven nutrient removal, several challenges persist in the field. The variability of influent wastewater composition remains a significant hurdle. Fluctuations in nutrient levels, presence of emerging contaminants, and changes in weather patterns can impact the performance of treatment processes [20]. Adapting AI models to handle such dynamic conditions requires continuous refinement and adaptation.

Another challenge lies in the need for extensive and high-quality data. Machine learning models heavily rely on robust datasets for training and validation. Obtaining sufficient and representative data, especially in diverse and complex wastewater treatment scenarios, can be challenging [51]. Incomplete or biased datasets may lead to inaccurate predictions and suboptimal performance of AI-driven systems.

Additionally, the scalability of AI solutions poses a challenge. While successful implementations have been demonstrated in pilot projects and small-scale facilities, scaling up AI-integrated wastewater treatment to larger plants with varying infrastructures and operational complexities requires careful consideration of technological, logistical, and economic factors [20].

Addressing these challenges necessitates collaboration between researchers, practitioners, and technology developers to refine AI models, enhance data collection methodologies, and ensure the scalability and adaptability of AI-driven nutrient removal strategies [51].

8.7.2 Ethical Considerations in AI-Integrated Wastewater Treatment

As AI becomes an integral part of wastewater treatment, ethical considerations come to the forefront. Privacy concerns, data security, and the responsible use of AI technologies must be carefully managed [11, 39]. The collection and utilization of sensitive data from treatment plants, especially in the context of real-time monitoring, require robust privacy protocols to safeguard against unauthorized access and potential misuse [39].

Transparency and explainability of AI models are essential ethical considerations. Understanding how AI algorithms make decisions is crucial for operators, regulators, and the public. Lack of transparency can lead to mistrust and hinder the widespread acceptance of AI-integrated wastewater treatment technologies [51].

Equity and access to AI technologies must also be considered. Smaller wastewater treatment facilities or those in economically disadvantaged regions may face challenges in adopting and implementing AI solutions due to financial constraints or lack of technical expertise [51]. Ensuring that the benefits of AI-driven nutrient removal are accessible to all communities is essential for fostering inclusivity and addressing environmental justice concerns [11].

Ethical guidelines and frameworks for AI in wastewater treatment should be developed, taking into account the unique challenges and responsibilities associated with managing critical infrastructure and sensitive environmental data [11, 52].

8.7.3 Future Trends and Potential Innovations

Looking ahead, several future trends and potential innovations are poised to shape the landscape of AI-integrated wastewater treatment. One key trend is the continued advancement of sensor technologies. Smaller, more affordable, and more sophisticated sensors will contribute to more comprehensive and real-time monitoring of wastewater treatment processes. These advancements will enhance the quality and quantity of data available for AI models, further improving their predictive capabilities [41].

The integration of decentralized and modular treatment systems is another emerging trend. These systems, coupled with AI, offer the potential for scalable and adaptable solutions. Decentralized treatment plants can respond more effectively to localized challenges, such as population growth or changes in industrial activities, providing a more resilient and flexible wastewater management infrastructure [20].

Innovations in explainable AI and interpretable machine learning models will address ethical concerns and enhance the trustworthiness of AI-integrated systems. Transparent models that provide clear insights into decision-making processes will be crucial for gaining acceptance among operators, regulators, and the public [52].

Furthermore, the convergence of AI with other emerging technologies, such as the Internet of Things (IoT) and edge computing, holds great promise. The integration of AI algorithms directly into sensors and devices at the edge of the wastewater treatment system enables faster decision-making and reduces the dependence on centralized computing resources [30].

As the field evolves, interdisciplinary collaboration between environmental engineers, data scientists, ethicists, and policymakers will become increasingly important. By fostering a holistic approach, future innovations can address challenges, uphold ethical standards, and contribute to the sustainable and efficient management of wastewater treatment processes.

Therefore, challenges and future directions in AI-integrated wastewater treatment are interconnected, requiring a balanced and collaborative approach. Ethical considerations must guide the responsible development and deployment of AI technologies, while innovations in sensor technologies, decentralized systems, and interdisciplinary collaborations will pave the way for a more resilient, adaptive, and sustainable future in wastewater treatment [11].

8.8 Conclusion

In navigating the landscape of artificial intelligence (AI) in nutrient removal and resource recovery within wastewater treatment, this exploration has unveiled key findings that underscore the transformative potential and challenges inherent in this dynamic field. As we summarize these findings, we also cast a gaze into the future, envisioning the evolving landscape of AI applications in wastewater treatment and the potential trajectory for nutrient removal and resource recovery.

8.9 Summary of Key Findings

The journey through this exploration began with an acknowledgment of the foundational aspects of nutrient removal, recognizing traditional methods and the importance of effective nutrient management. The integration of smart sensors and real-time monitoring systems emerged as a pivotal advancement, enabling dynamic and adaptive responses to fluctuations in nutrient levels. Machine learning models played a central role in nutrient optimization, providing predictive capabilities and adaptive control strategies for enhanced efficiency.

The exploration further delved into resource recovery strategies, emphasizing the paradigm shift from viewing wastewater as a disposal challenge to recognizing it as a valuable resource. AI-driven techniques were identified as catalysts for efficient nutrient recovery, offering innovative solutions to longstanding challenges. The economic and environmental impacts of these strategies were highlighted, showcasing how AI in resource recovery can contribute to both financial sustainability and environmental stewardship.

8.9.1 *The Future Landscape of AI in Nutrient Removal and Resource Recovery*

Looking ahead, the future landscape of AI in nutrient removal and resource recovery is marked by several promising trends and potential innovations. The integration of Internet of Things (IoT) devices and sensor networks is anticipated to enhance data connectivity and collection, providing a more comprehensive understanding of treatment processes. This interconnectedness will empower AI algorithms to make more informed decisions based on a broader range of real-time data, further optimizing nutrient removal.

Advancements in explainable AI will address the interpretability challenge, fostering transparency in AI-driven decisions. As stakeholders gain a better understanding of how these algorithms operate, trust in AI applications in wastewater treatment is likely to increase, facilitating broader adoption across the industry.

The incorporation of decentralized and modular treatment solutions represents another transformative trend. Decentralized systems, empowered by AI, can offer flexibility in nutrient removal and resource recovery, particularly in remote or underserved areas. These adaptable systems can optimize treatment processes based on specific environmental factors and influent characteristics, providing a tailored and efficient approach to wastewater management.

Ongoing research into novel AI-driven treatment methods, such as advanced oxidation processes and innovative filtration technologies, holds promise for breakthroughs in nutrient removal efficiency. These innovations have the potential to redefine wastewater treatment by offering more sustainable and cost-effective alternatives to traditional methods, contributing to the overall evolution of the field.

8.9.2 *Navigating the Future with AI in Wastewater Treatment*

So, the integration of artificial intelligence in nutrient removal and resource recovery has marked a significant leap forward in the evolution of wastewater treatment. The challenges faced today, including variability in influent composition and ethical considerations, are acknowledged as opportunities for refinement and advancement.

The key findings of this exploration affirm that AI-driven applications can revolutionize wastewater treatment, making it more adaptive, efficient, and aligned with sustainability goals. As we envision the future landscape, the trends of enhanced connectivity, explainable AI, decentralized systems, and novel treatment methods stand as beacons guiding the way forward.

The future of AI in nutrient removal and resource recovery within wastewater treatment promises a landscape where technology, ethics, and sustainability harmoniously converge. Through collaborative efforts among researchers, practitioners, and policymakers, the continued integration of AI will likely propel the field into new realms of innovation and efficiency, contributing to a more resilient and sustainable approach to wastewater management. As we navigate this future, the transformative journey driven by AI in wastewater treatment unfolds, promising a landscape that prioritizes both environmental stewardship and technological innovation.

References

1. Alcántara C, de Godos I, Muñoz R (2020) Wastewater treatment and biomass generation with algae. In: Wastewater treatment residues as resources for biorefinery products and biofuels. Elsevier, pp 229–254. https://doi.org/10.1016/B978-0-12-816204-0.00011-4
2. Altowayti WAH, Shahir S, Othman N, Eisa TAE, Yafooz WMS, Al-Dhaqm A, Soon CY, Yahya IB, Che Rahim NAN, binti, Abaker M, Ali A (2022) The role of conventional methods and artificial intelligence in the wastewater treatment: a comprehensive review. Processes 10(9):1832.https://doi.org/10.3390/pr10091832
3. Andrade Cruz I, Chuenchart W, Long F, Surendra KC, Renata Santos Andrade L, Bilal M, Liu H, Tavares Figueiredo R, Khanal SK, Fernando Romanholo Ferreira L (2022) Application of machine learning in anaerobic digestion: perspectives and challenges. Bioresour Technol 345:126433. https://doi.org/10.1016/j.biortech.2021.126433
4. Aparna KG, Swarnalatha R (2023) Dynamic optimization of a wastewater treatment process for sustainable operation using multi-objective genetic algorithm and non-dominated sorting cuckoo search algorithm. J Water Process Eng 53:103775. https://doi.org/10.1016/j.jwpe.2023.103775
5. Bairagi S, Ali SW (2020) Conventional and advanced technologies for wastewater treatment. In: Environmental nanotechnology for water purification. Wiley, pp 33–56. https://doi.org/10.1002/9781119641353.ch2
6. Bao H, Yin W, Wang H, Lu Y, Jiang S, Ajibade FO, Ouyang Q, Wang Y, Nie S, Bai Y, Gao H, Wang A (2023) Automated machine learning-based models for predicting and evaluating antibiotic removal in constructed wetlands. Biores Technol 385:129436. https://doi.org/10.1016/j.biortech.2023.129436
7. Ching PML, So RHY, Morck T (2021) Advances in soft sensors for wastewater treatment plants: a systematic review. J Water Process Eng 44:102367. https://doi.org/10.1016/j.jwpe.2021.102367
8. Chojnacka K, Chojnacki M (2024) Nutrient recovery from anaerobic digestate: fertilizer informatics for circular economy. Environ Res 245:117953. https://doi.org/10.1016/j.envres.2023.117953
9. Corominas L, Garrido-Baserba M, Villez K, Olsson G, Cortés U, Poch M (2018) Transforming data into knowledge for improved wastewater treatment operation: a critical review of techniques. Environ Model Softw 106:89–103. https://doi.org/10.1016/j.envsoft.2017.11.023
10. Diaz-Elsayed N, Xu X, Balaguer-Barbosa M, Zhang Q (2017) An evaluation of the sustainability of onsite wastewater treatment systems for nutrient management. Water Res 121:186–196. https://doi.org/10.1016/j.watres.2017.05.005
11. Doorn N (2021) Artificial intelligence in the water domain: opportunities for responsible use. Sci Total Environ 755:142561. https://doi.org/10.1016/j.scitotenv.2020.142561
12. El Alaoui El Fels A, Mandi L, Kammoun A, Ouazzani N, Monga O, Hbid ML (2023) Artificial intelligence and wastewater treatment: a global scientific perspective through text mining. Water 15(19):3487.https://doi.org/10.3390/w15193487
13. Emadi M, Taghizadeh-Mehrjardi R, Cherati A, Danesh M, Mosavi A, Scholten T (2020) Predicting and mapping of soil organic carbon using machine learning algorithms in Northern Iran. Remote Sensing 12(14):2234. https://doi.org/10.3390/rs12142234
14. Ennaji O, Vergütz L, El Allali A (2023) Machine learning in nutrient management: a review. Artif Intell Agric 9:1–11. https://doi.org/10.1016/j.aiia.2023.06.001

15. Garske B, Stubenrauch J, Ekardt F (2020) Sustainable phosphorus management in European agricultural and environmental law. Rev Eur Comp Int Environ Law 29(1):107–117. https://doi.org/10.1111/reel.12318
16. Gedda BDJS (2021) Introduction to conventional wastewater treatment technologies: limitations and recent advances. In: Advances in wastewater treatment I, vol 91. Materials Research Foundations, pp 1–36. https://doi.org/10.21741/9781644901144-1
17. Gizaw A, Zewge F, Kumar A, Mekonnen A, Tesfaye M (2021) A comprehensive review on nitrate and phosphate removal and recovery from aqueous solutions by adsorption. J Water Supply Res Technol AQUA 70(7):921–947. https://doi.org/10.2166/aqua.2021.146
18. Gowd SC, Kumar D, Lin R, Rajendran K (2022) Nutrient recovery from wastewater in India: a perspective from mass and energy balance for a sustainable circular economy. Bioresour Technol Rep 18:101079. https://doi.org/10.1016/j.biteb.2022.101079
19. Guo H, Jeong K, Lim J, Jo J, Kim YM, Park J, Kim JH, Cho KH (2015) Prediction of effluent concentration in a wastewater treatment plant using machine learning models. J Environ Sci 32:90–101. https://doi.org/10.1016/j.jes.2015.01.007
20. Hasan MN, Altaf MM, Khan NA, Khan AH, Khan AA, Ahmed S, Kumar PS, Naushad M, Rajapaksha AU, Iqbal J, Tirth V, Islam S (2021) Recent technologies for nutrient removal and recovery from wastewaters: a review. Chemosphere 277:130328. https://doi.org/10.1016/j.chemosphere.2021.130328
21. Ismail W, Niknejad N, Bahari M, Hendradi R, Zaizi NJM, Zulkifli MZ (2021) Water treatment and artificial intelligence techniques: a systematic literature review research. Environ Sci Pollut Res 30(28):71794–71812. https://doi.org/10.1007/s11356-021-16471-0
22. Kazadi Mbamba C, Batstone DJ, Flores-Alsina X, Tait S (2015) A generalised chemical precipitation modelling approach in wastewater treatment applied to calcite. Water Res 68:342–353. https://doi.org/10.1016/j.watres.2014.10.011
23. Korostynska O, Mason A, Al-Shamma'a A (2012) Monitoring of nitrates and phosphates in wastewater: current technologies and further challenges. Int J Smart Sens Intell Syst 5(1):149–176. https://doi.org/10.21307/ijssis-2017-475
24. Li J, Jin Q, Liang Y, Geng J, Xia J, Chen H, Yun M (2022) Highly efficient removal of nitrate and phosphate to control eutrophication by the dielectrophoresis-assisted adsorption method. Int J Environ Res Public Health 19(3):1890. https://doi.org/10.3390/ijerph19031890
25. Ludmerczki R, Mura S, Stagi L, Juhász T, Dettori M, Azara A, Innocenzi P, Malfatti L (2021) Fluorescence-based selective nitrite ion sensing by amino-capped carbon dots. Environ Nanotechnol Monit Manag 16:100573. https://doi.org/10.1016/j.enmm.2021.100573
26. Lv J, Du L, Lin H, Wang B, Yin W, Song Y, Chen J, Yang J, Wang A, Wang H (2024) Enhancing effluent quality prediction in wastewater treatment plants through the integration of factor analysis and machine learning. Biores Technol 393:130008. https://doi.org/10.1016/j.biortech.2023.130008
27. Mahmud MAP, Ejeian F, Azadi S, Myers M, Pejcic B, Abbassi R, Razmjou A, Asadnia M (2020) Recent progress in sensing nitrate, nitrite, phosphate, and ammonium in aquatic environment. Chemosphere 259:127492. https://doi.org/10.1016/j.chemosphere.2020.127492
28. Martínez R, Vela N, el Aatik A, Murray E, Roche P, Navarro JM (2020) On the use of an IoT integrated system for water quality monitoring and management in wastewater treatment plants. Water 12(4):1096. https://doi.org/10.3390/w12041096
29. Maúre E. de R., Terauchi G, Ishizaka J, Clinton N, DeWitt M (2021) Globally consistent assessment of coastal eutrophication. Nat Commun 12(1):6142.https://doi.org/10.1038/s41467-021-26391-9
30. Nair AM, Gonzalez-Silva BM, Haugen FA, Ratnaweera H, Østerhus SW (2020) Real-time monitoring of enhanced biological phosphorus removal in a multistage EBPR-MBBR using a soft-sensor for phosphates. J Water Process Eng 37:101494. https://doi.org/10.1016/j.jwpe.2020.101494
31. Neczaj E, Grosser A (2018) Circular economy in wastewater treatment plant–challenges and barriers. EWaS3 614. https://doi.org/10.3390/proceedings2110614

32. Oruganti RK, Biji AP, Lanuyanger T, Show PL, Sriariyanun M, Upadhyayula VKK, Gadhamshetty V, Bhattacharyya D (2023) Artificial intelligence and machine learning tools for high-performance microalgal wastewater treatment and algal biorefinery: a critical review. Sci Total Environ 876:162797. https://doi.org/10.1016/j.scitotenv.2023.162797
33. Paepae T, Bokoro P, Kyamakya K (2022) A virtual sensing concept for nitrogen and phosphorus monitoring using machine learning techniques. Sensors 22(19):7338. https://doi.org/10.3390/s22197338
34. Pellerin BA, Stauffer BA, Young DA, Sullivan DJ, Bricker SB, Walbridge MR, Clyde GA, Shaw DM (2016) Emerging tools for continuous nutrient monitoring networks: sensors advancing science and water resources protection. JAWRA J Am Water Resour Assoc 52(4):993–1008. https://doi.org/10.1111/1752-1688.12386
35. Petre E, Selisteanu D, Sulea-Iorgulescu C, Mehedinteanu S (2020) Multivariable adaptive control strategy for an activated sludge process inside a wastewater treatment plant. In: 2020 21th international carpathian control conference (ICCC), pp 1–6. https://doi.org/10.1109/ICCC49264.2020.9257214
36. Rout PR, Shahid MK, Dash RR, Bhunia P, Liu D, Varjani S, Zhang TC, Surampalli RY (2021) Nutrient removal from domestic wastewater: a comprehensive review on conventional and advanced technologies. J Environ Manage 296:113246. https://doi.org/10.1016/j.jenvman.2021.113246
37. Rugaika AM, Van Deun R, Njau KN, Van der Bruggen B (2019) Phosphorus recovery as calcium phosphate by a pellet reactor pre-treating domestic wastewater before entering a constructed wetland. Int J Environ Sci Technol 16(7):3851–3860. https://doi.org/10.1007/s13762-018-02179-1
38. Ryu H, Thompson D, Huang Y, Li B, Lei Y (2020) Electrochemical sensors for nitrogen species: a review. Sens Actuators Rep 2(1):100022. https://doi.org/10.1016/j.snr.2020.100022
39. Safeer S, Pandey RP, Rehman B, Safdar T, Ahmad I, Hasan SW, Ullah A (2022) A review of artificial intelligence in water purification and wastewater treatment: recent advancements. J Water Process Eng 49:102974. https://doi.org/10.1016/j.jwpe.2022.102974
40. Sangamnere R, Misra T, Bherwani H, Kapley A, Kumar R (2023) A critical review of conventional and emerging wastewater treatment technologies. Sustain Water Resour Manag 9(2):58. https://doi.org/10.1007/s40899-023-00829-y
41. Sharafati A, Asadollah SBHS, Hosseinzadeh M (2020) The potential of new ensemble machine learning models for effluent quality parameters prediction and related uncertainty. Process Saf Environ Prot 140:68–78. https://doi.org/10.1016/j.psep.2020.04.045
42. Sheik AG, Tejaswini ESS, Ambati SR (2022) Design of intelligent control strategies for full-scale wastewater treatment plants with struvite unit. J Water Process Eng 49:103104. https://doi.org/10.1016/j.jwpe.2022.103104
43. Singh P, Singh MK, Beg YR, Nishad GR (2019) A review on spectroscopic methods for determination of nitrite and nitrate in environmental samples. Talanta 191:364–381. https://doi.org/10.1016/j.talanta.2018.08.028
44. Singh V, Mishra V (2022) Evaluation of the effects of input variables on the growth of two microalgae classes during wastewater treatment. Water Res 213:118165. https://doi.org/10.1016/j.watres.2022.118165
45. Wang Y, Cheng Y, Liu H, Guo Q, Dai C, Zhao M, Liu D (2023) A review on applications of artificial intelligence in wastewater treatment. Sustainability 15(18):13557. https://doi.org/10.3390/su151813557
46. Wang Y-Q, Wang H-C, Song Y-P, Zhou S-Q, Li Q-N, Liang B, Liu W-Z, Zhao Y-W, Wang A-J (2023) Machine learning framework for intelligent aeration control in wastewater treatment plants: automatic feature engineering based on variation sliding layer. Water Res 246:120676. https://doi.org/10.1016/j.watres.2023.120676
47. Yang Y, Kim K-R, Kou R, Li Y, Fu J, Zhao L, Liu H (2022) Prediction of effluent quality in a wastewater treatment plant by dynamic neural network modeling. Process Saf Environ Prot 158:515–524. https://doi.org/10.1016/j.psep.2021.12.034

48. Ye Y, Ngo HH, Guo W, Chang SW, Nguyen DD, Zhang X, Zhang J, Liang S (2020) Nutrient recovery from wastewater: from technology to economy. Bioresour Technol Rep 11:100425. https://doi.org/10.1016/j.biteb.2020.100425
49. Zaghloul MS, Achari G (2022) Application of machine learning techniques to model a full-scale wastewater treatment plant with biological nutrient removal. J Environ Chem Eng 10(3):107430. https://doi.org/10.1016/j.jece.2022.107430
50. Zarei M (2020) Wastewater resources management for energy recovery from circular economy perspective. Water-Energy Nexus 3:170–185. https://doi.org/10.1016/j.wen.2020.11.001
51. Zhang S, Jin Y, Chen W, Wang J, Wang Y, Ren H (2023) Artificial intelligence in wastewater treatment: a data-driven analysis of status and trends. Chemosphere 336:139163. https://doi.org/10.1016/j.chemosphere.2023.139163
52. Zhang W, Tooker NB, Mueller AV (2020) Enabling wastewater treatment process automation: leveraging innovations in real-time sensing, data analysis, and online controls. Environ Sci Water Res Technol 6(11):2973–2992. https://doi.org/10.1039/D0EW00394H
53. Zhang Y, Wu H, Xu R, Wang Y, Chen L, Wei C (2024) Machine learning modeling for the prediction of phosphorus and nitrogen removal efficiency and screening of crucial microorganisms in wastewater treatment plants. Sci Total Environ 907:167730. https://doi.org/10.1016/j.scitotenv.2023.167730
54. Zhao L, Dai T, Qiao Z, Sun P, Hao J, Yang Y (2020) Application of artificial intelligence to wastewater treatment: a bibliometric analysis and systematic review of technology, economy, management, and wastewater reuse. Process Saf Environ Prot 133:169–182. https://doi.org/10.1016/j.psep.2019.11.014
55. Zhuang Y, Wen W, Ruan S, Zhuang F, Xia B, Li S, Liu H, Du Y, Zhang L (2022) Real-time measurement of total nitrogen for agricultural runoff based on multiparameter sensors and intelligent algorithms. Water Res 210:117992. https://doi.org/10.1016/j.watres.2021.117992

Chapter 9
Exploring Emerging Technologies for AI-Integrated Advancements in Wastewater Treatment

Harnessing Emerging Technologies for AI in Wastewater Treatment

Sheetal Kumari

Abstract Wastewater treatment has evolved significantly, from ancient methods of settling and filtration to sophisticated modern processes crucial for environmental and public health protection. The integration of artificial intelligence (AI) marks a transformative phase in this evolution, offering enhanced efficiency, precision, and sustainability. AI technologies, including machine learning and predictive analytics, empower treatment plants to optimize operations, predict system failures, and adapt to dynamic conditions. This chapter explores the historical context of wastewater treatment, highlighting the progression towards AI-integrated approaches. It discusses the foundational principles of conventional treatment, the limitations they face in addressing emerging contaminants, and the imperative for technological advancements. Key objectives include elucidating AI's role in revolutionizing wastewater treatment, examining applications and challenges, and exploring implications for sustainability and resource recovery. Case studies and examples illustrate AI's real-world impact, showcasing its potential to mitigate environmental challenges and shape the future of wastewater management practices.

Keywords Artificial intelligence (AI) · Wastewater treatment · Environmental management · Resource recovery · Sustainability

S. Kumari (✉)
Amity Institute of Environmental Science (AIES), Amity University Uttar Pradesh, Sector-125, Gautam Budh Nagar, Noida 201313, India
e-mail: Sheetalluhania99@gmail.com

M. C. Garg (ed.), *The AI Cleanse: Transforming Wastewater Treatment Through Artificial Intelligence*, Springer Water, https://doi.org/10.1007/978-3-031-67237-8_9

9.1 Introduction

Wastewater treatment has undergone a remarkable evolution over the years, transforming from rudimentary methods to sophisticated processes aimed at safeguarding our environment. The treatment of wastewater is an essential component of environmental management, ensuring the protection of ecosystems and public health [35]. As we delve into the twenty-first century, the convergence of traditional wastewater treatment practices with cutting-edge artificial intelligence (AI) technologies presents a new frontier of possibilities and advancements.

9.1.1 The Evolution of Wastewater Treatment

The journey of wastewater treatment can be traced back to ancient civilizations, where basic methods such as settling and natural filtration were employed to mitigate the adverse effects of wastewater on communities. Over time, the field has witnessed a gradual progression, marked by the development of sewage systems, primary and secondary treatment processes, and the advent of tertiary treatment for advanced pollutant removal [25]. In recent decades, the imperative to address emerging contaminants and the increasing complexity of industrial effluents has prompted a paradigm shift in wastewater treatment. Novel technologies, including membrane filtration, biological nutrient removal, and advanced oxidation processes, have become integral to modern treatment plants. The evolution is not merely technical, it also encompasses a broader perspective, incorporating sustainability and resource recovery in the treatment process [25].

9.1.2 Significance of Innovations in AI-Integrated Approaches

The integration of artificial intelligence (AI) into wastewater treatment processes represents a groundbreaking leap towards efficiency, precision, and sustainability. AI, with its machine learning algorithms and predictive analytics, empowers treatment plants to optimize operations, predict system failures, and enhance overall performance. By harnessing the power of AI, treatment facilities can adapt to dynamic conditions, improve energy efficiency, and minimize environmental impacts [4]. The significance of AI in wastewater treatment extends beyond operational enhancements. It enables real-time monitoring and analysis, facilitating proactive decision-making. AI algorithms can adapt to changing influent characteristics, weather patterns, and unforeseen events, ensuring that treatment processes remain robust and resilient [16]. This integration not only improves the treatment efficacy but also contributes to a more comprehensive understanding of the dynamics within wastewater treatment systems.

9.1.3 Objectives

The scope of this chapter is to explore the synergies between emerging AI technologies and wastewater treatment, providing an in-depth analysis of their integration, implications, and potential benefits. By delving into the historical context of wastewater treatment, we lay the foundation for understanding how AI has become a transformative force in an established field. The primary objective is to elucidate the specific ways in which AI-driven advancements are revolutionizing wastewater treatment. We will explore the key technologies underpinning this integration, examining their applications, successes, and challenges. Through case studies and examples, we aim to illustrate the real-world impact of AI in diverse wastewater treatment scenarios. Furthermore, this chapter seeks to address the broader implications of AI integration, including its environmental and economic ramifications. We will discuss the potential for sustainability, resource recovery, and the optimization of treatment processes, considering both current achievements and future possibilities. In essence, the overarching goal of this chapter is to provide a comprehensive overview of the evolving landscape where AI and wastewater treatment converge. By doing so, we hope to contribute to a deeper understanding of the transformative potential of AI in mitigating environmental challenges and shaping the future of wastewater treatment practices.

9.2 Foundations of Wastewater Treatment

Wastewater treatment, rooted in centuries of human ingenuity, has evolved into a complex and indispensable process for safeguarding public health and the environment. This section delves into the foundations of wastewater treatment, exploring the fundamental principles of conventional treatment, the inherent limitations of traditional approaches, and the imperative for ongoing technological progress.

9.2.1 Fundamental Principles of Conventional Treatment

At its core, conventional wastewater treatment relies on the principles of physical, chemical, and biological processes to remove contaminants from water before its discharge back into the environment. The primary stages typically include preliminary treatment, primary treatment, secondary treatment, and sometimes tertiary treatment. Preliminary treatment involves the removal of large debris, and grit, and

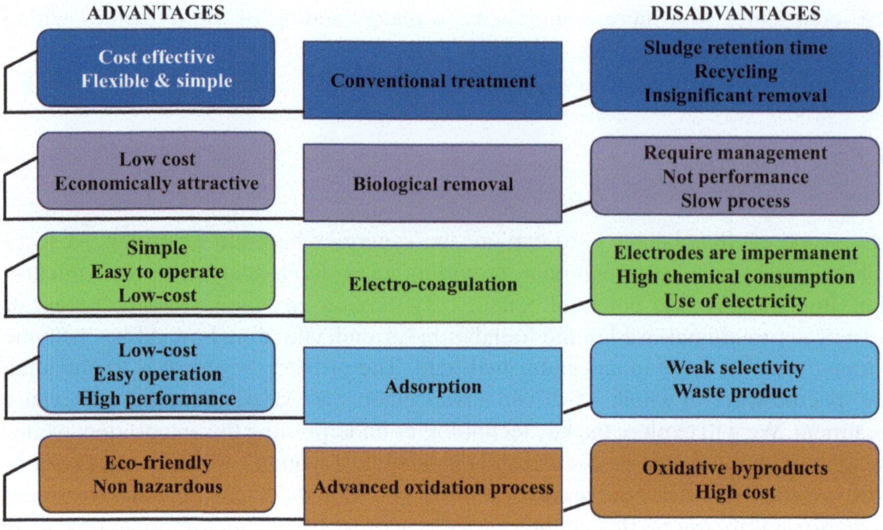

Fig. 9.1 Advantages and disadvantages of conventional methods

screens out objects that could damage equipment. Primary treatment utilizes physical processes like sedimentation to separate solids from the wastewater, while secondary treatment employs biological processes to further break down organic matter [38]. Conventional treatment methods have been instrumental in addressing conventional pollutants like suspended solids and organic compounds. However, they face limitations in effectively treating emerging contaminants such as pharmaceuticals, personal care products, and industrial chemicals [42]. Moreover, these methods often struggle to adapt to variations in influent characteristics and environmental conditions. The advantages and disadvantages of conventional methods are shown in Fig. 9.1.

9.2.2 Limitations and Evolving Challenges

While conventional treatment processes have been foundational in managing water pollution, they are not without their challenges and limitations. One of the prominent challenges is the inability to consistently address the diverse and evolving spectrum of contaminants present in wastewater. Emerging pollutants, including endocrine-disrupting chemicals and microplastics, pose new challenges that traditional treatment methods are ill-equipped to handle effectively [3]. Furthermore, the energy intensity of some conventional treatment processes, such as aeration in secondary treatment, raises concerns about the environmental footprint of wastewater treatment facilities. Ageing infrastructure, increased urbanization, and population growth amplify these challenges, placing additional stress on existing treatment systems.

Climate change adds another layer of complexity, with altered precipitation patterns and extreme weather events affecting the quantity and quality of wastewater entering treatment plants [8]. Conventional systems may struggle to cope with these dynamic conditions, leading to suboptimal performance and potential environmental risks.

9.2.3 Imperative for Technological Progress

Recognizing the limitations and evolving challenges of conventional wastewater treatment, there is a pressing need for continuous technological progress. This imperative extends beyond incremental improvements to embrace transformative technologies that can enhance treatment efficiency, address emerging contaminants, and promote sustainability. The integration of advanced technologies, particularly artificial intelligence (AI), into wastewater treatment processes holds immense promise [29]. AI can optimize the operation of treatment plants, predict system failures, and adapt to changing conditions in real time. This not only improves treatment efficacy but also contributes to energy savings and reduced environmental impact.

Technological progress also involves the development and implementation of innovative materials and processes. Membrane filtration, advanced oxidation processes, and biological nutrient removal are examples of technologies that push the boundaries of conventional treatment, offering more effective solutions for removing pollutants [43].

9.3 The Confluence of AI and Wastewater Treatment

In the intersection of artificial intelligence (AI) and wastewater treatment, a transformative synergy is unfolding, revolutionizing the way we approach environmental challenges. This section explores the dynamic confluence of AI and wastewater treatment, unveiling the synergy between these two domains and delving into the paradigm shifts that AI introduces to traditional treatment approaches.

9.3.1 Unveiling the Synergy: AI and Environmental Solutions

Artificial intelligence emerges as a powerful tool in the realm of environmental solutions, offering unprecedented capabilities to enhance the efficiency and sustainability of wastewater treatment processes. At its core, AI relies on machine learning algorithms and data analytics to make sense of vast datasets, adapt to changing conditions, and optimize complex systems. When applied to wastewater treatment, AI becomes a catalyst for precision and adaptability [41]. One of the primary contributions of

AI to environmental solutions is its ability to process and analyze real-time data. With sensors and monitoring systems providing a constant influx of information, AI algorithms can detect patterns, identify anomalies, and predict changes in wastewater characteristics. This real-time awareness enables treatment plants to respond proactively, adjusting operational parameters to optimize performance and mitigate potential issues.

9.3.2 *Paradigm Shifts in Wastewater Treatment Through AI*

The integration of AI into wastewater treatment herald's paradigm shifts that transcend traditional limitations. Conventional treatment methods often operate on predefined parameters and fixed processes, struggling to adapt to the dynamic nature of wastewater composition and environmental conditions. AI introduces a level of adaptability and intelligence that transforms treatment plants into agile and responsive systems [1]. Machine learning algorithms embedded in AI systems can learn from historical data, continuously improving their predictions and decision-making capabilities. This adaptability is particularly valuable in addressing the challenges posed by variable influent characteristics, climate-induced fluctuations, and unexpected events. AI empowers treatment plants to optimize processes in real time, ensuring consistent and efficient pollutant removal [24]. Furthermore, AI-driven automation reduces the burden on human operators and enhances the overall reliability of treatment facilities. Predictive maintenance algorithms can foresee equipment failures, allowing for proactive repairs and minimizing downtime. This shift towards automation not only improves operational efficiency but also contributes to a safer and more sustainable working environment for plant personnel [1].

9.3.3 *Unleashing the Potential: Transformative Benefits and Ongoing Challenges*

The transformative benefits of AI in wastewater treatment are multifaceted. Beyond optimizing operational efficiency, AI contributes to environmental sustainability by reducing energy consumption, minimizing chemical usage, and enhancing overall treatment efficacy. The potential for resource recovery, such as reclaiming valuable materials from wastewater, adds an extra layer of economic and ecological benefits [21]. However, the integration of AI in wastewater treatment is not without its challenges. The initial investment in AI infrastructure and training poses financial considerations for treatment plants. Moreover, ensuring the ethical use of AI, addressing data privacy concerns, and navigating regulatory frameworks present ongoing challenges that must be addressed for the widespread adoption of AI technologies [24]. Ongoing research and development are essential to unlocking the full potential of AI in wastewater treatment. Collaborations between technology developers, researchers, and wastewater treatment professionals play a crucial role in overcoming challenges and refining AI applications for specific treatment contexts. As AI continues to evolve, its potential to revolutionize wastewater treatment and contribute to a more sustainable and resilient future becomes increasingly evident.

9.4 Cutting-Edge AI Technologies

As wastewater treatment seeks to embrace the transformative potential of artificial intelligence (AI), cutting-edge technologies are at the forefront of this revolution. This section delves into three key AI technologies that are reshaping wastewater treatment: harnessing machine learning for optimization, exploring deep learning paradigms in treatment processes, and pioneering the use of neural networks for enhanced process efficiency (Fig. 9.2).

9.4.1 Harnessing Machine Learning for Wastewater Optimization

Machine learning (ML) stands as a cornerstone in the integration of AI into wastewater treatment, offering a dynamic optimization approach. ML algorithms excel in learning patterns from data, making them particularly well-suited for adapting to the

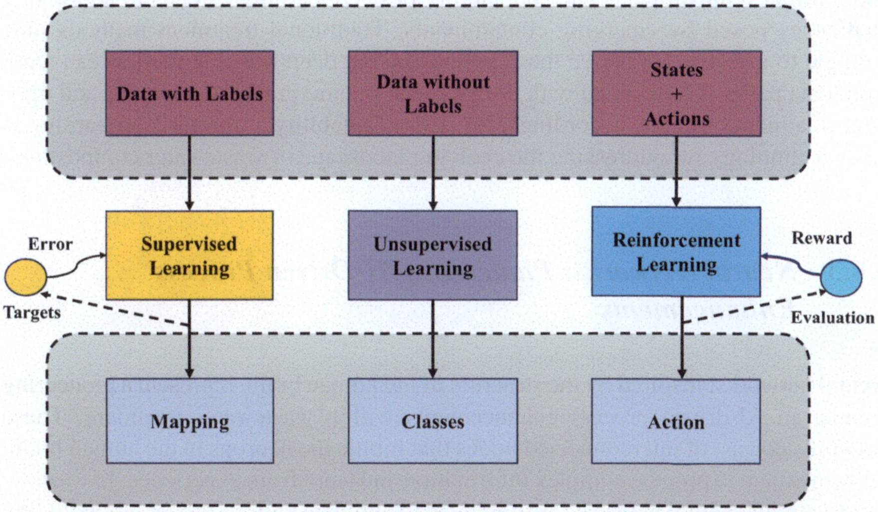

Fig. 9.2 Flow chart of different AI-driven approaches

variable and complex nature of wastewater treatment processes [8]. In the context of wastewater optimization, ML is employed to analyze historical data, sensor readings, and operational parameters. This analysis enables the development of predictive models that can anticipate changes in influent characteristics and optimize treatment processes accordingly. For instance, ML algorithms can predict peak flow periods, allowing treatment plants to adjust aeration rates, chemical dosages, and other operational parameters to enhance overall treatment efficiency [18]. Furthermore, ML algorithms contribute to predictive maintenance strategies by identifying patterns indicative of equipment failures. This proactive approach reduces downtime, extends the lifespan of critical components, and minimizes operational disruptions. The versatility of ML in optimizing various facets of wastewater treatment positions it as a powerful tool in the quest for sustainable and efficient operations [41].

9.4.2 Deep Learning Paradigms in Treatment Processes

Deep learning, a subset of machine learning, has gained prominence for its ability to process vast amounts of data and extract intricate patterns. In wastewater treatment, deep learning paradigms are proving instrumental in optimizing treatment processes and overcoming challenges associated with complex pollutant profiles [26]. Deep learning algorithms, particularly neural networks with multiple layers, excel in recognizing non-linear relationships within data. This capability is leveraged in wastewater treatment to enhance pollutant removal efficiency [27]. For example, in biological treatment processes, deep learning models can dynamically adjust aeration rates based on real-time sensor data, ensuring optimal conditions for microbial activity and nutrient removal [22]. The adaptability of deep learning extends to addressing challenges posed by emerging contaminants. Traditional treatment methods may struggle to effectively remove these pollutants, but deep learning models can learn from data patterns associated with the presence of emerging contaminants and optimize treatment strategies accordingly [5]. This adaptability positions deep learning as a key technology for addressing the evolving landscape of wastewater composition.

9.4.3 Neural Networks: Pioneering AI-Driven Process Enhancements

Neural networks, inspired by the structure of the human brain, represent a pioneering frontier in AI-driven process enhancements within wastewater treatment. These networks consist of interconnected nodes that mimic the neurons in the human brain, allowing them to process complex information and learn from experience. In wastewater treatment, neural networks are deployed to optimize various processes, including nutrient removal, sludge dewatering, and chemical dosing [30]. For instance, neural networks can be trained to recognize the relationship between influent characteristics and optimal chemical dosages, enabling precise and adaptive dosing strategies. This enhances treatment efficiency while minimizing the environmental impact associated with excessive chemical usage. Neural networks also play a crucial role in sensor fusion, where data from multiple sensors are integrated to provide a comprehensive understanding of treatment conditions. By processing diverse data inputs, neural networks enable a holistic view of the treatment system, allowing for more informed decision-making and adaptive control strategies [34]. The pioneering nature of neural networks in wastewater treatment extends to their ability to learn from real-time data, continuously improving their predictive capabilities. As neural networks adapt to changing conditions and learn from operational experiences, they contribute to a more resilient and efficient wastewater treatment process [15].

In conclusion, cutting-edge AI technologies, including machine learning, deep learning paradigms, and neural networks, are reshaping the landscape of wastewater treatment. These technologies bring adaptability, precision, and efficiency to treatment processes, addressing challenges posed by variable influent characteristics and the emergence of new contaminants [15]. As wastewater treatment facilities embrace these innovative AI-driven approaches, the potential for sustainable and effective environmental stewardship becomes increasingly attainable.

9.5 Revolutionary Sensor Technologies

In the era of advancing wastewater treatment, revolutionary sensor technologies are playing a pivotal role in reshaping monitoring and control strategies. This section explores three key aspects of sensor technologies that are driving innovation in wastewater treatment: the Internet of Things (IoT) revolutionizing wastewater monitoring, the centrality of real-time data analytics in AI-integrated systems, and the impact of remote sensing innovations on enhanced monitoring and control.

9.5.1 IoT Revolutionizing Wastewater Monitoring

The Internet of Things (IoT) has ushered in a new era of connectivity and data-driven insights in wastewater treatment. IoT-enabled sensor technologies have transformed traditional monitoring practices by providing real-time data on various parameters crucial to treatment efficiency. These sensors, strategically deployed throughout treatment facilities, continuously collect and transmit data, offering a comprehensive view of the treatment process [17]. In wastewater monitoring, IoT sensors measure parameters such as pH levels, dissolved oxygen, turbidity, and flow rates. This real-time data enables treatment plants to respond swiftly to fluctuations in influent characteristics, ensuring optimal operational conditions. For instance, if a sudden increase in organic load is detected, IoT sensors can trigger automated adjustments in aeration rates, optimizing the biological treatment process [23]. The integration of IoT in wastewater treatment not only enhances process control but also contributes to predictive analytics. By analyzing historical data trends, treatment plants can anticipate potential issues, implement proactive measures, and optimize resource utilization [17]. This shift towards data-driven decision-making is a cornerstone of the digital transformation in wastewater treatment.

9.5.2 Real-Time Data Analytics: The Heart of AI-Integrated Systems

At the heart of AI-integrated wastewater treatment systems lies the capability for real-time data analytics. While IoT sensors provide a continuous stream of data, it is the sophisticated analysis of this data that empowers AI algorithms to make informed decisions and optimize treatment processes [4]. Real-time data analytics involve the processing and interpretation of vast datasets generated by sensors in the treatment plant. These analytics provide valuable insights into influent variations, system performance, and potential issues [33]. By leveraging machine learning algorithms, treatment plants can identify patterns, correlations, and anomalies within the data, enabling predictive modelling and adaptive control. Furthermore, real-time data analytics facilitate the continuous improvement of AI models. As algorithms learn from operational data and outcomes, they can refine their predictions and decision-making capabilities [40]. This iterative learning process is fundamental to the adaptability and effectiveness of AI-integrated systems in wastewater treatment.

9.5.3 Remote Sensing Innovations for Enhanced Monitoring and Control

Remote sensing technologies are pushing the boundaries of wastewater monitoring and control, offering a broader perspective beyond the confines of treatment facilities. Satellites, drones, and other remote sensing platforms provide a bird's-eye view of watersheds, allowing for enhanced monitoring of environmental conditions and potential pollution sources [2]. Satellite-based remote sensing, for example, enables the monitoring of large-scale water bodies and can identify trends in water quality over time. This information is invaluable for watershed management and early detection of potential environmental threats. Drones equipped with sensors can be deployed to inspect difficult-to-reach areas, providing detailed data on water quality and ecosystem health [2]. In wastewater treatment, remote sensing innovations extend beyond traditional parameters to include thermal imaging, hyperspectral imaging,

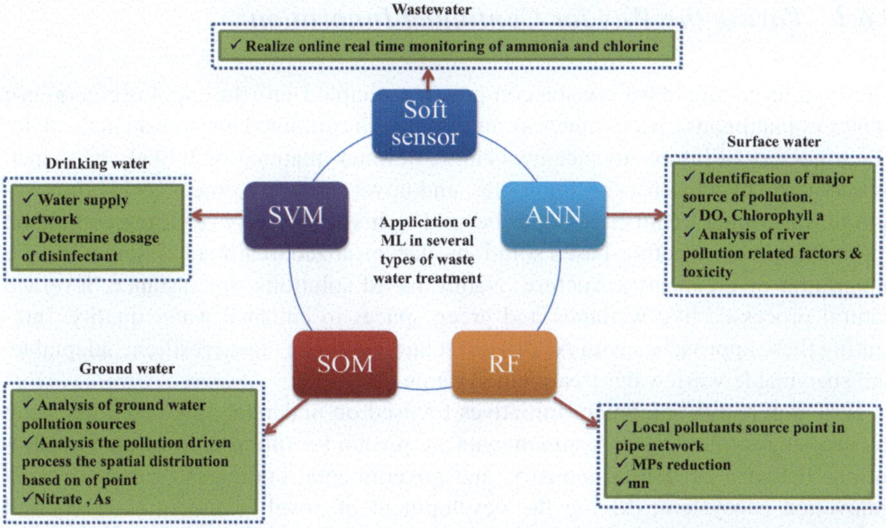

Fig. 9.3 Application of AI in several types of wastewater treatment

and other advanced techniques. These technologies can detect heat anomalies, identify specific pollutants, and assess vegetation health in the surrounding environment. Integrating these insights with AI-driven systems enhances the overall monitoring and control capabilities of wastewater treatment facilities [2]. The synergy between IoT, real-time data analytics, and remote sensing technologies marks a paradigm shift in wastewater monitoring and control. These revolutionary sensor technologies not only provide a deeper understanding of treatment processes but also enable proactive, data-driven decision-making [7]. As wastewater treatment facilities embrace these innovations, they pave the way for more efficient, sustainable, and resilient environmental management practices (Fig. 9.3).

9.6 Charting Future Trajectories and Addressing Challenges

As wastewater treatment undergoes a transformative journey driven by artificial intelligence (AI) and advanced sensor technologies, charting future trajectories becomes crucial for sustained innovation. This section explores the imperative of continued innovation in wastewater treatment, the need to navigate ethical and regulatory frontiers, and the strategies for tackling technological gaps and emerging challenges.

9.6.1 Paving the Way for Continued Innovation

The dynamic nature of wastewater composition, coupled with the ongoing emergence of new contaminants, necessitates a commitment to continued innovation in the field. The trajectory of future advancements in wastewater treatment will likely be shaped by a synergy of AI, sensor technologies, and novel treatment processes [4]. Innovation in wastewater treatment extends beyond technology itself to encompass holistic approaches such as nature-based solutions, decentralized treatment systems, and the integration of green infrastructure. Nature-based solutions, for instance, leverage natural processes like wetlands and green spaces to enhance water quality. Integrating these approaches with AI-driven technologies can create resilient, adaptable, and sustainable wastewater treatment systems [4].

Research and development initiatives focused on understanding and addressing the challenges of emerging contaminants are pivotal for future innovation. Collaborations between academia, industry, and governmental bodies can foster a robust innovation ecosystem, driving the development of novel treatment methods and technologies capable of addressing the evolving landscape of pollutants.

9.6.2 Navigating Ethical and Regulatory Frontiers

As AI and sensor technologies become integral to wastewater treatment, navigating ethical and regulatory frontiers is essential to ensure responsible and equitable use. Ethical considerations range from data privacy and security to the potential socio-economic impacts of technology adoption. Establishing clear ethical guidelines and frameworks is imperative to build public trust and address concerns associated with the use of advanced technologies [11]. Regulatory frameworks need to evolve in tandem with technological advancements to ensure that wastewater treatment facilities adhere to standards that safeguard the environment and public health. Regulatory bodies should actively engage with technology developers, researchers, and industry stakeholders to establish guidelines that balance innovation with responsible use [37].

Ensuring transparency in AI algorithms and decision-making processes is crucial for ethical wastewater treatment practices. Explainable AI methodologies, where the rationale behind AI-driven decisions is made understandable to human operators, can contribute to building trust and addressing concerns related to algorithmic opacity.

9.7 Integration with Smart Water Management Systems

The integration of wastewater treatment with smart water management systems represents a paradigm shift towards a more interconnected and sustainable approach to water resources. This section explores the crucial role of integrating wastewater treatment with water distribution networks, bridging the gap between these traditionally distinct domains, and emphasizes the significance of adopting a holistic, sustainable approach to water infrastructure.

9.7.1 Bridging the Gap: Wastewater Treatment and Water Distribution Networks

Historically, wastewater treatment and water distribution networks have operated as separate entities within the broader water management landscape. However, the integration with smart water management systems seeks to bridge the gap between these two critical components [20]. By fostering seamless communication and coordination, this integration offers a host of benefits that extend beyond traditional siloed approaches. Smart water management systems leverage real-time data from both wastewater treatment plants and water distribution networks, creating a symbiotic relationship between these essential elements of the water cycle [14]. The integration allows for a more dynamic response to fluctuations in water demand, quality, and environmental conditions. For example, suppose a water treatment plant detects a sudden increase in pollutant levels. In that case, this information can be relayed to the water distribution network, prompting adjustments to water supply sources or treatment strategies to maintain water quality standards.

Furthermore, integrating wastewater treatment with water distribution networks facilitates improved resource utilization. Treated wastewater, often regarded as a valuable resource, can be strategically repurposed for non-potable uses such as irrigation, industrial processes, or even replenishing aquifers [10]. This circular water management approach enhances overall water sustainability and resilience in the face of growing water scarcity challenges. In smart water management systems, the integration also extends to predictive modelling, where AI algorithms analyze data from both wastewater and water distribution networks to anticipate future trends and optimize operational strategies. This proactive approach enhances the efficiency of water systems, reduces energy consumption, and contributes to a more sustainable water management ecosystem [20].

9.7.2 *Sustainable Water Infrastructure: A Holistic Approach*

The integration of wastewater treatment with smart water management systems underscores the importance of adopting a holistic and sustainable approach to water infrastructure. Sustainability in this context encompasses environmental, economic, and social considerations, aiming to balance the needs of current and future generations while preserving the health of ecosystems [31]. From an environmental perspective, a holistic approach involves minimizing the environmental impact of wastewater treatment and water distribution processes. By leveraging smart water management systems, treatment plants can optimize energy consumption, reduce carbon emissions, and enhance overall environmental stewardship. Additionally, the strategic reuse of treated wastewater contributes to reducing the strain on freshwater resources, promoting a more sustainable water cycle [32]. Economically, a holistic approach focuses on optimizing resource utilization and minimizing operational costs. The integration of smart technologies allows for more efficient use of energy, chemicals, and infrastructure, resulting in cost savings for water utilities. Moreover, the repurposing of treated wastewater for non-potable uses adds economic value to a resource that was traditionally considered a waste product [32]. Social sustainability in water infrastructure involves considering the impact on communities and ensuring equitable access to clean water. The integration with smart water management systems can enhance water quality monitoring, enabling quicker response to potential contamination events and safeguarding public health. The repurposing of treated wastewater for non-potable uses can also contribute to ensuring reliable water sources for communities, particularly in regions facing water scarcity [31]. Furthermore, a holistic approach considers the resilience of water infrastructure to the impacts of climate change and extreme weather events. By integrating climate data and predictive modelling into smart water management systems, water utilities can better prepare for and respond to changing conditions, ensuring the continued availability of water resources.

9.8 Fostering Collaborations and Strategic Partnerships

In the dynamic landscape of wastewater treatment augmented by artificial intelligence (AI) and advanced sensor technologies, fostering collaborations and strategic partnerships is instrumental in driving innovation, sharing knowledge, and addressing global water challenges. This section explores the synergies in industry-academia collaborations, the pivotal role of government initiatives in driving AI integration, and the significance of international cooperation for a global impact on wastewater treatment.

9.8.1 Synergies in Industry-Academia Collaborations

Collaborations between industry and academia play a crucial role in advancing the frontiers of wastewater treatment technologies. The synergy between academic research institutions and industry stakeholders fosters an exchange of expertise, resources, and perspectives that accelerates innovation [28]. In the realm of AI-integrated wastewater treatment, industry-academia collaborations can facilitate the development and validation of cutting-edge technologies. Academic researchers bring a deep understanding of fundamental principles and theoretical frameworks, while industry partners contribute practical insights, real-world data, and operational considerations. This collaborative approach ensures that AI technologies are not only scientifically sound but also viable and effective in the complex environments of wastewater treatment plants [13]. Joint research projects enable the testing and refinement of AI algorithms, sensor technologies, and innovative treatment processes in real-world scenarios. This iterative feedback loop between academia and industry ensures that advancements are not confined to theoretical frameworks but are translated into tangible solutions that address the evolving challenges of wastewater treatment [12]. Moreover, industry-academia collaborations contribute to the education and training of the next generation of water professionals. Students and researchers gain valuable insights into the practical applications of their work, and industry professionals benefit from exposure to the latest scientific developments. This collaborative ecosystem nurtures a skilled workforce capable of navigating the complexities of AI-integrated wastewater treatment.

9.8.2 Government Initiatives Driving AI Integration

Government initiatives play a pivotal role in driving the integration of AI into wastewater treatment by providing funding, regulatory frameworks, and strategic guidance (Table 9.1). Recognizing the potential of AI to enhance efficiency, sustainability, and resilience in water management, governments worldwide are increasingly investing in initiatives that propel the adoption of these technologies. Funding programs support research and development efforts, encouraging collaboration between academic institutions, research organizations, and technology developers [12]. By incentivizing innovation, governments stimulate the creation of novel AI algorithms, sensor technologies, and treatment processes that can address specific challenges in wastewater treatment. Regulatory frameworks established by governments provide a roadmap for the ethical and responsible use of AI in water management. These frameworks address issues related to data privacy, transparency, and accountability, ensuring that the integration of AI aligns with societal values and environmental goals. Clear regulations also offer certainty to industry stakeholders, encouraging investment and adoption of AI technologies [39]. Government initiatives

Table 9.1 Summary of government initiative driving AI integration

Government initiative	Description	Goals/objectives
National AI strategy	A comprehensive plan outlining the government's approach to AI development, research, and integration across various sectors	Foster AI innovation and research. Develop AI talent and expertise. Ensure ethical and responsible AI deployment
AI research funding	Financial support provided by the government to academic institutions, research labs, and private companies for AI research	Accelerate advancements in AI technology. Address key challenges and gaps in AI development. Promote collaboration
Data sharing mandates	Policies require government agencies to share datasets for AI research and development purposes	Expand access to diverse datasets for AI training and testing. Enhance transparency and accountability in AI applications
Regulatory frameworks	Establishment of legal frameworks and standards governing the development and deployment of AI technologies	Ensure AI systems adhere to ethical and safety standards. Address privacy concerns and data protection. Manage AI risks
AI education programs	Initiatives aimed at promoting AI education and literacy at various levels of education, from schools to professional training	Equip individuals with AI skills for future job markets. Foster a workforce capable of leveraging AI technologies effectively
Public–private partnerships	Collaborative efforts between government agencies, industry players, and academic institutions to drive AI innovation	Pool resources and expertise for AI research and development. Accelerate the adoption of AI solutions in public services and industries

extend beyond funding and regulations to include strategic guidance and knowledge-sharing platforms. By convening stakeholders through conferences, workshops, and collaborative platforms, governments facilitate the exchange of best practices, lessons learned, and emerging trends in AI-integrated wastewater treatment. This collective knowledge-sharing approach accelerates the pace of adoption and ensures that advancements are disseminated throughout the water management community [9].

9.8.3 International Cooperation for Global Impact

The global nature of water challenges necessitates international cooperation to address shared concerns and leverage collective expertise. Collaborative efforts between countries, international organizations, and research institutions contribute to a more holistic and inclusive approach to AI-integrated wastewater treatment.

International partnerships enable the pooling of resources and expertise, fostering innovation that transcends geographical boundaries. By sharing data, research findings, and technological advancements, countries can collectively tackle global water challenges, such as the impact of climate change on water resources and the presence of emerging contaminants in wastewater [6]. Joint research initiatives on an international scale enhance the robustness and applicability of AI technologies in diverse contexts. Different regions face unique water quality issues, and collaborative projects ensure that solutions are adaptable and effective across varied environmental conditions. For example, a successful AI model developed for wastewater treatment in one country may serve as a blueprint for addressing similar challenges in another region [36].

Moreover, international cooperation contributes to capacity-building efforts, particularly in developing regions where access to advanced technologies may be limited. Collaborative initiatives can include knowledge transfer, training programs, and technology transfer to empower local communities with the tools and knowledge needed to implement AI-integrated wastewater treatment solutions [19]. In conclusion, fostering collaborations and strategic partnerships is paramount in driving the adoption of AI in wastewater treatment on a global scale. The synergies between industry and academia, the support of government initiatives, and international cooperation contribute to a thriving ecosystem where innovation flourishes, knowledge is shared, and collective efforts address the complex water challenges facing the world. As the water community continues to navigate the frontiers of AI-integrated wastewater treatment, collaboration remains a cornerstone for achieving sustainable and resilient water management solutions worldwide.

9.9 Conclusion: A Glimpse into the AI-Enhanced Future

As we navigate the intersection of AI and wastewater treatment, the synthesis of key findings offers a glimpse into a future where innovation, collaboration, and sustainability converge to redefine the landscape of water management. This concluding section encapsulates the key insights from the exploration of AI in wastewater treatment and envisions the role of AI in shaping the future of water stewardship.

9.9.1 Synthesis of Key Findings

The exploration of AI in wastewater treatment has unveiled a multifaceted landscape where technological advancements, data-driven decision-making, and collaborative frameworks converge to redefine traditional approaches. Several key findings emerge from the comprehensive journey through the chapters:

(a) **Transformative Potential of AI**: AI technologies, including machine learning, deep learning, and neural networks, offer transformative potential in optimizing wastewater treatment processes. These technologies enhance efficiency, adaptability, and sustainability, addressing the challenges posed by variable influent characteristics and emerging contaminants.
(b) **Integration with Smart Water Management**: The integration of wastewater treatment with smart water management systems represents a paradigm shift towards a more interconnected and sustainable approach. Bridging the gap between wastewater treatment and water distribution networks fosters dynamic responses to fluctuations in water demand, quality, and environmental conditions.
(c) **Collaborations and Partnerships**: Industry-academia collaborations and strategic partnerships play a pivotal role in driving innovation and knowledge exchange. Government initiatives provide crucial support through funding, regulatory frameworks, and knowledge-sharing platforms. International cooperation enhances the collective capacity to address global water challenges.
(d) **Holistic Approach to Sustainability**: A holistic approach to sustainability encompasses environmental, economic, and social considerations. AI-integrated wastewater treatment contributes to environmental stewardship by optimizing resource utilization, reducing energy consumption, and enhancing water quality. Economic benefits arise from cost savings and resource recovery, while social sustainability involves ensuring equitable access to clean water and fostering community resilience.

9.9.2 *Future Horizons: Envisioning AI's Role in Wastewater Treatment*

As we peer into the future, the role of AI in wastewater treatment holds tremendous promise for reshaping water management practices globally. Envisioning the future horizons involves embracing continuous innovation, ethical governance, and a commitment to addressing emerging challenges:

(a) **Continuous Innovation**: The trajectory of AI in wastewater treatment points towards a future of continuous innovation. Advances in machine learning, sensor technologies, and treatment processes will refine the precision and adaptability of wastewater treatment systems. Nature-based solutions, decentralized treatment approaches, and green infrastructure will complement AI technologies, creating a holistic and resilient water management ecosystem.
(b) **Ethical Governance and Responsible AI**: The future of AI in wastewater treatment necessitates a robust ethical framework and responsible governance. Transparent and explainable AI methodologies, coupled with clear regulatory guidelines, will ensure the ethical use of AI technologies. Privacy, security, and socio-economic impacts must be carefully considered to build public trust and foster responsible AI integration.

(c) **Addressing Emerging Challenges**: Anticipating and addressing emerging challenges will be crucial for the sustained success of AI-integrated wastewater treatment. This involves tackling technological gaps, ensuring interoperability of sensor systems, and developing energy-efficient AI algorithms. International collaborations will play a vital role in sharing insights and collectively addressing global challenges such as climate change impacts on water resources and the presence of novel contaminants.

(d) **Inclusive and Equitable Implementation**: The future of AI in wastewater treatment envisions inclusive and equitable implementation. Community engagement, capacity-building, and tailored solutions for diverse regions will ensure that the benefits of AI technologies are accessible to all. Empowering local communities with the knowledge and tools to manage their water resources will contribute to a more sustainable and resilient water future.

In conclusion, the journey through the exploration of AI in wastewater treatment unveils a landscape where technology, collaboration, and sustainability converge to redefine the future of water management. As AI continues to evolve and integrate with water treatment practices, it holds the potential to enhance efficiency, optimize resource utilization, and contribute to a more sustainable and equitable water future for generations to come. The synthesis of key findings and the envisioning of future horizons provide a roadmap for navigating this transformative journey towards a water-resilient and AI-enhanced future.

References

1. Alprol AE, Mansour AT, Ibrahim MEE-D, Ashour M (2024) Artificial intelligence technologies revolutionizing wastewater treatment: current trends and future prospective. Water 16(2):314. https://doi.org/10.3390/w16020314
2. Andres L, Boateng K, Borja-Vega C, Thomas E (2018) A review of in-situ and remote sensing technologies to monitor water and sanitation interventions. Water 10(6):756. https://doi.org/10.3390/w10060756
3. Anthony ET, Ojemaye MO, Okoh OO, Okoh AI (2020) A critical review on the occurrence of resistomes in the environment and their removal from wastewater using apposite treatment technologies: limitations, successes and future improvement. Environ Pollut 263:113791. https://doi.org/10.1016/j.envpol.2019.113791
4. Bahramian M, Dereli RK, Zhao W, Giberti M, Casey E (2023) Data to intelligence: the role of data-driven models in wastewater treatment. Expert Syst Appl 217:119453. https://doi.org/10.1016/j.eswa.2022.119453
5. Bhangale KB, Mohanaprasad K (2021) A review on speech processing using machine learning paradigm. Int J Speech Technol 24(2):367–388. https://doi.org/10.1007/s10772-021-09808-0
6. Bibri SE, Krogstie J, Kaboli A, Alahi A (2024) Smarter eco-cities and their leading-edge artificial intelligence of things solutions for environmental sustainability: a comprehensive systematic review. Environ Scie Ecotechnol 19:100330. https://doi.org/10.1016/j.ese.2023.100330
7. Chen J, Chen S, Fu R, Li D, Jiang H, Wang C, Peng Y, Jia K, Hicks BJ (2022) Remote sensing big data for water environment monitoring: current status, challenges, and future prospects. Earth's Future 10(2). https://doi.org/10.1029/2021EF002289

8. Croll HC, Ikuma K, Ong SK, Sarkar S (2023) Reinforcement learning applied to wastewater treatment process control optimization: approaches, challenges, and path forward. Crit Rev Environ Sci Technol 53(20):1775–1794. https://doi.org/10.1080/10643389.2023.2183699
9. Dwivedi YK, Hughes L, Ismagilova E, Aarts G, Coombs C, Crick T, Duan Y, Dwivedi R, Edwards J, Eirug A, Galanos V, Ilavarasan PV, Janssen M, Jones P, Kar AK, Kizgin H, Kronemann B, Lal B, Lucini B et al (2021) Artificial intelligence (AI): multidisciplinary perspectives on emerging challenges, opportunities, and agenda for research, practice and policy. Int J Inf Manage 57:101994. https://doi.org/10.1016/j.ijinfomgt.2019.08.002
10. Foglia A, González-Camejo J, Radini S, Sgroi M, Li K, Eusebi AL, Fatone F (2023) Transforming wastewater treatment plants into reclaimed water facilities in water-unbalanced regions. An overview of possibilities and recommendations focusing on the Italian case. J Cleaner Prod 410:137264. https://doi.org/10.1016/j.jclepro.2023.137264
11. Gajjar SP, Singh C, Deshpande T (2019) Tracing back to move ahead: a review of development pathways that constrain adaptation futures. Climate Dev 11(3):223–237. https://doi.org/10.1080/17565529.2018.1442793
12. Garousi V, Herkiloglu K (2016) Selecting the right topics for industry-academia collaborations in software testing: an experience report. In: 2016 IEEE international conference on software testing, verification and validation (ICST), pp 213–222. https://doi.org/10.1109/ICST.2016.7
13. Hamrani A, Bouarab FZ, Agarwal A, Ju K, Akbarzadeh H (2023) Advancements and applications of multiple wire processes in additive manufacturing: a comprehensive systematic review. Virtual Phys Prototyping 18(1). https://doi.org/10.1080/17452759.2023.2273303
14. Herrera-Navarrete R, Colín-Cruz A, Arellano-Wences HJ, Sampedro-Rosas ML, Rosas-Acevedo JL, Rodríguez-Herrera AL (2022) Municipal wastewater treatment plants: gap, challenges, and opportunities in environmental management. Environ Manage 69(1):75–88. https://doi.org/10.1007/s00267-021-01562-y
15. Jawad J, Hawari AH, Javaid Zaidi S (2021) Artificial neural network modeling of wastewater treatment and desalination using membrane processes: a review. Chem Eng J 419:129540. https://doi.org/10.1016/j.cej.2021.129540
16. Malviya A, Jaspal D (2021) Artificial intelligence as an upcoming technology in wastewater treatment: a comprehensive review. Environ Technol Rev 10(1):177–187. https://doi.org/10.1080/21622515.2021.1913242
17. Michael AD, Majemite MT, Obaigbena A, Daraojimba OH, Oliha JS, Nwokediegwu ZQS (2024) Review of smart water management: IoT and AI in water and wastewater treatment. World J Adv Res Rev 21(1):1373–1382. https://doi.org/10.30574/wjarr.2024.21.1.0171
18. Mondal PP, Galodha A, Verma VK, Singh V, Show PL, Awasthi MK, Lall B, Anees S, Pollmann K, Jain R (2023) Review on machine learning-based bioprocess optimization, monitoring, and control systems. Biores Technol 370:128523. https://doi.org/10.1016/j.biortech.2022.128523
19. Ngeno EC, Mbuci KE, Necibi MC, Shikuku VO, Olisah C, Ongulu R, Matovu H, Ssebugere P, Abushaban A, Sillanpää M (2022) Sustainable re-utilization of waste materials as adsorbents for water and wastewater treatment in Africa: recent studies, research gaps, and way forward for emerging economies. Environ Adv 9:100282. https://doi.org/10.1016/j.envadv.2022.100282
20. Oberascher M, Rauch W, Sitzenfrei R (2022) Towards a smart water city: a comprehensive review of applications, data requirements, and communication technologies for integrated management. Sustain Cities Soc 76:103442. https://doi.org/10.1016/j.scs.2021.103442
21. Pamungkas MRSP, Asyivadibrata A, Susilawati T, Huda MN (2023) Unleashing the potentials of artificial intelligence for micro, small, and medium enterprises: a systematic literature review. Jurnal Teknologi Dan Sistem Informasi Bisnis 5(3):303–310. https://doi.org/10.47233/jteksis.v5i3.860
22. Pérez-Beltrán CH, Robles AD, Rodriguez NA, Ortega-Gavilán F, Jiménez-Carvelo AM (2024) Artificial intelligence and water quality: from drinking water to wastewater. TrAC Trends Anal Chem 172:117597. https://doi.org/10.1016/j.trac.2024.117597
23. Prabha C, Kaur S, Singh J, Malik M (2023) An overview of IoT and smart application environments: research and challenges. In: Communication and intelligent systems, pp 111–124. https://doi.org/10.1007/978-981-99-2100-3_10

24. Rowan NJ, Murray N, Qiao Y, O'Neill E, Clifford E, Barceló D, Power DM (2022) Digital transformation of peatland eco-innovations ('Paludiculture'): enabling a paradigm shift towards the real-time sustainable production of 'green-friendly' products and services. Sci Total Environ 838:156328. https://doi.org/10.1016/j.scitotenv.2022.156328
25. Rudi NN, Muhamad MS, Te Chuan L, Alipal J, Omar S, Hamidon N, Abdul Hamid NH, Mohamed Sunar N, Ali R, Harun H (2020) Evolution of adsorption process for manganese removal in water via agricultural waste adsorbents. Heliyon 6(9):e05049. https://doi.org/10.1016/j.heliyon.2020.e05049
26. Sahu S, Kaur A, Singh G, Kumar Arya S (2023) Harnessing the potential of microalgae-bacteria interaction for eco-friendly wastewater treatment: a review on new strategies involving machine learning and artificial intelligence. J Environ Manage 346:119004. https://doi.org/10.1016/j.jenvman.2023.119004
27. Sarker IH (2021) Machine learning: algorithms, real-world applications and research directions. SN Comput Sci 2(3):160. https://doi.org/10.1007/s42979-021-00592-x
28. Segovia-Hernández JG, Hernández S, Cossío-Vargas E, Sánchez-Ramírez E (2023) Challenges and opportunities in process intensification to achieve the UN's 2030 agenda: goals 6, 7, 9, 12 and 13. Chem Eng Process Process Intensification 192:109507. https://doi.org/10.1016/j.cep.2023.109507
29. Shah AI, Din Dar MU, Bhat RA, Singh JP, Singh K, Bhat SA (2020) Prospectives and challenges of wastewater treatment technologies to combat contaminants of emerging concerns. Ecol Eng 152:105882. https://doi.org/10.1016/j.ecoleng.2020.105882
30. Sheik AG, Kumar A, Patnaik R, Kumari S, Bux F (2024) Machine learning-based design and monitoring of algae blooms: recent trends and future perspectives—a short review. Crit Rev Environ Sci Technol 54(7):509–532. https://doi.org/10.1080/10643389.2023.2252313
31. Silva JA (2023) Wastewater treatment and reuse for sustainable water resources management: a systematic literature review. Sustainability 15(14):10940. https://doi.org/10.3390/su151410940
32. Starkl M, Brunner N, Das S, Singh A (2022) Sustainability assessment for wastewater treatment systems in developing countries. Water 14(2):241. https://doi.org/10.3390/w14020241
33. Thayyib PV, Mamilla R, Khan M, Fatima H, Asim M, Anwar I, Shamsudheen MK, Khan MA (2023) State-of-the-art of artificial intelligence and big data analytics reviews in five different domains: a bibliometric summary. Sustainability 15(5):4026. https://doi.org/10.3390/su15054026
34. Tsang YP, Lee CKM (2022) Artificial intelligence in industrial design: a semi-automated literature survey. Eng Appl Artif Intell 112:104884. https://doi.org/10.1016/j.engappai.2022.104884
35. Wang J, Wang S (2021) Toxicity changes of wastewater during various advanced oxidation processes treatment: an overview. J Clean Prod 315:128202. https://doi.org/10.1016/j.jclepro.2021.128202
36. Wang L, Xu Y, Qin T, Wu M, Chen Z, Zhang Y, Liu W, Xie X (2023) Global trends in the research and development of medical/pharmaceutical wastewater treatment over the half-century. Chemosphere 331:138775. https://doi.org/10.1016/j.chemosphere.2023.138775
37. Wani MA, Din A, Nazki IT, Rehman TU, Al-Khayri JM, Jain SM, Lone RA, Bhat ZA, Mushtaq M (2023) Navigating the future: exploring technological advancements and emerging trends in the sustainable ornamental industry. Front Environ Sci 11.https://doi.org/10.3389/fenvs.2023.1188643
38. Xu A, Wu Y-H, Chen Z, Wu G, Wu Q, Ling F, Huang WE, Hu H-Y (2020) Towards the new era of wastewater treatment of China: development history, current status, and future directions. Water Cycle 1:80–87. https://doi.org/10.1016/j.watcyc.2020.06.004
39. Yigitcanlar T, Corchado JM, Mehmood R, Li RYM, Mossberger K, Desouza K (2021) Responsible urban innovation with local government artificial intelligence (AI): a conceptual framework and research agenda. J Open Innov Technol Market Complex 7(1):71. https://doi.org/10.3390/joitmc7010071

40. Zhang W, Tooker NB, Mueller AV (2020) Enabling wastewater treatment process automation: leveraging innovations in real-time sensing, data analysis, and online controls. Environ Sci Water Res Technol 6(11):2973–2992. https://doi.org/10.1039/D0EW00394H
41. Zhao L, Dai T, Qiao Z, Sun P, Hao J, Yang Y (2020) Application of artificial intelligence to wastewater treatment: a bibliometric analysis and systematic review of technology, economy, management, and wastewater reuse. Process Saf Environ Prot 133:169–182. https://doi.org/10.1016/j.psep.2019.11.014
42. Zhao Q, Peng Y, Li J, Gao R, Jia T, Deng L, Du R (2022) Sustainable upgrading of biological municipal wastewater treatment based on anammox: from microbial understanding to engineering application. Sci Total Environ 813:152468. https://doi.org/10.1016/j.scitotenv.2021.152468
43. Zieliński M, Kazimierowicz J, Dębowski M (2022) Advantages and limitations of anaerobic wastewater treatment—technological basics, development directions, and technological innovations. Energies 16(1):83. https://doi.org/10.3390/en16010083

Chapter 10
The AI Cleanse: Revolutionizing the Future of Wastewater Treatment with AI and Machine Learning

R. Sanjeevi, Prashantkumar B. Sathvara, Sandeep Tripathi, and J. Anuradha

Abstract Artificial intelligence and machine learning (AIML) are now popular trends rapidly transforming wastewater treatment. These technologies are being used to improve wastewater treatment facilities' efficiency, efficacy, and long-term viability. One of the ways AI is being used in wastewater treatment is to automate tasks. AI may be used to monitor and regulate treatment operations, detect leaks and spills, and optimize chemical dosage, for example. This can free up operators to focus on other tasks, such as troubleshooting and maintenance. Artificial intelligence is also being utilized to create novel therapeutic methods. AI, for example, may be used to create designs more efficient membranes for desalination, or to develop new methods for treating wastewater from industrial sources. In addition to improving the efficiency and effectiveness of wastewater treatment, AI can also help to make it more sustainable. For example, AI can be used to optimize energy use, reduce the release of greenhouse gases, and recover valuable resources from wastewater. Some of the most common techniques like Machine learning algorithms may be trained to recognize patterns in data and to predict when problems are likely to occur. Deep learning algorithms can learn more complex patterns in data than machine learning algorithms. This makes them well-suited for tasks such as identifying rare or subtle problems. NLP (natural language processing) approaches: NLP techniques can be used to analyze text data since WWTPs, such as sensor readings or operator logs. This can help to identify problems that are not easily detected by other methods. AI-enhanced real-time monitoring and fault detection systems can be used to monitor a variety of parameters in WWTPs, like Flow rate, Temperature, Color, pH, Turbidity, Electrical conductivity, Biological oxygen demand (BOD), and Chemical oxygen demand (COD). Collecting data on these parameters, AI-enhanced systems can help operators to identify problems early on, such as Overflows or spills, Equipment failures, Process upsets, Changes in the quality of the wastewater. Early identification

Present Address:
R. Sanjeevi · P. B. Sathvara · S. Tripathi · J. Anuradha (✉)
Department of Environmental Science, Nims Institute of Allied Medical Science and Technology, NIET, Nims University Rajasthan, Jaipur, India
e-mail: j.anuradha@nimsuniversity.org

M. C. Garg (ed.), *The AI Cleanse: Transforming Wastewater Treatment Through Artificial Intelligence*, Springer Water, https://doi.org/10.1007/978-3-031-67237-8_10

of problems can help operators to take corrective action before they cause major disruptions to the treatment process. This can save time, money, and environmental damage. In addition to monitoring parameters, AI-enhanced systems can also be used to optimize the treatment process. For example, AI-enhanced systems can be used to determine the optimal dosage of chemicals to be used in the treatment process. Otherwise, AI-enhanced systems can be used to identify opportunities to reduce energy consumption or water usage. AI-enhanced real-time monitoring and fault detection systems are a powerful tool that can help to improve the efficiency and sustainability of WWTPs. The application of artificial intelligence in the treatment of wastewater is still in its earliest phases, but it has a chance to revolutionize the sector. Artificial intelligence can help to make wastewater treatment more efficient, effective, and sustainable. This is essential as we face the challenges of climate change and water scarcity. In addition to the applications mentioned above, AI is also being used in wastewater treatment for a variety of other tasks, such as Predicting the impact of climate change on wastewater treatment systems, developing new methods for treating emerging pollutants, Improving the safety and security of wastewater treatment plants. As advances in artificial intelligence (AI), we should expect to see even more novel applications of artificial intelligence in wastewater treatment. This is an exciting area of research with the potential to make a significant positive impact on the environment.

Keywords Real-time monitoring · Fault detection · Artificial intelligence · Wastewater treatment

10.1 Introduction

One of the most significant trends is the emergence of artificial intelligence (AI) astonishing developments popular science that has captured the interest of practically all study disciplines. Since machine learning took over the majority of mainstream research AI technology expanded rapidly in the 1990s, and many novel AI techniques are developing. Artificially intelligent (AI) technologies include artificial neural networks, also known as (ANN), neural networks with support vector machines (SVM) [1], algorithms based on genetics (GA), fuzzy reasoning (FL), and others used in a variety of fields including nanotechnology, finance, engineering, education, and medical. These techniques, which are regarded as effective and cost-effective alternatives to traditional methods and mathematics, have shown to offer a high degree of competence in order to cope with the complexities of unpredictable, interactive, and dynamic concerns [2].

The need for further advancement in this field has been sparked by the fact that environmental pollution is becoming the society's top concern and that treatment of wastewater, contaminants in the air, and solid waste is now subject to stricter rules and laws. However, the majority of Controlling pollution in the environment is related to a number of nonlinear, time-dependent, multisource, and multiple-purpose aspects, making it challenging to accomplish the required system performance and optimization of significant components [3]. MLR (multiple linear regression), RSM (response surface technique), and PCA (principal component analysis), partially least-squares regression (PLS), the algorithm for clustering k-means (K-means), and other statistical and multivariate data analysis techniques have all been used in Several research have underlined the utility of these techniques to this topic [4]. RSM, which is outside the purview of machine learning algorithms, is the approach that has received the most attention. The findings of research contrasting These data analysis approaches combined with AI models shown that the former provides more accurate information that is supported by statistical analysis and uncertainty calculations [5]. However, AI technologies often perform more accurately in the majority of works. For instance, showed that the predicted DO in riverine habitats was more in line with the actual experimental values than MLR [6]. Pointed out in their evaluation that for the removal efficiencies estimations of various contaminants, RSM was outperformed by the GA-ANN and PSO-ANN models, which had higher R2 values and lower average errors. However, AI technologies differ from traditional mechanism modelling techniques in that They permit the omission of complex mathematical formulas and system details, including the relationships between inputs and corresponding outputs, without sacrificing accuracy [7]. Numerous research have shown that artificial intelligence (AI) technologies are useful helpers for contamination of the environment controls in wastewater, air pollution, and solid waste. Meanwhile, it was stated that some research has been done employing AI technologies in the areas of climate change, natural water pollution, and other environmental sectors including simulating vegetation dynamics and natural catastrophe early warning. Because of the extensive use of artificial intelligence in contamination cleanups, this study will focus on the modelling, forecasting, optimization, and autonomous control of eliminating contaminants in the environment [8].

To the best of our knowledge, prior review papers have summarized AI-assisted environmental cleanup applications, focused on wastewater treatment techniques such as contaminants removal efficiency and soft measurement [9–11]. Due to the review articles' primary attention on ANNs rather than other types of AI approaches, there was a dearth of thorough investigate the applications of various AI-based technologies for environmental control [12].

The purpose of this study is to provide an up-to-date overview of artificial intelligence (such as ANN, SVM, FL, GA, and others) utilized in the treatment of wastewater, air pollution control, and solid waste treatment. An overview of the fundamental concepts behind various AI approaches is followed by a study of the most common single or blended AI strategies. Comprehensive research of AI approaches is going to be carried out in the field of green controls, notably intelligent oversight of the treatment of wastewater, early warning and evaluation of contaminants in the air, and the handling of solid waste. Each of them adds new and crucial elements to this assessment. Future challenges with AI approaches will then be investigated.

10.2 Literature Review

10.2.1 Methodological Strategy

This section contains the state-of-the art knowledge on the research publications reported on the AI monitored wastewater treatment processes. Assessed methods for using data generated by WWTPs to produce relevant knowledge and information that may be used to enhance wastewater treatment operations or assist decision-making. The approaches that have been examined are organized into three levels: basic information extraction, complex data extraction, or human interpretable data extraction [13]. Keep in mind that the complexity levels (time required to arrive at a solution) gradually increase from the first (analyzing single dataset or parameters) to the third (taking into account numerous variables and many distinct forms of information). We have organized the approaches into the following categories under the advanced information extraction level: Dimension-reduction-based techniques make up the first group, followed by methods for deriving In the second group, linear functions derived from labelled training information are used, methods for deriving Clustering of nonlinear functions in the subsequent group methods in the fourth group, and qualitative feature detection methods in the fifth group. In the human comprehensible information extraction level, we used rules, CBR, fuzzy logic, EDSS, and ontologies. As a result, we are particularly interested in data mining methodologies that mix machine learning, mathematical modelling, and visualization disciplines to find patterns in massive data sets. A detailed analysis of machine learning techniques used in the water and wastewater industry. Mechanistic approaches or models weren't included in this analysis [14, 15].

The searches were conducted using the terms wastewater treatment and the names of the approaches (and any pertinent modifications), with a restriction on articles published up to 2015. In order to exclude studies on emerging pollutants, Papers dealing with classic systems of activated sludge or anaerobic digestion were also excluded from the search, as well as publications dealing with traditional chemicals (nutrients, organic stuff, and so forth). Prior to 2010, publications with less than five citations were not included. The accompanying material includes details from all publications that have been examined including the software platforms used to assess the methodologies, technologies, difficulties that the approaches have to overcome, the amount the number of variables involved, as well as the types of data used (pilot-plant, full-scale, synthetic) [16].

10.2.2 Overview's Generalities

Following a reduction in scope for significance, the SCOPUS search yielded 340 documents. In the bulk of these studies, fuzzy logic (12%), PCA (13%) and ANN (21%) are discussed (Fig. 10.1). In the literature, ANN has been used to control, soft sensing, and process performance prediction (Table 10.1). Fuzzy reasoning has been used in prediction and control applications whereas PCA has mostly been used for defect identification and process comprehension. The majority of the articles (7% and 4%, respectively) used partial least squares (PLS) and multiple linear regression to predict water quality indicators such as using information from UV–VIS observations for oxygen chemical demand, total solids in suspension, nitrate, and other parameters. 8% of the publications employed clustering to improve process comprehension. In order to manage the duration between the anoxic and aerobic phases, qualitative characteristics have been utilized (in 6% of the articles) to identify the bending points of inexpensive sensors (pH, ORP) installed in alternating systems or sequencing batch reactors (SBRs). EDSSs used to select wastewater treatment technology or EDSSs that give controllers a managerial position have been presented in 9% of the research [17, 18]. The knowledge gathered from the procedure has mostly been included into these EDSS's rules or CBR. Ontologies (2%), rule induction (less than 1%), and decision trees have all been utilized in the search for knowledge. SVM has been used for soft sensing and chemo metrics, whereas Process knowledge has been improved primarily through the use of SOM (unsupervised dimension reduction) [19].

European Union (EU) is the area that has contributed the most to this topic, appearing in 61% of the publications, The United States and Canada (12% of the studies) and Asia-Oceania (34% of the publications) came next. Not even 4% of investigations were carried out by research teams from South America or Africa. The EU has made the most contributions to each strategy. Nonetheless, the Asia-Oceania region has contributed significantly to PCA (20 studies), fuzzy logic seventeen studies), and ANN (38 study). Furthermore, 37 papers (12%) are the result of international collaboration between research teams [21].

Regarding the academic backgrounds of the authors participated in the research, engineers (i.e., environmental, civil, and water engineers) have contributed the most, with involvement in 60% of the papers that have been examined, followed by electricians 10%) and 3% are computer engineers. Communities with different ancestries

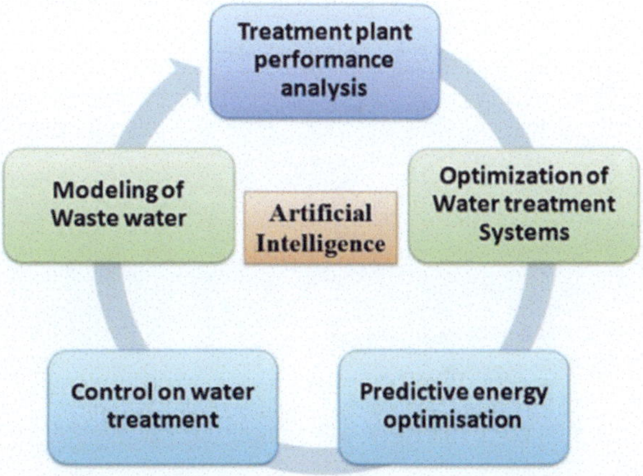

Fig. 10.1 Application of artificial intelligence in wastewater treatment [20]

Table 10.1 Techniques are arranged by level and are accompanied by a brief description

Level	Technique	Short description
Basic information extraction	Control charts with a single variable	To keep track of a particular process attribute
	Balancing techniques	Mass, power and stoichiometric balances for large mistake detection and data reconciliation
Advanced information extraction	Dimensionality reduction techniques include PCA, ICA, and SOM	Reduce the number of variables assessed by creating a fewer number of new variables that describe the most significant elements of the original data
	Regression, PLS, and PCR are all examples of linear supervised learning	Linear functions should be fitted to labelled training data
	Nonlinear supervised learning techniques include artificial neural networks (ANN, ANFIS, WNN) and support vector machines (SVM)	Non-linear functions are fitted to labelled training data
	Clustering methods include k-means, fuzzy C-means (FCM), and Gustafsonkessel., GK-FCM (fuzzy C-means)	Sort data samples into understandable categories
	Detection of qualitative features (qualitative trend analysis, QTA)	Automated detection of understood and visually recognisable time series features
Human-interpretable information extraction	Decision trees and rule induction	To derive rules from a collection of observations. The retrieved rules may represent local trends or a whole academic model of the data
	Case-based cognition	Attempts to solve new issues by adapting existing solutions to one or more previous comparable issues in the system's memory
	Fuzzy logic	Methods for approaching reasoning in an imprecise and coarse-grained rather than precise and accurate manner

(continued)

Table 10.1 (continued)

Level	Technique	Short description
	Environmental decision-making systems	Computer-based systems of information that integrate models and organisational frameworks to aid in making choices
	Ontologies	Methods for structuring knowledge by compartmentalising the variables required for a specific set of calculations and establishing the links between the pieces of information that exist in a given domain

work together. To develop EDSS, which consists of decision trees and rules that are closely related to EDSS, engineers (water, civil, and environmental) worked with computer engineers in 7% of the publications or with electrical professionals in 3% of the publications [22, 23]. Wastewater-related research and data, information, and expertise are published in 93 publications. Water Sci. Techno has published the most articles on this issue (21% of all published papers), Water Res., and Environment Model came next. Gentle. With seven percent and five percent of all published publications, in that order. Journal of Hazardous Materials, Journal for Environmental Management, Journal of Process Control, the Chemical Engineering The Journal, Engineering Applications of Machine Learning, The field of chemo and Intelligent Scientific Systems, Bioprocess and Bio systems Engineering, etc.. are other journals with smaller contributions [24–27].

10.3 Specifics for Each Approach

10.3.1 Basic Data Extraction

The aforementioned category refers to a range of approaches targeted at extracting critical data from wastewater treatment facilities (WWTPs). These strategies are critical in guaranteeing the efficient and dependable functioning of WWTPs. Specific variables critical for control of processes and error detection techniques to identify substantial deviations may be included in the retrieved data. This study additionally discusses the use of balances of mass and multivariate control graphs for data analysis, particularly with a small number of variables. Such technologies are critical for monitoring, sending timely alerts and notifications in the event of a deviation, recognizing outliers, and recognizing potential calibration concerns [28]. In this area, control graphs are an essential tool. They function as an illustration of data, enabling for real-time trend tracking and visualization. Control graphs identify drift, rapid changes, and the existence of outliers, all of which indicate possible issues with the WWTP processes. They also give a visual depiction of the curve's quality, making it possible to identify curves which may be inadequate for reliable measurements [28–30]. Another important part of analysis of data in WWTPs is mass balance calculations. These computations are used to evaluate the overall dependability of the treatment plant's operating metrics. Mass balances can uncover anomalies or mistakes in data by comparing the inputs and outputs flows for substances and chemicals. Control graphs may be used in the setting of mass balances in order to assess historical data and systematically uncover problems in WWTP datasets [31]. This method can be quite useful in identifying systematic faults that would otherwise go undiscovered [32]. Redundancy is established by mass balances across several system boundaries to improve data quality and ensure resilience. This redundancy facilitates in the evaluation of systemic flaws and the proper reconciliation of data. WWTPs may reduce the effect of mistakes and inconsistencies by adopting mass balances across all conceivable system boundaries, resulting in more accurate data for process optimization and regulatory compliance. In order to guarantee that the data used for handle control and decision-making is accurate and dependable, this category discusses an all-encompassing approach that offers a comprehensive methodology for evaluating data quality inside WWTPs. It combines a number of data analysis techniques, such as control graphs, weight balances, and data reconciliation. The category also emphasizes how crucial it is to use fundamental descriptive statistics. Although it acknowledges their importance, it also highlights a significant vacuum in the literature: there are no academic works that particularly address fundamental descriptive statistics in the context of WWTPs [33]. This discovery suggests that there are few scholarly resources that concentrate on this facet of wastewater treatment plant data management, even though descriptive statistics are useful in data analysis and interpretation. However, the category subtly draws attention to the possibility for additional study and advancement in this field by recommending that basic descriptive statistics be applied more often in WWTPs in order to improve the quality and dependability of data. A comprehensive approach to analyzing data and quality assurance in WWTPs is presented by this category, which includes methods like mass balances, control graphs, data reconciliation, and—most importantly—the requirement for fundamental descriptive statistics. This latter point may be the subject of further study and development. When combined, these techniques guarantee the accuracy and dependability of the data utilized in wastewater treatment facility operations as well as decision-making processes, hence promoting the effective and ecologically conscious administration of water treatment plants [34, 35].

10.3.2 Advanced Information Extraction

The techniques of data science covered in this section have their origins in artificial intelligence, machine learning, and statistics. These methods are essential for comprehending data and drawing insightful conclusions from it. Approaches for handling both continuous and discrete variables can be roughly classified into the techniques mentioned in this area, each with its own methodology and goal. Data points with comparable properties are grouped together using independent variables, such as clustering. In unsupervised learning, an algorithm can identify structures and patterns in the data without the requirement for labelled results by using a basic approach called clustering. When data is divided into meaningful categories, clustering is very helpful in revealing hidden patterns or underlying structures in the dataset. Conversely, continuous variables—such as Principal Component Analysis, or PCA—are intended to examine the correlations and variations present in a dataset. Using PCA, a dimensionality reduction approach, one may reduce complexity while maintaining key information by finding the most important dimensions or parts of the data. This is helpful in streamlining data visualization and facilitating comprehension of the main factors behind dataset variances. In this area, the difference among supervised and unsupervised methods for modelling is quite important. When it is necessary to predict an explicit output variable, supervised techniques like regression and classification are used. Regression, for instance, aims to predict an ongoing outcome variable, whereas classification divides data points into pre-established classes or categories according to their qualities. For applications like predictive modelling, where the links between variables that are input and output are investigated and utilized, these supervised approaches are crucial [36–38]. On the other hand, unsupervised methods such as PCA and clustering do not try to find input–output correlations. Rather, they are employed to investigate and evaluate the intrinsic connections among various variables or measures in the dataset. While PCA lowers the complexity of the data and reveals underlying structures and trends among the variables, clustering aids in the identification of natural clusters or groups of data points. These unsupervised methods are useful for anomaly identification, exploratory data analysis, and finding concealed trends in the data. Finding qualitative characteristics in a data set is another topic of interest in this category. Finding important locations within a dataset, such as minima (lowest values) and points of inflection (points of notable change), may be necessary to do this. Finding these characteristics can be important for a number of uses, such as analysis of time series, where it's crucial to comprehend the temporal patterns and pivotal moments in the data. To sum up, this section's discussion of data science methodologies covers a broad range of methods with their roots in artificial intelligence, machine learning, and statistics. When using supervised or unsupervised modelling techniques, these techniques are crucial for handling discrete and continuous data. This area covers a wide range of tools to extract useful insights from data, including clustering to uncover hidden patterns, PCA for decreasing dimensionality, and supervised approaches like classification and regression for predictive modeling [39, 40]. Furthermore, particular attention is paid to identifying qualitative characteristics such as minima and points of inflection, given their special significance in the analysis and interpretation of data [41].

10.3.2.1 Techniques for Reducing Dimensionality

Many attempts have been made in the field of multidimensional data analysis, specifically to address whether the data in multidimensional data sets might be compressed into a compact yet meaningful collection of variables. The majority of approaches employ methods of unsupervised learning with constant data representation, sometimes known as scores and latent variables for short. PCA is the most common usually used [42]. PCA is often used in its original, exploratory form within the framework of the bacterial community research, which means that its major function is to visualize the main causes of variances in gathered information. Analyzing the physical–chemical makeup of wastewaters employs PCA in a similar way. As an alternative, PCA is widely used to characterize the distribution of data that corresponds to typical operating circumstances. In this instance, A PCA model is used as a normative or predictive framework for potential samples, and the resulting distribution is used to generate statistics that have theoretical or empirical support bounds are provided. Therefore, in addition to data visualization, PCA may be utilized for the live identification of aberrant data samples and process behavior. In later modelling processes, it may also be applied as a dimension reduction approach for diagnostic or predictive reasons. To handle the unpredictable, multi-model, non-normal, and quadratic process parameters for which PCA is known to be useful suboptimal, several approaches have been put forth. Alternative models with latent variables like ICA and SOM are among these tactics. Despite the fact that SOMs have many similarities to the neural networks that are employed in predictions (i.e. supervision learning) discussed below, SOM has been included here owing to its utility for dimension reduction. By using appropriate extensions based on adaptive updating, data reorganization, unfolding, missing data imputation, mixture modelling, nonlinear transformation, or wavelet decomposition, the existing fundamental data models are frequently enhanced. As long as the data are almost steady, using the PCA and ICA algorithms is quite straightforward. However, choosing from the aforementioned extensions presents challenges. There are no rules for making a methodical choice among the numerous potential extensions. Additionally, the stated approaches suffer from a severe lack of benchmarking. Claims of advancement are typically based on case-specific statistics that have not been validated thoroughly examined or that have only been treated to a handful of the preprocessing methods that are available. Similar findings were reached after looking at soft-sensors.

10.4 Discussion

10.4.1 Putting the Techniques into Action

In around 16% of the articles considered for the review, it was shown how the research efforts led to a commercial product. Even however, only 9% of all papers explicitly mentioned that these solutions had been evaluated at full-scale, such as a Controller that operates in real time. The remaining 7% are commodities that have not been subjected to substantial real-world testing and hence are less developed. Even if full-scale information were used, as 46% of the research do, the remaining publications (84% of the total papers) continue to be considered academic exercises. In around 25% of these academic tasks, different approaches were examined for the same aim. Other academic activities covered enhancements or modifications to traditional approaches to account for the unique characteristics of WWTPs [43]. We think that techniques that become increasingly prominent in academics frequently attract interest for commercial application. In order to improve the usefulness of the approaches, software development businesses are essential (for example, through specific processes and user-friendly interfaces). Even though a sizable number of businesses provide wastewater data management software, just a small portion state that data quality is their primary concern. Even still, it is challenging to keep track of which techniques were really used because This is beyond the scope of scientific journals and requires a targeted search for commercial items [16, 44].

According to our observations, in order to assure the implementation of a technique, a high degree of end-user engagement (such as from WWTP operators) must be guaranteed. Decision trees and fuzzy logic are two examples of effective methods that are used in daily life [45]. Such end-user involvement increases trust significantly and ensures that the strategies are used and tuned going forward. Furthermore, academics with little real-world experience—often employing fictitious or benchmark examples—have embraced the majority of statistical methodologies [46]. Today, it has proved challenging to introduce such sophisticated procedures since practitioners do not engage in learning during their growth. Future decisions will be significantly impacted by this, thus we must take this into account. Therefore, practitioners should either receive training in contemporary techniques or WWTPs should hire qualified consultants to provide services. To assure its implementation, researchers should look at alternative data analysis approaches that are more intelligible to operators. For instance, qualitative reasoning and analysis show promise but have only been used in a small number of wastewater treatment-related applications. The insufficient compensation for another challenge in academia is the transfer of technology from research results to application.

10.4.2 What Are the Present Technique Development Constraints?

Although technique justification is a crucial stage, obtaining high-quality data can be difficult, making it particularly tough. This is in part because efforts to install sensors and maintain them are insufficient to ensure the accuracy of the data. Additionally, wastewater processes exhibit variations on a wide range of timescales, from seconds to decades (changes in plant architecture and construction stages) [47]. As a result, it can be challenging to find datasets that are sufficiently high quality, cover all of the important dynamics, and span substantial amounts of time. Therefore, building a database a collection of data sets that may be used for validation and comparisons would be quite beneficial. In actuality, only 7 of the publications that were examined made use of repository data [48].

The absence of a connection between the domains of engineering and statistics is another constraint we found. Environmental, Chemical and Civil Engineering and computer engineering (including AI experts), especially for the development of EDSS (7% of papers), and Environmental/Civil/Civil Engineer and Electrical Engineering (3% of publications), have collaborated most frequently. When designing techniques, engineers and other disciplines must work more closely together to ensure the methods' effective implementation. The ideal situation would involve researchers or practitioners who have received interdisciplinary training in areas such as statistics, computer science, and engineering in addition to their engineering education. We believe that method development has little direction. Each of the techniques has many alternative implementations, but it is still unclear how to define what makes an acceptable solution implementation. At the very least, for those techniques that have been around for a while, there should be guidelines for good data mining and information extraction practices [49]. An excellent example is the manual on effective Modelling practice for mechanistic the wastewater treatment plant model identification.

Although it has received little attention in the field of wastewater treatment, knowledge management discipline is the one we recognized as having significant promise. Ontologies are the logical method for sharing information across many players and integrating knowledge from various sources [50]. Ontologies are used to create systems that seamlessly combine many forms of information and reasoning techniques; however, the rate of development is accelerated by faster processing infrastructure, less expensive Memory storage, as well as more reliable sensors. In the ecosystem of the World Wide Web of Things, where WWTPs are merely one type of element, ontologies are already a consumable. The coming generations of EDSS must be capable of storing and disseminating extremely important knowledge, as well as organizing memory and incorporating lessons learned. These systems ought to be able to leverage ontologies to give the correct language and techniques for categorizing material and mapping to the form of databases, databases, and knowledge bases as a mechanism for collecting implicit and expert knowledge. They ought to offer a setting for documenting and recognizing human skill, if it exists [51].

10.4.3 What Are the Present Method Selection Constraints?

Last but not least, there is insufficient advice for technique selection, which makes it challenging for practitioners to choose the appropriate approach to solve a certain problem. A single method may use several implementations, or multiple methods may do the same thing. An accurate comparison is difficult as a result. Furthermore, different methodologies draw attention to various data aspects or patterns, making comparison in certain circumstances pointless [49]. In fact, 21% of the publications that were assessed compared the new procedures to other ones that already existed and highlighted their benefits. Even so, only a tiny fraction of strategies are often evaluated (on average, just two are), necessitating a straightforward comparison of many different ways that perform the same purpose in order to offer advice on which method to choose. In general, there are no standardized procedures for selecting and using data analytic methodologies. The endeavor to define and map the area has proven to be difficult, even for academics. We believe that rather than creating new approaches, there is now a greater need than ever for instructions and guides [52].

This standard the wastewater treatment plant plant-wide simulation model has been created to make it possible to compare process monitoring, diagnostic, and control methods in an efficient manner. Analyses of the cost–benefit of specialized process tracking, diagnostics, and control algorithms are also made easier by this platform. In reality, that platform was previously utilized for technique development and comparison in roughly 9% of the investigations. Additionally, as 38% of the articles did not mention the software platform utilized, researchers should benefit from the information that Matlab was used for method implementation in at least 25% of the articles published. This number may even be higher. A group of people that distributes method representations (at a minimum in Matlab or through Git Hub) greatly aid in method comparison and provide appropriate direction for method choosing [53].

The future of wastewater treatment is being revolutionized in a number of ways using artificial intelligence or algorithmic learning [54]. Improve the effectiveness and effectiveness of the methods used to treat wastewater. Treatment procedures may be improved using AI and ML, which can also help save energy and raise effluent quality. On the basis of real-time data, AI-powered control systems, for instance, may automatically alter the dose of chemicals and the flow of effluent through treatment units. This can support ensuring that treatment procedures are effective and that effluent complies with legal requirements. Predict issues and prevent them. In order to forecast future issues, such as equipment breakdowns or increases in pollution concentrations, AI and ML can analyse previous data and find trends [55]. Utilizing this knowledge, one may plan maintenance or modify treatment procedures as a preventive measure. Find spills and other events, then take action. Real-time monitoring systems that can rapidly and reliably identify spills and other issues may be created using AI and ML. Once operators are informed, they can take corrective measures to lessen the impact on the environment. Create fresh, cutting-edge wastewater treatment methods. New wastewater treatment methods that are more effective, efficient, and sustainable may be designed and developed using AI and ML. AI-driven systems, for instance, may be utilized to create novel membrane materials for adsorption and filtration. In several cities across the world, wastewater treatment procedures are being optimized using AI-powered control systems. For instance, AI is being utilized in San Francisco to manage the city's wastewater treatment facility, resulting in an energy consumption decrease of 15–20% [56]. At wastewater treatment facilities, ML models are used to forecast and stop issues before they arise. For instance, scientists at the University of California, Berkeley have created a machine learning model that, with up to 90% accuracy, can forecast equipment breakdowns in wastewater treatment plants. Spills and other issues are being detected and handled at wastewater treatment plants using AI-powered monitoring systems. For instance, the business Enviro-Incite has created an AI-powered system that uses satellite photography to instantly identify spills and other issues. The development of cutting-edge wastewater treatment systems makes use of AI and ML. For instance, scientists at the University of Minnesota have created an AI-powered device that uses magnetic nanoparticles to filter out microplastics from wastewater [57]. In general, wastewater treatment might be completely changed by AI and ML. These technologies may enhance the efficacy and efficiency of treatment procedures, forecast and avoid issues, identify and address spills and other accidents, and provide fresh and cutting-edge approaches to wastewater treatment.

In addition to the advantages already discussed, AI and ML can also contribute to more sustainable and inexpensive wastewater treatment. For instance, AI and ML can assist to lessen the environmental effect of wastewater treatment by optimizing treatment operations and lowering energy use. Additionally, AI and ML can lower the cost of wastewater treatment by foreseeing issues and avoiding them. We may anticipate even more cutting-edge and efficient methods to employ AI and ML to remediate wastewater as these technologies advance.

10.5 Summary

The wastewater treatment sector is undergoing a rapid transformation due to the increasing popularity of the use of AI/ML. The long-term sustainability, efficacy, and efficiency of wastewater treatment facilities are only a few benefits that these technologies provide. The importance of AI/ML in wastewater treatment will be concluded and essential topics will be highlighted in this summary. AI is being utilized in wastewater treatment to automate a number of functions, including chemical dose optimization, leak and spill detection, and treatment operation monitoring. By enabling operators to devote more time to maintenance and troubleshooting, automation enhances the overall performance of the plant. AI is used to create new kinds of treatment, such as treating wastewater from industrial sources or inventing membranes for desalination that are more effective. These developments improve the treatment procedure and deal with certain environmental issues. With its capacity to maximize energy efficiency, lower greenhouse gas emissions, and recover valuable materials from wastewater, artificial intelligence (AI) enhances the sustainability of wastewater treatment. It complies with legal criteria and environmental aims. While deep learning algorithms are superior at managing delicate and difficult problems, machine learning algorithms are better at identifying patterns in data and forecasting possible problems. Text data from operator logs or sensor measurements is analyzed using language processing, which helps identify issues. Artificial intelligence (AI)-enhanced systems keep an eye on a number of characteristics in wastewater treatment facilities, such as pH, temperature, and flow rate. Early identification of problems, such as overflows, equipment malfunctions, or alterations in the quality of wastewater, enables operators to take fast corrective action. The entire efficiency of wastewater treatment is increased by AI-enhanced systems, which also minimize water usage, cut energy consumption, and select the best chemical doses. Although the use of AI in wastewater treatment is still in its infancy, it has the potential to completely transform the industry. Artificial Intelligence (AI) has the potential to improve safety and security in wastewater treatment plants and tackle issues pertaining to climate change, water shortages, and new contaminants.

10.6 Conclusion

One exciting and revolutionary development in wastewater treatment is the application of AI and machine learning. It has the potential to greatly improve treatment facilities' efficacy, sustainability, and efficiency. We may anticipate even more cutting-edge uses for AI technology to appear in this important field of study as it develops. Artificial Intelligence has the potential to be a key component in guaranteeing the efficient and responsible management of wastewater resources at a time of environmental difficulties.

Ethical Acceptance

Not Applicable

Participant Consent

Not Applicable

Contributions of Authors

The study's inception and design were contributed to by all authors. Prashantkumar B. Sathvara, J. Anuradha, Sandeep Tripathi, and R. Sanjeevi prepared the materials, collected data, and analyzed the results. J. Anuradha and R. Sanjeevi wrote the initial draft of the paper, and other writers provided feedback on prior drafts. The final manuscript was reviewed and approved by all writers.

Funding

No Research grants from funding agencies

Competing Interests Declaration

The authors state that they do not have any known competing financial interests or personal ties that may seem to have influenced the work disclosed in this study.

Data and resources are available

Not applicable

References

1. Chambers DM, Reese CM, Thornburg LG, Sanchez E, Rafson JP, Blount BC et al (2018) Distinguishing petroleum (crude oil and fuel) from smoke exposure within populations based on the relative blood levels of benzene, toluene, ethylbenzene, and xylenes (BTEX), styrene and 2,5-dimethylfuran by pattern recognition using artificial neural networks. Environ Sci Technol 52(1):308–316
2. Hong H, Tsangaratos P, Ilia I, Liu J, Zhu AX, Chen W (2018) Application of fuzzy weight of evidence and data mining techniques in construction of flood susceptibility map of Poyang County, China. Sci Total Environ 1(625):575–588
3. Offenberg JH, Lewandowski M, Kleindienst TE, Docherty KS (2017) Predicting thermal behavior of secondary organic aerosols. Environ Sci Technol 51(17):9911–9919
4. Pearce AR, Rizzo DM, Watzin MC, Druschel GK (2013) Unraveling associations between cyanobacteria blooms and in-lake environmental conditions in Missisquoi Bay, Lake Champlain, USA, using a modified self-organizing map. Environ Sci Technol 47(24):14267–14274
5. da Rocha SJSS, Torres CMME, Jacovine LAG, Leite HG, Gelcer EM, Neves KM et al (2018) Artificial neural networks: modeling tree survival and mortality in the Atlantic Forest biome in Brazil. Sci Total Environ 15(645):655–661
6. Wang C, Ye Z, Yu Y, Gong W (2018) Estimation of bus emission models for different fuel types of buses under real conditions. Sci Total Environ 1(640–641):965–972
7. Zhang H, Wu P, Yin A, Yang X, Zhang M, Gao C (2017) Prediction of soil organic carbon in an intensively managed reclamation zone of eastern China: a comparison of multiple linear regressions and the random forest model. Sci Total Environ 15(592):704–713
8. Fan M, Hu J, Cao R, Xiong K, Wei X (2017) Modeling and prediction of copper removal from aqueous solutions by nZVI/rGO magnetic nanocomposites using ANN-GA and ANN-PSO. Sci Rep 7(1):18040
9. Ghaedi AM, Ghaedi M, Pouranfard AR, Ansari A, Avazzadeh Z, Vafaei A et al (2016) Adsorption of Triamterene on multi-walled and single-walled carbon nanotubes: Artificial neural network modeling and genetic algorithm optimization. J Mol Liq 1(216):654–665
10. Khataee AR, Kasiri MB (2010) Artificial neural networks modeling of contaminated water treatment processes by homogeneous and heterogeneous nanocatalysis. J Mol Catal A: Chem 331(1):86–100
11. Impact of societal development and infrastructure on biodiversity decline methodology and approaches for biodiversity conservation. IGI Global, pp. 271–299
12. Haimi H, Corona F, Mulas M, Sundell L, Heinonen M, Vahala R (2015) Shall we use hardware sensor measurements or soft-sensor estimates? Case study in a full-scale WWTP. Environ Model Softw 1(72):215–229
13. Haimi H, Mulas M, Corona F, Vahala R (2013) Data-derived soft-sensors for biological wastewater treatment plants: an overview. Environ Model Softw 1(47):88–107
14. Hadjimichael A, Comas J, Corominas L (2016) Do machine learning methods used in data mining enhance the potential of decision support systems? A review for the urban water sector. AI Commun 29(6):747–756

15. Impact of societal development and infrastructure on biodiversity decline current scenario of biodiversity loss due to developmental activities. IGI Global pp. 50–69
16. Corominas Ll, Garrido-Baserba M, Villez K, Olsson G, Cortés U, Poch M (2018) Transforming data into knowledge for improved wastewater treatment operation: a critical review of techniques. Environ Modell Softw 106:89–103
17. Thomann M (2008) Quality evaluation methods for wastewater treatment plant data. Water Sci Technol 57(10):1601–1609
18. Sensors for environmental monitoring identification and assessment water pollutants sensor types and their advantages and challenges. IGI Global, pp. 78–101
19. Berthouex PM, Lai W, Darjatmoko A (1989) Statistics-based approach to wastewater treatment plant operations. J Environ Eng 115(3):650–671
20. Malviya A, Jaspal D (2021) Artificial intelligence as an upcoming technology in wastewater treatment: a comprehensive review. Environ Technol Rev 10(1):177–187
21. Villez K, Rieger L, Keser B, Venkatasubramanian V (2012) Probabilistic qualitative analysis for fault detection and identification of an on-line phosphate analyzer. Int J Adv Eng Sci Appl Math 4(1):67–77
22. Rieger L, Thomann M, Joss A, Gujer W, Siegrist H (2004) Computer-aided monitoring and operation of continuous measuring devices. Water Sci Technol 50(11):31–39
23. Sensors for environmental monitoring identification and assessment air sensors and their capabilities. IGI Global, pp. 10–24
24. Ray SS, Verma RK, Singh A, Ganesapillai M, Kwon YN (2023) A holistic review on how artificial intelligence has redefined water treatment and seawater desalination processes. Desalination 15(546):116221
25. Chan H, Nai-He Y (2020) A pretreatment method of wastewater based on artificial intelligence and fuzzy neural network system. J Intell Fuzzy Syst 39(2):1711–1720
26. Li K, Duan H, Liu L, Qiu R, van den Akker B, Ni BJ et al (2022) An integrated first principal and deep learning approach for modeling nitrous oxide emissions from wastewater treatment plants. Environ Sci Technol 56(4):2816–2826
27. Yan J, Xu Z, Yu Y, Xu H, Gao K (2019) Application of a hybrid optimized BP network model to estimate water quality parameters of Beihai Lake in Beijing. Appl Sci 9(9):1863
28. Bhardwaj A, Kishore S, Pandey DK (2022) Artificial intelligence in biological sciences. Life 12(9):1430
29. Sanjeevi R (2011) Studies on the treatment of low-strength wastewaters with upflow anaerobic sludge blanket (UASB) reactor: with emphasis on granulation studies. Centre for Pollution Control and Environmental Engineering, Pondicherry
30. Biodiversity loss assessment for ecosystem protection biodiversity and its importance an introduction. IGI Global, pp. 1–17
31. Ramakrishnan S, Jayaraman A (2019) Pesticide contaminated drinking water and health effects on pregnant women and children. In: Handbook of research on the adverse effects of pesticide pollution in aquatic ecosystems [Internet]. IGI Global; 2019 [cited 2023 Oct 27], pp 123–136. Available from: https://www.igi-global.com/chapter/pesticide-contaminated-drinking-water-and-health-effects-on-pregnant-women-and-children/213500
32. Altowayti WAH, Shahir S, Othman N, Eisa TAE, Yafooz WMS, Al-Dhaqm A et al (2022) The role of conventional methods and artificial intelligence in the wastewater treatment: a comprehensive review. Processes 10(9):1832
33. Abbasi T, Sanjeevi R, Makhija M, Abbasi SA (2012) Role of vitamins B-3 and C in the fashioning of granules in UASB reactor sludge. Appl Biochem Biotechnol 167(2):348–357
34. Narayanan CM, Narayan V (2019) Biological wastewater treatment and bioreactor design: a review. Sustain Environ Res 29(1):33
35. Ramakrishnan S, Jayaraman A (2019) Pesticide contaminated drinking water and health effects on pregnant women and children [internet], vol 123–136. Available from: https://doi.org/10.4018/978-1-5225-6111-8.ch007
36. Mamandipoor B, Majd M, Sheikhalishahi S, Modena C, Osmani V (2020) Monitoring and detecting faults in wastewater treatment plants using deep learning. Environ Monit Assess 192(2):148

37. Sathvara PB, Anuradha J, Sanjeevi R, Tripathi S, Rathod AB (2023) Spatial analysis of carbon sequestration mapping using remote sensing and satellite image processing. In: Multimodal biometric and machine learning technologies [internet]. [cited 2023 Oct 27], pp 71–83. Available from: https://doi.org/10.1002/9781119785491.ch4
38. Impact of climate change and its importance on human performance. In: Insights on impact of climate change and adaptation of biodiversity (1st ed., pp. 1–9). KD Publication.
39. Rathod AB, Sathvara PB, Tripathi A, Anuradha J, Tripathi S, Sanjeevi R (2022) Vegetational cartography analysis utilizing multi-temporal ndvi data series: a case study from Rajkot District (Gujarat), India. J Tianjin Univ Sci Technol 55(4):490–497
40. Rathod AB, Anuradha J, Sathvara PB, Tripathi S, Sanjeevi R (2023) Vegetational change detection using machine learning in Gis technique: a case study from Jamnagar (Gujarat). J Data Acquisition Process 38(1):1046–1061
41. Samuel AL (1959) Some studies in machine learning using the game of checkers. IBM J Res Dev 3(3):210–229
42. Yekkehkhany B, Safari A, Homayouni S, Hasanlou M (2014) A comparison study of different Kernel functions for Svm-based classification of multi-temporal polarimetry Sar data. Int Arch Photogrammetry Remote Sensing Spatial Inf Sci XL-2-W3:281–285
43. Martín de la Vega PT, Jaramillo-Morán MA (2018) Obtaining key parameters and working conditions of wastewater biological nutrient removal by means of artificial intelligence tools. Water 10(6):685
44. Prashantkumar Sathvara et al. (2021) Chlorination treatment in common effluent treatment plant: evaluation of effectiveness and acute toxicity effects. Mendeley Data. https://doi.org/10.17632/f7jctrd8jj.3
45. Bahramian M, Dereli RK, Zhao W, Giberti M, Casey E (2023) Data to intelligence: the role of data-driven models in wastewater treatment. Expert Syst Appl 1(217):119453
46. Shivaram KB, Bhatt P, Applegate B, Simsek H (2023) Bacteriophage-based biocontrol technology to enhance the efficiency of wastewater treatment and reduce targeted bacterial biofilms. Sci Total Environ 1(862):160723
47. Sahu S, Kaur A, Singh G, Kumar AS (2023) Harnessing the potential of microalgae-bacteria interaction for eco-friendly wastewater treatment: a review on new strategies involving machine learning and artificial intelligence. J Environ Manage 15(346):119004
48. Oruganti RK, Biji AP, Lanuyanger T, Show PL, Sriariyanun M, Upadhyayula VKK et al (2023) Artificial intelligence and machine learning tools for high-performance microalgal wastewater treatment and algal biorefinery: a critical review. Sci Total Environ 10(876):162797
49. Nourani V, Zonouz RS, Dini M (2023) Estimation of prediction intervals for uncertainty assessment of artificial neural network based wastewater treatment plant effluent modeling. J Water Process Eng 1(55):104145
50. Yang Y, Kim KR, Kou R, Li Y, Fu J, Zhao L et al (2022) Prediction of effluent quality in a wastewater treatment plant by dynamic neural network modeling. Process Saf Environ Prot 1(158):515–524
51. Haenlein M, Kaplan A (2019) A brief history of artificial intelligence: on the past, present, and future of artificial intelligence. Calif Manage Rev 61(4):5–14
52. Alvi M, Batstone D, Mbamba CK, Keymer P, French T, Ward A et al (2023) Deep learning in wastewater treatment: a critical review. Water Res 15(245):120518
53. Nguyen XC, Nguyen TTH, Tran QB, Bui XT, Ngo HH, Nguyen DD (2022) Chapter 21—artificial intelligence for wastewater treatment. In: Bui XT, Nguyen DD, Nguyen PD, Ngo HH, Pandey A (eds) Current developments in biotechnology and bioengineering [internet]. Elsevier [cited 2023 Oct 10], pp 587–608. Available from: https://www.sciencedirect.com/science/article/pii/B9780323998741000087
54. Wang Y, Cheng Y, Liu H, Guo Q, Dai C, Zhao M et al (2023) A review on applications of artificial intelligence in wastewater treatment. Sustainability 15(18):13557
55. El Alaoui El Fels A, Mandi L, Kammoun A, Ouazzani N, Monga O, Hbid ML (2023) Artificial intelligence and wastewater treatment: a global scientific perspective through text mining. Water (19):3487

56. Nourani V, Elkiran G, Abba SI (2018) Wastewater treatment plant performance analysis using artificial intelligence—an ensemble approach. Water Sci Technol 78(10):2064–2076
57. Lowe M, Qin R, Mao X (2022) A review on machine learning, artificial intelligence, and smart technology in water treatment and monitoring. Water 14(9):1384

Chapter 11
Predictive Modeling and Machine Learning for Optimal Wastewater Treatment Performance

Salim Heddam

Abstract This paper proposes new wastewater treatment performances predictive model based on a novel Outlier-robust extreme learning machine (ORELM). The new proposed ORELM model is proposed to resolve some limitations of the standard extreme learning machine (ELM) model and to enhance their performances. The ORELM is than developed for predicting influent five days biochemical oxygen demand (BOD_5) measured at the inlet of the wastewater treatment plant (WWTP). The ORELM is developed using several influent wastewater quality variables namely, wastewater temperature (T_w), pH, specific conductance (SC), suspended solids (MES), and wastewater chemical oxygen demand (COD). The performances of the ORELM were compared with those of the regularized extreme learning machine (RELM), weighted regularized extreme learning machine (WRELM), and the standard ELM models. The models were evaluated and compared using Pearson correlation coefficient (R), Nash–Sutcliffe efficiency (NSE), root-mean-square error (RMSE), and mean absolute error (MAE), and the historical wastewater dataset are used for training and testing the proposed models using a ratio of 70% and 30%, respectively. The comparison demonstrates the superior performance of the proposed ORELM model compared to the ELM, RELM, and WRELM, exhibiting an R, NSE, RMSE and MAE values of 0.807, 0.649, 63.83, and 45.21. Moreover, obtained results in the present study have demonstrated the potential benefits of the proposed models.

Keywords Wastewater · BOD_5 · Modelling · ORELM · WRELM · RELM · ELM

S. Heddam (✉)
Hydraulics Division, Agronomy Department, Faculty of Science, University 20 Août 1955 Skikda, Route El Hadaik, BP 26, Skikda, Algeria
e-mail: heddamsalim@yahoo.fr

M. C. Garg (ed.), *The AI Cleanse: Transforming Wastewater Treatment Through Artificial Intelligence*, Springer Water, https://doi.org/10.1007/978-3-031-67237-8_11

11.1 Introduction

Nowadays, the wastewater (WT) rejected into natural watercourse has become a serious concerns for the overall environmental management system [7]. The presence of several undesirable WT component pose serious problems and among these harmful components, the presence of organic and inorganic pollutants, various nutrients and more precisely the presence of heavy metals [8]. High concentration of WT with some elements, i.e., Nitrogen (N) and phosphorus (P) is the major responsible of the freshwater eutrophication and can cause the emission of greenhouse gases. Taking into account theses serious problems related to the presence of WT in the naturel watercourse, the needs for wastewater treatment and the possibility to impose conditions on the reuse of WT for several purposes, has become a critical element for the preservation of the water resources [6]. Furthermore, making an efficient WT treatment can be guaranteed by the implementation of small and large-scale wastewater treatment plant (WWTP). The WWTP play an increasingly important role in the collect of WT and they are mainly employed for treating the effluent and ensuring an acceptable wastewater quality.

In recent years, the control of effluent and influent WT has gained increasing attention and the biochemical oxygen demand (BOD_5) is one of the most important variable that needs to be continuously controlled. A parts form the direct in situ measures, the use of various modelling approaches was also broadly reported in the literature. While various models have been reported to be good approximators, the needs for more works in this subject is always acknowledged. The use machine learning models (ML) based on artificial intelligence techniques (AI) has been increasing in the last decades, and several authors have highlighted the importance of these ML algorithms. Yadav et al. [18] compared between gradient boosting regression (GBR), random forest regression (RFR), artificial neural network (ANN), and the k-nearest neighbour (KNN) models for predicting and forecasting the influent and effluent BOD_5 at the Bharwara WWTP in Lucknow, India. For developing the models, they used various wastewater variables namely, pH, dissolved oxygen (DO), chemical oxygen demand (COD), total suspended solids (TSS), biochemical oxygen demand (BOD) and myeloproliferative neoplasms (MPN). Acceptable numerical performances were obtained and the ANN was the most accurate with correlation coefficient (R) of ≈ 0.74. Rizal and Hayder [13] compared between the ANN and support vector machine (SVR) for predicting the effluent BOD_5 using various wastewater variables namely pH, COD, TSS, temperature (T_w), nitrogen nitrate (NN, mg/L), ammonia nitrogen (AN, mg/L), oil and grease (OG, mg/L). Based on various numerical performances, the ANN was found to be more accurate with root mean square error (RMSE), R, coefficient of determination (R^2) of approximately, ≈ 0.02, ≈ 0.99, and ≈ 0.99, respectively. Sahu et al. [14] used the SVM model with various kernel function namely, normalised poly kernels (Npoly), poly kernel (Poly), the radial basis function (RBF) kernel, and the new Pearson VII universal kernel (Puk). By combining various water variables namely, DO, COD, pH, TSS, electrical conductivity (EC), NO_3–NO_2, NO_3–N, and alkalinity (Alk). The

SVM with Puk kernel function was found to be more accurate exhibiting and R and RMSE values of approximately ≈ 0.97 and ≈ 2.38, respectively. Sheikh Khozani et al. [15] introduced a new improved version of the convolutional neural network (CNN) with multi-kernel RBFNN (MKRBFNN) for predicting the effluent BOD_5. The hybrid model was called the CNN-MKRBFNN and it is optimized using the salp swarm algorithm (SSA). The models was developed using influents BOD_5, COD, TSS, volatile suspended solids (VSS), and sediment (SED). High predictive accuracy was obtained with Nash–Sutcliffe efficiencies of ≈ 0.97.

Qambar and Al Khalidy [12] compared between various ML models namely decision tree (DT), RFR, adaptive boosting (AB), and gradient boost algorithms (GBA), for predicting influent BOD_5 at Madinat Salman WWTP located in Bahrain. The authors used various WT quality variables, namely pH, EC, TSS, NH_3, COD, total Phosphorus (TP), and total Nitrogen (TN). Local and global modelling strategy were analyzed and obtained results revealed high predictive accuracies with $R^2 \approx 0.92$, MAE ≈ 6.56, and $R^2 \approx 0.95$, MAE ≈ 3.93 for local and global modelling, respectively. Qambar and Al Khalidy [11] compared between DT, RFR, AdaBoost, GBA, and extreme gradient boosting (XGBoost) for modelling BOD_5 using various WT quality variables. They reported acceptable prediction accuracy. Asami et al. [1] compared between ANN and M5Tree models for modelling BOD_5 using various WT variables namely, T_w, pH, TSS, DO, turbidity (TU), EC, TDS and COD. The ANN was found to be more accurate with $R^2 \approx 0.95$, RMSE ≈ 3.52, respectively. Finally, Azimi et al. [2] applied an optimized ANN model using genetic algorithm (ANN-GA) for modelling BOD_5 measured at the Zarghandeh WWTP located in Tehran, Iran. Three scenarios were analyzed and good numerical performances were obtained with NSE, RMSE, and MAE ranged from ≈ 0.882 to ≈ 0.915, ≈ 0.689 to ≈ 0.812, and ≈ 0.546 to ≈ 0.599, respectively.

The main objective of this chapter is to develop a framework for improving the wastewater treatment performance by modelling the influent BOD_5 using historical measured time series data of various wastewater quality variables. We propose an improved extreme learning machine (ELM) models called outlier-robust extreme learning machine (ORELM), and we perform a comparison of the ORELM with those of ELM, regularized extreme learning machine (RELM), and weighted regularized extreme learning machine (WRELM). The novelty of our study is that, the proposed models were introduced for the first time for modelling WWTP performances. The proposed models were calibrated and validates using real in situ measured dataset.

11.2 Study Area and Data Used

11.2.1 Description of the Wastewater Treatment Plant Process

Batna is located in the North-Est of Algeria country with semi-arid climate. It is bound to the south by the Biskra Province, to the east by Khenchela and Oum El Bouaghi, to the north by Setif, and to the west by M'sila. Wastewater is collected by the collecting systems and rejected in Oued El Gourzi. Thus, it is transported to the Batna wastewater treatment plant (Fig. 11.1). Batna WWTP adopts a conventional activated sludge process with the treatment capacity of about 19,875 m^3/days. The WWTP is composed from two major's lines: (*i*) water line and (*ii*) sludge line, and it is mainly strutured into primary treatment, secondary treatment, and tertiary treatment. The WWTP sections included: (*i*) bar screen workshop, (*ii*) sewer lift station, (*iii*) mechanical screening, (*iv*) degritting/deoiling (grit chambers and oil–water separators), (*v*) the distribution box or flow splitter, (*vi*) two primary sedimentation tank, (*vii*) sludge pumping station, (*viii*) floating room, (*ix*) aeration tank, (*x*) sludge pumping station, (*xi*) secondary sedimentation tank, and (*xii*) disinfectant dip. In the present study, the wastewater characteristics, including five days biochemical oxygen demand (BOD_5), chemical oxygen demand (COD), wastewater temperature (T_w), pH, specific conductance (SC), and total suspended solids (TSS), measured at daily time scale from 01 March 2006 to 07 April 2009, with a total of 1134 patterns of input–output were collected for model development. The measured dataset was partitioned into two subsets: a training set with a total 794 samples (794 days) and a validation set consisting of 340 samples (340 days). The correlation heatmaps between all variables is depicted in Fig. 11.2 showing the correlation coefficient between all variables for the training, validation and all data. Furthermore, this study employed five variables as input, i.e., COD, TSS, pH, T_w, and SC in the model development for predicting BOD_5. Descriptive statistic for all data is depicted in Table 11.1. In the present chapter, we have compared the performances of the models taking into count nine input combinations, with five, four, three and two input variables (Table 11.2).

11.2.2 Performance Assessment of the Models

In the present study, four performances metrics were selected for models comparison and evaluation: the root mean square error (RMSE), mean absolute error (MAE), correlation coefficient (R), and Nash–Sutcliffe efficiency (NSE).

$$MAE = \frac{\sum_{i=1}^{N} \left| BOD_{5pre,i} - BOD_{5obs,i} \right|}{N} \tag{11.1}$$

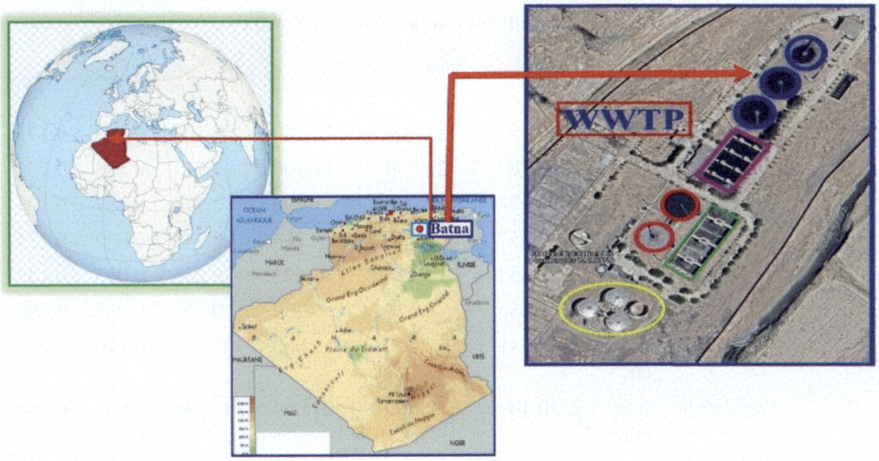

Fig. 11.1 Map showing the location of Batna wastewater treatment plant

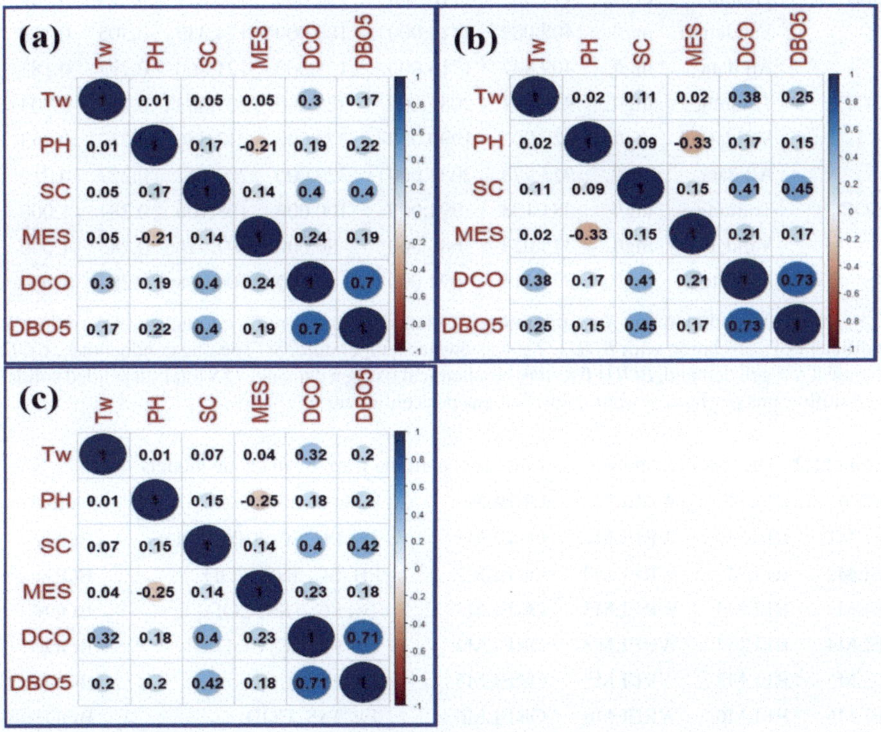

Fig. 11.2 The correlation heatmaps between all variables for the training, validation and all data

Table 11.1 Summary statistics of the influent Batna municipal WWTP

Variables	Subset	Unit	X_{mean}	X_{max}	X_{min}	S_x	C_v	*R*
Batna municipal WWTP								
Tw	Training	°C	20.055	27.000	11.300	3.804	0.190	0.171
	Validation	°C	19.782	26.500	11.200	3.717	0.188	0.252
	All data	°C	19.973	27.000	11.200	3.778	0.189	0.195
pH	Training	//	7.657	8.276	6.259	0.248	0.032	0.224
	Validation	//	7.653	8.120	6.493	0.255	0.033	0.147
	All data	//	7.656	8.276	6.259	0.250	0.033	0.200
SC	Training	μ.s/cm	2341.541	3330.000	1016.000	327.923	0.140	0.404
	Validation	μ.s/cm	2330.407	3330.000	1004.000	332.948	0.143	0.454
	All data	μ.s/cm	2338.203	3330.000	1004.000	329.330	0.141	0.419
TSS	Training	mg/L	410.602	1716.000	155.000	120.854	0.294	0.189
	Validation	mg/L	408.082	1716.000	180.000	124.339	0.305	0.170
	All data	mg/L	409.847	1716.000	155.000	121.860	0.297	0.183
COD	Training	mg/L	925.681	2084.000	320.000	207.652	0.224	0.703
	Validation	mg/L	920.002	1990.000	378.000	213.886	0.232	0.733
	All data	mg/L	923.978	2084.000	320.000	209.462	0.227	0.712
BOD_5	Training	mg/L	381.153	994.000	100.000	108.106	0.284	1.000
	Validation	mg/L	375.023	681.000	86.000	107.916	0.288	1.000
	All data	mg/L	379.315	994.000	86.000	108.038	0.285	1.000

X_{mean} mean; X_{max} maximum; X_{min} minimum; S_x standard deviation; C_v coefficient of variation; *R* coefficient of correlation with BOD_5, T_w wastewater temperature, *SC* specific conductance, *COD* chemical oxygen demand, BOD_5 five-day biochemical oxygen demand, *TSS* total suspended solids, *mg/l* milligrams per liter, μ.s/cm microsiemens per centimeter

Table 11.2 The input combinations of different extreme learning machine models

ELM	RELM	WRELM	ORELM		Input combination		Output
ELM1	RELM1	WRELM1	ORELM1		Tw, PH, SC, TSS, COD		BOD_5
ELM2	RELM2	WRELM2	ORELM2		PH, SC, TSS, COD		BOD_5
ELM3	RELM3	WRELM3	ORELM3		Tw, SC, TSS, COD		BOD_5
ELM4	RELM4	WRELM4	ORELM4		Tw, PH, SC, COD		BOD_5
ELM5	RELM5	WRELM5	ORELM5		SC, TSS, COD		BOD_5
ELM6	RELM6	WRELM6	ORELM6		Tw, TSS, COD		BOD_5
ELM7	RELM7	WRELM7	ORELM7		PH, TSS, COD		BOD_5
ELM8	RELM8	WRELM8	ORELM8		TSS, COD		BOD_5
ELM9	RELM9	WRELM9	ORELM9		SC, COD		BOD_5

$$RMSE = \sqrt{\frac{\sum_{i=1}^{N} \left(BOD_{5obs,i} - BOD_{5pre,i}\right)^2}{N}} \tag{11.2}$$

$$NSE = 1 - \left[\frac{\sum_{i=1}^{N} \left(BOD_{5obs,i} - BOD_{5pre,i}\right)^2}{\sum_{i=1}^{N} \left(BOD_{5obs,i} - \overline{BOD_{5obs}}\right)^2}\right] \tag{11.3}$$

$$R = \frac{\sum_{i=1}^{N} \left(BOD_{5obs,i} - \overline{BOD_{5obs}}\right)\left(BOD_{5pre,i} - \overline{BOD_{5pre}}\right)}{\sqrt{\sum_{i=1}^{N} \left(BOD_{5obs,i} - \overline{BOD_{5obs}}\right)^2 \sum_{i=1}^{N} \left(BOD_{5pre,i} - \overline{BOD_{5pre}}\right)^2}} \tag{11.4}$$

$\overline{BOD_{5obs}}$ and $\overline{BOD_{5pre}}$ are the mean measured, and mean predicted five days biochemical oxygen demand, respectively, BOD_{obs} and BOD_{pre} specifies the observed and predicted five days biochemical oxygen demand, and N shows the number of data points.

11.3 Methodology

11.3.1 Extreme Learning Machine (ELM)

Given a training dataset $D_l = \{(x_i, y_i)\}^l$, the extreme learning machine (ELM) can be formulated as an approximation function as follow (Fig. 11.3):

$$f(x) = \sum_{j=1}^{L} \beta_j G(w_j, b_j, x) \tag{11.5}$$

where β_j is the weight between the hidden layer and the output layer, L is the number of hidden neurons, w_j and b_j are the weight and bias linking the input to the hidden layer, and G is the activation function, generally the sigmoidal function. Training the ELM model is achieved in two distinguished stages. First, from the input to the hidden layer, the parameters, i.e., the weights and biases are generated randomly based on a uniform distribution, while the weights from the hidden to the output layer are analytically determined, making the training of the ELM very faster. The estimation of the optimal weight vector β values is the major objective of the training algorithm of the ELM proposed by Huang et al. [5, 4], which is formulated as an optimization problem as follow:

$$\min_{(\beta,\xi)} \frac{1}{2} \|\beta\|^2 + \frac{C}{2} \sum_{i=1}^{M} \xi_i^2 \tag{11.6}$$

where C is the regularization coefficient and ξ is the training error.

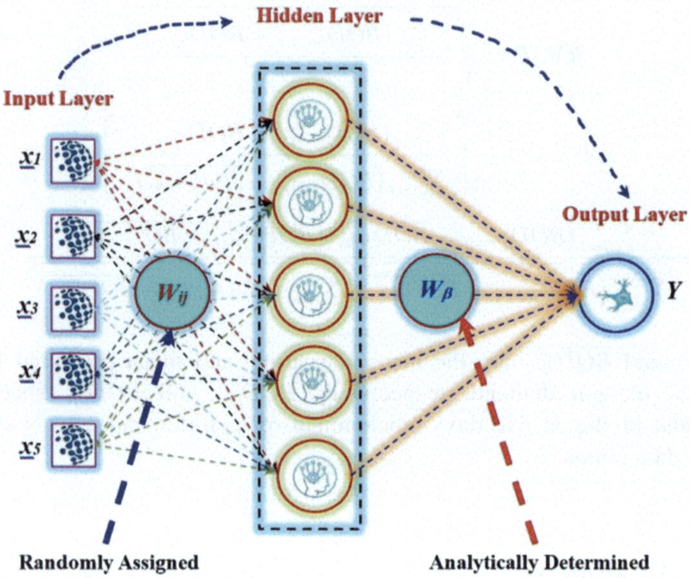

Fig. 11.3 The ELM architecture

Equation 11.6 is subject to:

$$h(x_i)\beta - y_i = \xi_i, i = (1, 2, \ldots, M) \tag{11.7}$$

Equation 11.6 can be converted to:

$$\min_{(\beta)} \frac{1}{2}\beta^2 + \frac{C}{2}\sum_{i=1}^{M}(h(x_i)\beta - y_i)^2 \tag{11.8}$$

The output weight β, is computed using the linear system as follow:

$$\beta = \beta^{\dagger}\boldsymbol{T} \tag{11.9}$$

where $\boldsymbol{H}\dagger$ is the Moore–Penrose generalized inverse of $\boldsymbol{H}$.

According to Zhang and Luo [20], it is necessary to have an ELM algorithm with high level of generalisation, which can be achieved by finding the pertinent tradeoff between the "*training error*" and the norm of output weight. This led to the introduction of a regularization coefficient and the introduction of the regularized extreme learning machine (RELM) algorithm [20], formulated as follow:

$$\min_{(\beta)} C\|y - H\beta\|_2^2 + \|\beta\|_2^2 \tag{11.10}$$

Another improved ELM algorithm was also proposed to handling the presence of sample with outlier, the weighted ELM (WELM) was proposed by adding a weight coefficient "*w*" to the calculated error [3, 16]:

$$\min_{(\beta)} \frac{1}{2}\|\beta\|^2 + \frac{C}{2}\sum_{i=1}^{M} w_i \xi_i^2 \tag{11.11}$$

Zhang and Luo [20] introduced another improved ELM algorithm called outlier robust extreme learning machine (ORELM). The ORELM is constructed by combining the ELM and the "*l*1*-norm loss function*" for reducing the interference of outliers and it can be summarized as follow [17]:

$$\min_{(\beta)} Ce_0 + \frac{1}{C}\beta_2^2 \tag{11.12}$$

$$y_i - H\beta = \xi_i \tag{11.13}$$

$$L_\mu(\xi, \beta, \lambda) = \xi_1 + \frac{1}{C}\beta_2^2 + \lambda^T(y, H\beta - \xi) + \frac{\mu}{2}y - H\beta - \xi_2^2 \tag{11.14}$$

where ξ is the estimated training error; β represents the output weight; y represents the target output; H represents the output of hidden layer; $L\mu$ corresponds to the augmented Lagrangian function; μ represents the penalty parameter [17, 20]. During the last few years, the ORELM has been applied successfully for solving various problems, i.e., multi-step ship motion forecasting [17], wind speed forecasting [9], solar radiation prediction [19], and multi-step air quality index forecasting [10].

11.4 Results and Discussion

The historical wastewater quality variables are used to verify the performances of the proposed extreme learning models. Data was divided into training (70%) and validation (30%), and in total 9 input combination having different water quality variables were evaluated and compared. Obtained results are depicted in Table 11.3. According to Table 11.3, the results depend on the number of input variables and it is clear that the best results, i.e., the smallest RMSE and MAE and the largest R and NSE values can be attained by considering the structure having five input variables, i.e., the ELM1, RELM1, WRELM1 and ORELM1, respectively. This section compares the performances of different models taking into account different numerical indices. Using the ELM models, it is clear that, the performances did not change significantly between different input combinations, and the numerical performances does not decreased with increasing number of input from two to five. Table 11.3 reveals that the proposed ELM models always yielded best performances. In particular, the R and NSE values were in excess of ≈ 0.780 and ≈ 0.600, respectively. Comparison of the results obtained using the ELM models we can see that, during the validation stage, the RMSE values were ranged from ≈ 64.732 to ≈ 67.486 with mean value of ≈ 66.354. Additionally, the MAE values were ranged from ≈ 47.417 to ≈ 50.822 with mean value of ≈ 49.036. The best performances were achieved using the ELM1 with R, NSE, RMSE and MAE values of approximately ≈ 0.799, ≈ 0.639, ≈ 64.732, and ≈ 47.417, respectively. Furthermore, obtained results revealed that, ELM1 model was found to be slightly higher than the ELM9 having only two inputs variables (i.e., SC and COD), and comparison of numerical performances shows that the ELM1 improves the ELM9 by $\approx 2.25\%$, $\approx 4.85\%$, $\approx 4.05\%$, $\approx 6.70\%$, respectively.

In order to investigate the predictive ability of the standard ELM, it is compared with three other improved ELM models, namely the RELM, WRELM, and ORELM. The comparisons are made based on the four numerical performances for which the values are listed in Table 11.3. It can be seen from Table 11.3 that, the performances of the RELM were slightly superior to those of the ELM models and there is not a remarkable improvement in terms of numerical performances. The mean R, NSE, RMSE and MAE values were equal to ≈ 0.788, ≈ 0.620, ≈ 66.43, $\approx$ and 48.951, respectively, which negligible difference compared to the values obtained using the ELM models. The best performances were achieved using the RELM1 with R, NSE, RMSE, and MAE of ≈ 0.803, ≈ 0.644, ≈ 64.32, $\approx$ and 46.945 equally with the RELM6 having only three input variables, i.e., the Tw, TSS, and COD as input variables. Furthermore, using the RELM models, the lowest performances were obtained using the RELM8 with R, NSE, RMSE, and MAE of ≈ 0.775, ≈ 0.600, ≈ 68.189, $\approx$ and 49.929, respectively. For comparison, the RELM1 improves the performances of RELM8 in terms of R, NSE, RMSE, and MAE by $\approx 3.48\%$, $\approx 6.83\%$, $\approx 5.67\%$, $\approx 5.97\%$, respectively.

The results used the WRELM are also depicted in Table 11.3. Evidently, when each model is considered individually (i.e., models WRELM1-9), the WRELM1 model yielded the highest correlation coefficient ($R \approx 0.814$) and correspondingly the highest NSE value of ≈ 0.660, and the low RMSE (≈ 62.80) and the low MAE value (≈ 44.83), more than any other WRELM models. This is followed by (in the order) WRELM4, WRELM3 and the last rank with see the WRELM5 having the poorest numerical performances of $R \approx 0.776$, $NSE \approx 0.602$, $RMSE \approx 67.963$, and $MAE \approx 49.191$, respectively. The best results were achieved when all input variables were combined (Tw, pH, SC, TSS, COD, Table 11.2). Comparison between the best and the poorest models revealed that the WRELM1 improves the performances of the WRELM5 by $\approx 4.668\%$, $\approx 8.788\%$, $\approx 7.588\%$, $\approx 8.857\%$, by in terms of R, NSE, RMSE and MAE, respectively.

Further analysis on the effect of input variables shows that, beyond the best models, i.e., the WRELM1 and WRELM4, increasing the number of input variables did reduce the error between measured and predicted data only marginally, and between the last five models, i.e., WRELM5 to WRELM9, these reductions are even lower, or negligible. More importantly, the performances of the WRELM models are superior to those of the RELM and ELM models.

Table 11.3 Performances of different machine learning models for BOD_5 prediction

Models	Training				Validation			
	R	NSE	RMSE	MAE	R	NSE	RMSE	MAE
ELM1	0.811	0.657	63.277	46.407	0.799	0.639	64.732	47.417
ELM2	0.816	0.666	62.467	46.386	0.797	0.635	65.067	47.900
ELM3	0.799	0.638	65.031	47.876	0.787	0.619	66.481	48.717
ELM4	0.799	0.638	65.031	46.678	0.791	0.625	65.989	48.532
ELM5	0.803	0.645	64.349	47.387	0.786	0.617	66.715	49.861
ELM6	0.813	0.660	62.955	45.796	0.789	0.621	66.361	48.301
ELM7	0.755	0.570	70.855	51.440	0.781	0.608	67.486	50.378
ELM8	0.741	0.550	72.495	51.287	0.786	0.615	66.885	49.396
ELM9	0.740	0.547	72.697	51.510	0.781	0.608	67.468	50.822
RELM1	0.820	0.672	61.861	45.680	0.803	0.644	64.323	46.945
RELM2	0.807	0.652	63.757	47.098	0.776	0.603	67.917	50.541
RELM3	0.808	0.653	63.605	47.287	0.788	0.620	66.465	48.508
RELM4	0.804	0.647	64.192	46.174	0.795	0.633	65.314	47.559
RELM5	0.798	0.636	65.147	47.438	0.788	0.619	66.556	49.276
RELM6	0.815	0.665	62.555	45.964	0.802	0.640	64.662	47.478
RELM7	0.802	0.643	64.579	47.531	0.781	0.608	67.430	50.551
RELM8	0.789	0.622	66.415	48.713	0.775	0.600	68.189	49.929
RELM9	0.752	0.565	71.227	49.927	0.783	0.612	67.083	49.772
WRELM1	0.800	0.639	64.870	45.951	0.814	0.660	62.806	44.834
WRELM2	0.781	0.610	67.499	47.201	0.797	0.636	65.042	47.909
WRELM3	0.788	0.621	66.494	46.852	0.800	0.640	64.691	46.148
WRELM4	0.797	0.636	65.218	46.189	0.801	0.641	64.561	46.802
WRELM5	0.794	0.628	65.913	46.390	0.776	0.602	67.963	49.191
WRELM6	0.804	0.646	64.270	45.760	0.785	0.615	66.830	48.423
WRELM7	0.803	0.645	64.405	46.324	0.779	0.607	67.573	50.088
WRELM8	0.786	0.615	67.041	47.361	0.781	0.608	67.453	48.235
WRELM9	0.749	0.559	71.765	49.728	0.783	0.614	66.991	49.065
ORELM1	0.787	0.620	66.636	45.030	0.817	0.666	62.233	44.818
ORELM2	0.781	0.609	67.539	47.024	0.802	0.642	64.484	48.905
ORELM3	0.800	0.640	64.855	45.151	0.794	0.629	65.591	46.329
ORELM4	0.792	0.627	65.958	44.776	0.801	0.642	64.509	46.743
ORELM5	0.787	0.615	67.024	46.365	0.786	0.616	66.774	48.779
ORELM6	0.807	0.651	63.863	44.351	0.792	0.625	66.005	47.645
ORELM7	0.801	0.640	64.861	45.641	0.787	0.618	66.579	49.076
ORELM8	0.782	0.606	67.782	46.824	0.787	0.617	66.699	47.610
ORELM9	0.748	0.557	71.901	49.402	0.778	0.604	67.819	49.438

Finally, we discuss the results obtained using the ORELM models. According to Table 11.3, there has been a marked improvement in the predictive accuracies in comparison to the all previous models, and the ORELM1 exhibited the best performances compared to the ELM1, RELM1, and WRELM1. The R, NSE, RMSE and MAE values obtained suing the ORELM1 were respectively ≈ 0.817, ≈ 0.666, ≈ 62.233, and ≈ 44.918, respectively. The ORELM1 improves the performances of the ELM1 in terms of R, NSE, RMSE and MAE by $\approx 2.203\%$, $\approx 4.054\%$, $\approx 3.861\%$, $\approx 5.481\%$, respectively. Similarly, the ORELM1 improves the performances of the RELM1 in terms of R, NSE, RMSE and MAE by $\approx 2.203\%$, $\approx 4.054\%$, $\approx 3.861\%$, $\approx 5.481\%$, respectively. More importantly, the difference between the WRELM1 and ORELM1, there is a marginal difference and the improvement are even lower, or negligible, all below the rate of 01%.

Beyond the numerical comparison of the models performances, we provide a graphical comparison between all models for evaluation and or highlighting the difference between the models in terms of forecasting accuracy. The scatterplot of measured and calculated BOD5 using the best models are depicted in Fig. 11.4. According to Fig. 11.4, the difference between the ORELM, ELM and RELM is more obvious, while the ORELM and the WRELM were approximately similar showing less scattered data. The obtained results are also depicted in Fig. 11.5 as curve of comparison between measured and predicted data, for which we can see that the ORELM and the WRELM were the models having the superior capability in predicting the extreme (i.e., the minimal and maximal) values. The boxplot and violinplot are depicted in Fig. 11.6, while the Taylor diagram is provided in Fig. 11.7. By analysing the Taylor diagram of Fig. 11.7, we can see that the high correlation coefficient was achieved using the ORELM model, while the lowest value was obtained using the ELM model.

11.5 Conclusion

The feasibility of the application of various extreme learning machine methods and its potential for real predictive application could have a significant impact on the improvement of the wastewater treatment performance. Our results support the obtained models performances reported by various authors worldwide, and it contributes to the improvement of our understanding about the possible prediction of BOD_5 based on the combination of various other wastewater quality variables. Indeed, while we have tested various input combination, it was found that the ORELM model exhibited the high predictive accuracy compared to the all other models by using all five input variables. For a future studies, some important points need to be taken into consideration for improving the models performances. First, it is highly recommended to introduce the optimization algorithms base on the various metaheuristic algorithms for improving the performances of the different ELM models.

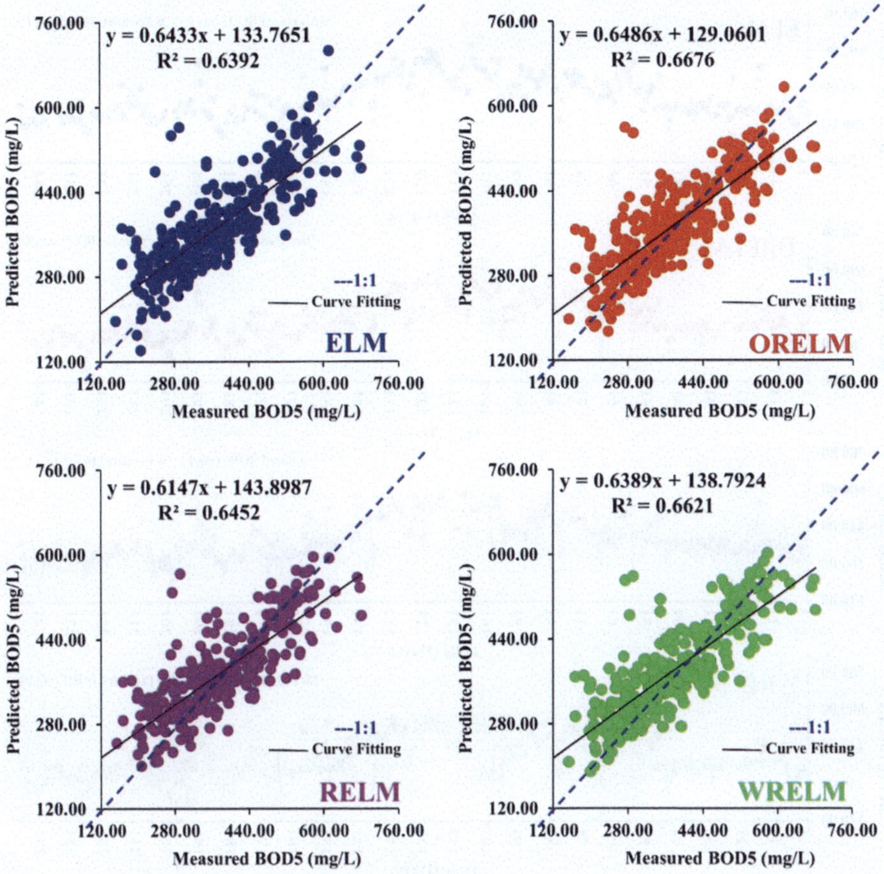

Fig. 11.4 Scatterplot of measured against predicted five days biochemical oxygen demand (BOD_5) using the best machine learning models: validation stage

Second, the possibility of including more dataset and other wastewater quality variables will certainly increase the performances of the models and the predictive accuracy. Third, the possibility of testing the modelling algorithms at different wastewater treatment process can help in providing more comprehensive about the variation and fluctuation of the water quality.

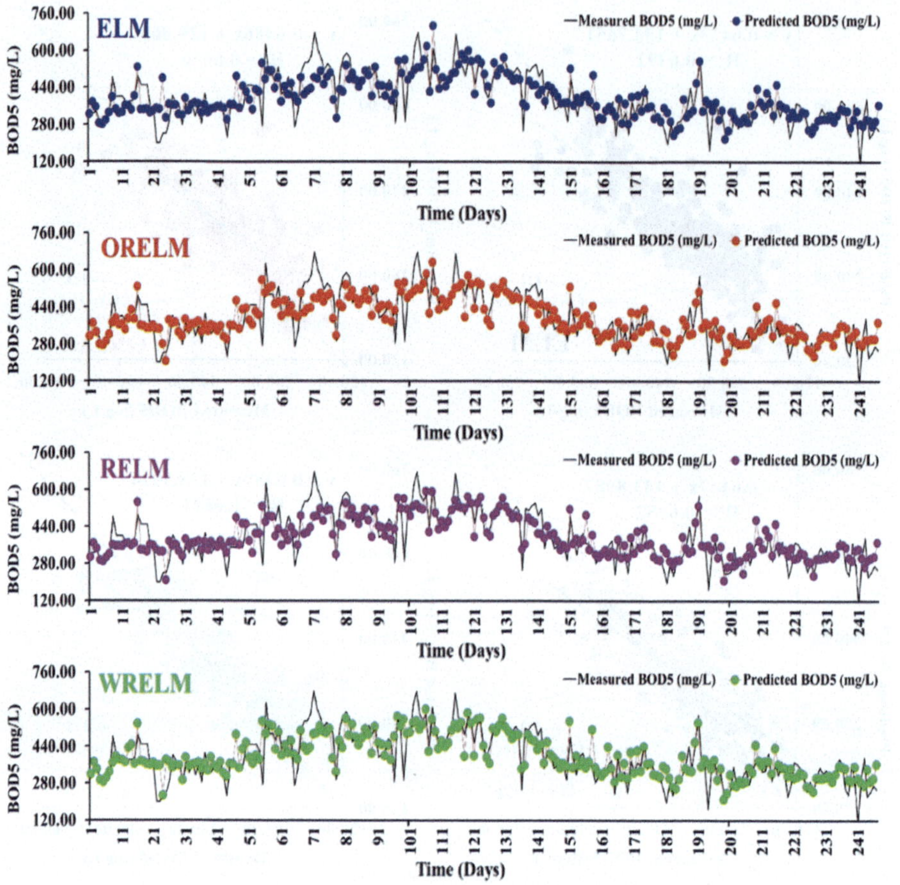

Fig. 11.5 Comparison between measured and predicted five days biochemical oxygen demand (BOD_5) using the best machine learning models: validation stage

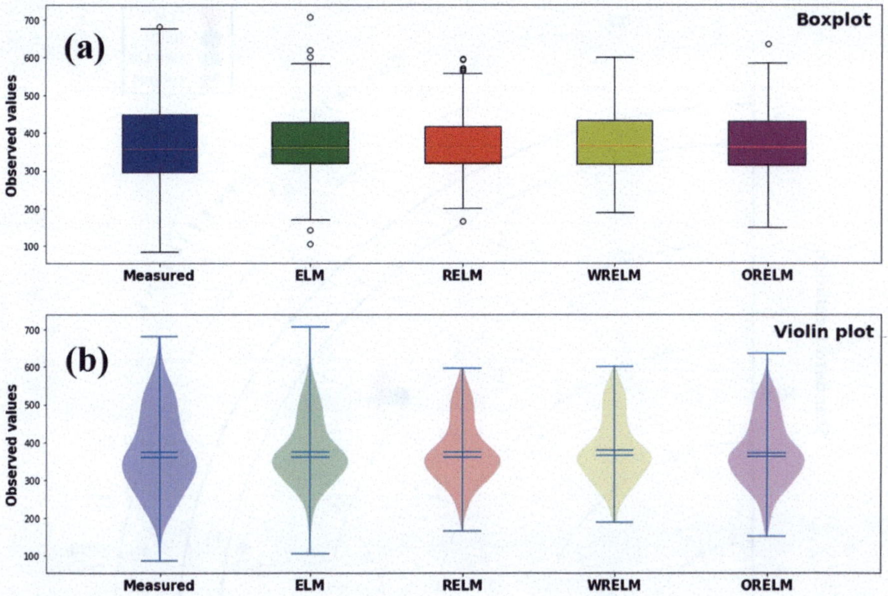

Fig. 11.6 **a** Boxplots and **b** Violin plots of the measured and predicted five days biochemical oxygen demand (BOD_5) during validation stage for all machine leaning models

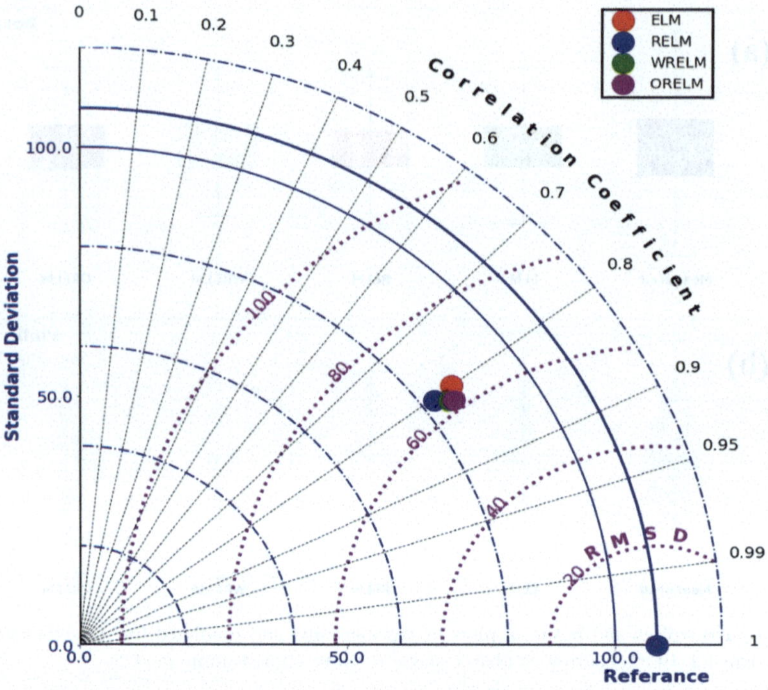

Fig. 11.7 Taylor diagram showing the performances of the best models

References

1. Asami H, Golabi M, Albaji M (2021) Simulation of the biochemical and chemical oxygen demand and total suspended solids in wastewater treatment plants: data-mining approach. J Clean Prod 296:126533. https://doi.org/10.1016/j.jclepro.2021.126533
2. Azimi Y, Talaeian M, Sarkheil H, Hashemi R, Shirdam R (2022) Developing an evolving multi-layer perceptron network by genetic algorithm to predict full-scale municipal wastewater treatment plant effluent. J Environ Chem Eng 10(5):108398. https://doi.org/10.1016/j.jece.2022.108398
3. Deng W, Zheng Q, Chen L (2009) Regularized extreme learning machine. In: 2009 IEEE symposium on computational intelligence and data mining. IEEE, pp 389–395
4. Huang, G.B., Zhu, Q.Y., Siew, C.K. (2006b). Extreme learning machine: theory and applications. Neurocomputing 70 (1-3), 489-501. https://doi.org/10.1016/j.neucom.2005.12.126
5. Huang GB, Chen L, Siew CK (2006a) Universal approximation using incremental constructive feedforward networks with random hidden nodes. IEEE Trans. Neural Network 17(4):879–892. https://doi.org/10.1109/TNN.2006.875977
6. Ismail W, Niknejad N, Bahari M, Hendradi R, Zaizi NJM, Zulkifli MZ (2023) Water treatment and artificial intelligence techniques: a systematic literature review research. Environ Sci Pollut Res 30:71794–71812. https://doi.org/10.1007/s11356-021-16471-0
7. Ismail R (2023). Improving wastewater treatment plant performance: an ANN-based predictive model for managing average daily overflow and resource allocation optimization using Tabu search. Asian J Civ Eng:1–15. https://doi.org/10.1007/s42107-023-00853-5

8. Jadhav AR, Pathak PD, Raut RY (2023) Water and wastewater quality prediction: current trends and challenges in the implementation of artificial neural network. Environ Monit Assess 195(2):321. https://doi.org/10.1007/s10661-022-10904-0
9. Liu H, Chen C (2019) Multi-objective data-ensemble wind speed forecasting model with stacked sparse autoencoder and adaptive decomposition-based error correction. Appl Energy 254:113686. https://doi.org/10.1016/j.apenergy.2019.113686
10. Liu H, Yang R (2021) A spatial multi-resolution multi-objective data-driven ensemble model for multi-step air quality index forecasting based on real-time decomposition. Comput Ind 125:103387. https://doi.org/10.1016/j.compind.2020.103387
11. Qambar AS, Al Khalidy MMM (2022) Prediction of municipal wastewater biochemical oxygen demand using machine-learning techniques: a sustainable approach. Process Saf Environ Prot 168:833–845. https://doi.org/10.1016/j.psep.2022.10.033
12. Qambar AS, Al Khalidy MMM (2023) Development of local and global wastewater biochemical oxygen demand real-time prediction models using supervised machine learning algorithms. Eng Appl Artif Intell 118:105709. https://doi.org/10.1016/j.engappai.2022.105709
13. Rizal NNM, Hayder G (2023) Forecasting effluent biochemical oxygen demand in sewage treatment plants using machine learning and user-friendly interface. Int J Environ Res 17(1):4. https://doi.org/10.1007/s41742-022-00493-8
14. Sahu P, Londhe SN, Kulkarni PS (2023) Prediction of water quality parameters using support vector regression. Inno Infrastructure Sol 8(9):229. https://doi.org/10.1007/s41062-023-011 95-6
15. Sheikh Khozani Z, Ehteram M, Mohtar WHMW, Achite M, Chau KW (2023) Convolutional neural network-multi-kernel radial basis function neural network-SALP swarm algorithm: a new machine-learning model for predicting effluent quality parameters. Environ Sci Poll Res:1–18
16. Wang K, Pei H, Cao J, Zhong P (2020) Robust regularized extreme learning machine for regression with non-convex loss function via DC program. J Franklin Inst 357(11):7069–7091. https://doi.org/10.1016/j.jfranklin.2020.05.027
17. Wei Y, Chen Z, Zhao C, Chen X, Tu Y, Zhang C (2022) Big multi-step ship motion forecasting using a novel hybrid model based on real-time decomposition, boosting algorithm and error correction framework. Ocean Eng 256:111471. https://doi.org/10.1016/j.oceaneng.2022.111471
18. Yadav P, Chandra M, Fatima N, Sarwar S, Chaudhary A, Saurabh K, Yadav BS (2023) Predicting influent and effluent quality parameters for a UASB-based wastewater treatment plant in Asia covering data variations during COVID-19: a machine learning approach. Water 15(4):710. https://doi.org/10.3390/w15040710
19. Zhang C, Hua L, Ji C, Nazir MS, Peng T (2022) An evolutionary robust solar radiation prediction model based on WT-CEEMDAN and IASO-optimized outlier robust extreme learning machine. Appl Energy 322:119518. https://doi.org/10.1016/j.apenergy.2022.119518
20. Zhang K, Luo M (2015) Outlier-robust extreme learning machine for regression problems. Neurocomputing 151:1519–1527. https://doi.org/10.1016/j.neucom.2014.09.022

Chapter 12
Novel Security Mechanism for AI Enabled Wastewater Treatment Systems

Manas Kumar Yogi and A. S. N. Chakravarthy

Abstract As wastewater treatment systems increasingly leverage artificial intelligence (AI) technologies to enhance efficiency and environmental impact, the need for robust security mechanisms becomes paramount. This chapter presents a novel security framework designed specifically for AI-enabled wastewater treatment systems. The integration of AI in these critical infrastructures brings forth unique challenges and vulnerabilities, necessitating innovative approaches to safeguard against potential threats. Our proposed security mechanism combines traditional cybersecurity measures with AI-driven anomaly detection and adaptive response mechanisms. By harnessing the power of AI, we aim to proactively identify and mitigate security risks while ensuring uninterrupted wastewater treatment operations. The framework incorporates machine learning algorithms for real-time monitoring of system behavior, anomaly detection, and predictive threat analysis. Additionally, it employs variants of bio-inspired algorithm for optimization of measures of cyber security like protection of data integrity and confidentiality, ensuring sensitive operational data remains secure. The chapter also discusses the importance of comprehensive security policies and user training to create a culture of awareness and vigilance within wastewater treatment facilities. Furthermore, it highlights the role of regulatory compliance and standards adherence in enhancing the overall security posture of AI-enabled wastewater treatment systems. Through a thorough analysis of potential threats and vulnerabilities specific to these systems, this research contributes to the evolving field of AI-enabled critical infrastructure security. The proposed security mechanism aims to strike a balance between the advancements in AI technology and the imperative to safeguard against cyber threats, ultimately ensuring the resilience and reliability of wastewater treatment systems in an increasingly connected world.

Keywords AI · Wastewater · Bio-inspired · Cyber · Threat · Security · Privacy

M. K. Yogi (✉)
CSE Department, Pragati Engineering College (A), Surampalem, India
e-mail: manas.yogi@gmail.com

A. S. N. Chakravarthy
CSE Department, JNTUK, Kakinada, Andhra Pradesh, India

M. C. Garg (ed.), *The AI Cleanse: Transforming Wastewater Treatment Through Artificial Intelligence*, Springer Water, https://doi.org/10.1007/978-3-031-67237-8_12

12.1 Introduction

In the twenty-first century, the world is witnessing a rapid urbanization trend with more than half of the global population now living in cities. This unprecedented growth of urban areas has given rise to the concept of "smart cities". Smart cities are characterized by their innovative use of technology and data to enhance the quality of life for their residents and improve the overall efficiency of urban services [1]. Amid this transformative urban evolution, wastewater treatment systems play a pivotal role. This article explores the importance of wastewater treatment systems in smart cities, highlighting their role in environmental sustainability, public health, economic development, and the realization of a truly smart urban future.

1. The Growing Challenge of Urbanization

Urbanization is an inevitable global trend, and its pace is accelerating. The United Nations predicts that by 2050, nearly 70% of the world's population will live in cities [2]. This influx of people into urban areas places immense pressure on city infrastructure, including water and wastewater management systems. As cities grow, so does the volume of wastewater produced, and this poses significant challenges if not managed efficiently.

1.1. Wastewater Generation and Its Consequences

Urban centers generate vast quantities of wastewater, a mixture of water from various sources, including residential, industrial, and commercial activities. Without proper management, untreated wastewater can have dire consequences for both human health and the environment. It can lead to the contamination of water bodies, the spread of waterborne diseases, and ecological damage.

1.2. Urban Wastewater Management Challenges

In the context of smart cities, managing wastewater becomes even more complex. Smart cities aim to optimize resource usage, enhance sustainability, and provide a high quality of life for their residents. Therefore, the efficient management of wastewater is a fundamental component of this vision, as it affects the environment, public health, and the overall liveability of the city [3].

2. Environmental Sustainability

Wastewater treatment systems play a critical role in ensuring environmental sustainability within smart cities. Several key aspects underscore their importance in this regard.

2.1. Water Resource Conservation

One of the primary environmental benefits of wastewater treatment is the conservation of water resources. In smart cities, advanced treatment technologies can recycle and purify wastewater, making it safe for reuse in various applications such as irrigation, industrial processes, and even potable water supply. This approach not only reduces the burden on freshwater sources but also helps in water resource management, particularly in regions facing water scarcity.

2.2. Pollution Mitigation

Wastewater treatment systems are essential for mitigating pollution. Untreated wastewater often contains harmful substances, including heavy metals, pathogens, and organic pollutants. When discharged into the environment, these contaminants can degrade ecosystems, harm aquatic life, and pose health risks to both humans and wildlife [4]. Proper treatment ensures that these pollutants are removed or reduced to safe levels before discharge, thus preserving the quality of natural water bodies.

2.3. Ecosystem Health

Smart cities recognize the importance of maintaining a balanced and healthy environment. Wastewater treatment contributes to ecosystem health by reducing the impacts of pollution on aquatic ecosystems. When wastewater is treated to remove harmful pollutants, the receiving water bodies become less stressed, supporting the diversity of aquatic life and maintaining a healthier balance within the ecosystem.

2.4. Energy Efficiency

Sustainability in smart cities extends to energy efficiency. Wastewater treatment plants often require substantial energy inputs to operate, but innovations in energy-efficient treatment technologies can reduce the carbon footprint associated with wastewater treatment. This aligns with the broader goals of smart cities to minimize their environmental impact and optimize resource use.

3. Public Health

Effective wastewater treatment is a cornerstone of public health in smart cities. Ensuring that the water supply remains free from contaminants is crucial for the well-being of the city's residents.

3.1. Disease Prevention

Inadequate wastewater treatment can lead to the contamination of water sources with disease-causing microorganisms. This, in turn, can result in the spread of waterborne diseases such as cholera, typhoid, and dysentery. Smart cities prioritize public health and recognize that the provision of safe water and the proper treatment of wastewater are essential for preventing these diseases.

3.2. Access to Safe Water

The treated wastewater, once recycled and purified, can contribute to the provision of safe drinking water. Smart cities leverage advanced treatment methods to ensure the quality of treated water, enabling it to meet the highest standards for potable water. This not only reduces the strain on traditional water sources but also ensures a reliable supply of clean drinking water for residents.

3.3. Healthcare Cost Reduction

By reducing the occurrence of waterborne diseases through effective wastewater treatment, smart cities can significantly reduce the burden on their healthcare systems. Fewer cases of water-related illnesses lead to decreased medical expenses, fewer hospitalizations, and improved overall public health.

3.4. Quality of Life

Incorporating proper wastewater treatment into the urban infrastructure enhances the overall quality of life for residents. Access to clean and safe water contributes to better health, sanitation, and well-being. This is a core goal of smart cities: to create environments that promote the happiness and prosperity of their citizens.

4. Economic Development

Wastewater treatment systems not only contribute to environmental sustainability and public health but also play a pivotal role in economic development within smart cities [5].

4.1. Job Creation

The construction, operation, and maintenance of wastewater treatment plants create employment opportunities, which can have a significant impact on the local economy. Skilled and unskilled labor is required for the day-to-day operation of treatment facilities, and these jobs provide a source of stable income for many city residents.

4.2. Industrial Growth

Effective wastewater treatment can attract industries to smart cities. Industries often require large quantities of water for their processes, and having a reliable and sustainable water supply, including recycled water; can make a city more attractive for investment. As a result, industrial growth is fostered, contributing to the economic vitality of the city.

4.3. Property Values

Investment in wastewater treatment infrastructure can positively impact property values. A city with efficient wastewater treatment is seen as a more desirable place to live, work, and invest. This increased desirability can lead to higher property values and, in turn, increased tax revenues for the city.

4.4. Innovation and Research

Smart cities often invest in research and development to improve their wastewater treatment systems. These innovations can lead to the creation of new technologies and the growth of local businesses focused on water and environmental technologies. Research and innovation in this field not only enhance the city's reputation but also drive economic growth.

5. The Path to a Truly Smart City

Wastewater treatment is not just an isolated component of a smart city; it is a fundamental piece of the broader vision of urban intelligence and sustainability. Here's how wastewater treatment systems contribute to the realization of a truly smart city.

5.1. Data-Driven Insights

Smart cities are characterized by their use of data and technology to optimize urban services. Wastewater treatment systems generate vast amounts of data that can be harnessed for predictive maintenance, real-time monitoring, and system optimization. These data-driven insights lead to cost savings, operational efficiency, and improved service delivery (Table 12.1).

Table 12.1 Optimization aspects in wastewater treatment systems [6]

Sl. No.	Aspect	Description
1	Sensor technology	Integration of sensors for real-time water quality and level monitoring
2	Data collection and analytics	Gathering and processing data for better system management
3	Predictive maintenance	Using AI and analytics to predict and prevent system failures
4	Energy efficiency	Optimizing energy consumption through automation and control
5	Smart pumping systems	Intelligent pump control for efficient water distribution
6	Resource recovery	Extracting energy or reusable resources from wastewater
7	Water quality prediction	AI-driven models to forecast water quality and pollution levels
8	Remote monitoring and control	Real-time control and management of treatment processes
9	Integration with IoT	Connecting wastewater systems to the Internet of Things (IoT)
10	GIS and mapping	Geographic Information Systems (GIS) for spatial analysis
11	Water reuse and recycling	Treating wastewater for reuse in various applications
12	Compliance and regulation	Ensuring adherence to environmental standards and regulations
13	Resilience and disaster management	Preparing for and responding to emergencies and natural disasters
14	Public engagement and education	Involving citizens in water conservation and awareness programs
15	Infrastructure sustainability	Building eco-friendly and sustainable treatment infrastructure
16	Cost reduction and efficiency	Lowering operational costs while maintaining or improving performance
17	Integration with smart grids	Coordinating water and energy systems for greater efficiency
18	Cybersecurity and data privacy	Protecting data and control systems from cyber threats
19	Green technologies	Implementing eco-friendly and low-impact treatment methods

5.2. Resource Optimization

Wastewater treatment in smart cities goes beyond the treatment of wastewater itself. It encompasses the optimization of resource usage, from energy to water. Smart cities integrate wastewater treatment with water recycling, energy recovery, and nutrient harvesting, creating a closed-loop system that minimizes

5.3. Smart Infrastructure

Wastewater treatment infrastructure can be equipped with sensors, automation, and remote monitoring capabilities, making it part of the broader smart infrastructure network. This connectivity allows for real-time adjustments, remote operation, and rapid response to issues, enhancing the reliability and resilience of the wastewater treatment system.

5.4. Resilience and Disaster Preparedness

Smart cities are designed to be resilient in the face of natural disasters and emergencies. Wastewater treatment systems are a critical component of disaster preparedness. Advanced treatment technologies and redundancy measures ensure that even in the event of a crisis, the city can continue to manage its wastewater, preventing environmental and public health disasters.

5.5. Community Engagement

Wastewater treatment can also be a focal point for community engagement and education. Smart cities promote awareness of the importance of wastewater treatment and involve residents in the decision-making process. This fosters a sense of shared responsibility for the city's well-being and its sustainable development.

6. Challenges and Barriers

While the importance of wastewater treatment systems in smart cities is clear, there are several challenges and barriers that must be addressed to ensure their successful implementation and operation. In Table 12.2, a list of some of the existing challenges in the design of a smart wastewater management system is presented. These challenges highlight the complexity and multidisciplinary nature of designing a smart wastewater management system, which must consider technical, regulatory, environmental, and societal factors to be successful.

7. Case Studies

Several smart cities around the world are already demonstrating the crucial role of wastewater treatment systems in building a sustainable and technologically advanced urban future. Let's look at a few examples [7–9].

7.1. Singapore—NEWater

Singapore, known for its limited freshwater resources, has invested heavily in wastewater treatment and recycling. The NEWater project treats used water to ultra-pure standards, making it safe for drinking. This initiative not only addresses the city's

Table 12.2 Inherent challenges in design of a smart waste water management system

Sl. No.	Challenge	Description
1	Data collection and integration	Efficiently gathering and integrating data from various sources, including sensors, SCADA systems, and public input
2	Real-time monitoring	Ensuring continuous, real-time monitoring of wastewater systems to detect issues promptly
3	Data quality and accuracy	Maintaining high data quality and accuracy for informed decision-making
4	Sensor calibration and maintenance	Regular calibration and maintenance of sensors to avoid drift and inaccuracies
5	Infrastructure aging and maintenance	Addressing aging wastewater infrastructure and managing necessary maintenance and upgrades
6	Energy efficiency	Optimizing energy consumption and reducing the environmental footprint of wastewater treatment
7	Cybersecurity and data privacy	Protecting data and control systems from cyber threats and ensuring data privacy
8	Big data management	Efficiently handling and processing the large volumes of data generated by smart systems
9	Interoperability and standardization	Ensuring compatibility and data exchange between different systems and technologies
10	Scalability and flexibility	Designing systems that can scale and adapt to changing city needs and growth
11	Resilience and disaster preparedness	Preparing for emergencies, natural disasters, and infrastructure failures
12	Regulatory compliance	Meeting and adhering to water quality and environmental regulations and standards
13	Public engagement and education	Engaging citizens in wastewater management and raising awareness about water conservation
14	Environmental impact	Minimizing the environmental impact of wastewater treatment and resource recovery
15	Cost efficiency and ROI	Balancing the costs of implementing smart systems with their long-term benefits
16	Reducing non-revenue water	Addressing water losses and non-revenue water to enhance water efficiency
17	Green technologies and sustainability	Implementing eco-friendly and sustainable wastewater treatment methods
18	Human resource and training	Ensuring that staff have the necessary skills to manage and operate smart wastewater systems
19	Funding and investment	Securing the necessary funding for the design, deployment, and maintenance of smart wastewater systems
20	Data visualization and decision support	Developing effective data visualization tools and decision support systems for city officials and operators

water scarcity issues but also serves as a model for sustainable water management in smart cities.

7.2. Vienna—The Danube Island Project

Vienna, Austria, has transformed a former sewage treatment plant into a recreational area known as the Danube Island. The plant uses cutting-edge technology to treat wastewater and is a prime example of how wastewater treatment can contribute to the revitalization of urban spaces, enhancing the quality of life for residents.

7.3. San Francisco—Recycled Water

San Francisco, California, is actively pursuing the use of recycled water to offset the demand for freshwater. The city's wastewater treatment facilities produce high-quality recycled water that is used for irrigation, industrial processes, and cooling systems. This initiative showcases the economic and environmental benefits of wastewater recycling in a smart city context.

Wastewater treatment systems are a linchpin of smart cities, offering numerous benefits in terms of environmental sustainability, public health, economic development, and the realization of a truly smart urban future. These systems not only address the challenges associated with rapid urbanization but also provide a pathway to creating more resilient, efficient, and liveable cities.

As the world continues to urbanize, the importance of wastewater treatment in smart cities cannot be overstated. It is not merely a utility but an essential pillar upon which the entire concept of smart cities is built. Addressing the challenges and barriers associated with wastewater treatment is crucial to unlocking the full potential of these systems and ensuring their long-term success in the urban environments of the future. Smart cities that prioritize and invest in wastewater treatment will undoubtedly reap the rewards of a healthier, more sustainable, and economically vibrant urban future.

In today's rapidly evolving technological landscape, the integration of artificial intelligence (AI) has permeated numerous sectors, and wastewater treatment systems are no exception. This convergence of cutting-edge AI technologies with critical infrastructure promises significant advancements in terms of efficiency and environmental impact. However, as we venture into this new era, where AI plays an instrumental role in the optimization of wastewater treatment processes, a critical concern looms large on the horizon: the imperative need for robust security mechanisms. The deployment of AI in wastewater treatment plants offers immense promise, presenting opportunities for cost reduction, resource optimization, and a reduced ecological footprint. Yet, this promise is tempered by an equally significant challenge—the emergence of unique security vulnerabilities and potential threats that come hand-in-hand with the incorporation of AI. The consequences of a breach or cyber-attack on wastewater treatment systems can be catastrophic, with far-reaching repercussions for the environment and public health [9].

The deployment of AI in wastewater treatment plants offers immense promise by revolutionizing the way these essential facilities operate. Here's how AI contributes to this promise:

1. Cost Reduction: AI systems can optimize and automate various aspects of wastewater treatment. They can predict equipment failures, detect inefficiencies, and dynamically adjust processes to reduce energy and chemical consumption. By doing so, AI can significantly lower operational costs. For instance, AI-driven predictive maintenance can help avoid costly breakdowns, and real-time data analysis can optimize chemical dosing, resulting in substantial savings in operational expenses.
2. Resource Optimization: AI excels in data analysis and can handle vast amounts of data from sensors, instruments, and other sources in real-time. By processing and analyzing this data, AI systems can optimize resource allocation, such as water flow, chemical dosing, and energy consumption. This leads to improved resource efficiency and ensures that resources are used exactly where and when they are needed, reducing waste and increasing overall system performance.
3. Environmental Impact: Traditional wastewater treatment processes can be resource-intensive and can have a significant ecological footprint. AI enables smarter decision-making that can lead to the reduction of harmful emissions and the overall environmental impact. By optimizing treatment processes, AI can help decrease the release of pollutants and minimize the energy footprint of wastewater treatment, thereby reducing the environmental burden.
4. Real-time Monitoring and Response: AI can provide real-time monitoring and control of wastewater treatment processes. It can quickly identify anomalies or issues, allowing for rapid response to potential problems. This proactive approach can help prevent environmental disasters and ensure that treated water meets stringent regulatory standards, further reducing the ecological impact of wastewater treatment.
5. Data-Driven Decision-Making: AI systems make decisions based on data, patterns, and predictive analytics. This data-driven approach can result in more informed, precise, and efficient decision-making within wastewater treatment plants. It can lead to better process control, which, in turn, can improve water quality and reduce the release of contaminants into natural water bodies.
6. Innovation and Research: AI can stimulate innovation in wastewater treatment. Researchers and engineers can use AI to develop and test new treatment methods, conduct simulations, and explore alternative approaches that can further reduce costs and environmental impact.

In this context, this paper explores the intersection of AI and variant of bio-inspired system for cyber threat detection in wastewater treatment systems, shedding light on the challenges and vulnerabilities that necessitate innovative security approaches. As we delve into this domain, it becomes evident that the safeguarding of these vital infrastructures against potential threats is a paramount concern, demanding our attention and innovation to ensure the sustainable and secure future of water treatment technologies.

Bio-inspired algorithms are particularly suitable for AI-based threat detection systems in wastewater treatment for several compelling reasons [10, 11]:

1. Resilience and Adaptability: Many bio-inspired algorithms are modeled after biological systems, which are known for their adaptability and resilience. Wastewater treatment systems face varying input conditions and dynamic environmental factors. Bio-inspired algorithms can adapt to these changes, ensuring that the threat detection system remains effective under different circumstances.
2. Anomaly Detection: Biological systems excel at identifying anomalies and threats. In a wastewater treatment context, detecting abnormal patterns or deviations in sensor data is essential for identifying potential cyber threats or equipment malfunctions. Bio-inspired algorithms can excel in recognizing unusual data patterns.
3. Parallel Processing: Some bio-inspired algorithms, such as genetic algorithms and swarm intelligence, can perform parallel processing, allowing for efficient analysis of vast amounts of sensor data in real-time. Wastewater treatment facilities generate large datasets that need to be processed promptly, making bio-inspired algorithms advantageous.
4. Adaptive Learning: Many bio-inspired algorithms incorporate adaptive learning, which enables AI agents to learn from past experiences. This is particularly valuable in improving threat detection accuracy over time and recognizing evolving cyber threats.
5. Robustness: Biological systems are robust and can often tolerate minor perturbations or noise. Bio-inspired algorithms can be designed to filter out irrelevant information or noise in sensor data, reducing the likelihood of false alarms.
6. Self-Organization: Some bio-inspired algorithms, like ant colony optimization and neural networks, demonstrate self-organization, allowing AI agents to work together efficiently in a distributed system. In a wastewater treatment facility, sensors and AI agents must collaborate to monitor and protect the system effectively.
7. Resource Efficiency: Bio-inspired algorithms, such as swarm intelligence, can optimize resource utilization and energy efficiency, which is crucial for the sustainable operation of wastewater treatment facilities.
8. Parallel to Natural Systems: Wastewater treatment facilities aim to mimic the natural processes of biodegradation and purification. Bio-inspired AI aligns with the principles of these natural systems, making it well-suited to enhance the overall functionality of the treatment process.
9. Integration with Biological Systems: Bio-inspired AI can be seamlessly integrated with biological processes used in wastewater treatment. This integration enhances the overall efficiency and resilience of the treatment system.
10. Interdisciplinary Approach: Threat detection in wastewater treatment is a multidisciplinary challenge. Bio-inspired AI encourages collaboration between experts in AI, environmental science, and cybersecurity to design and implement effective solutions.

The motivation for the suitability of bio-inspired algorithms in AI-based threat detection systems for wastewater treatment lies in their adaptability, efficiency, and the capacity to mimic the principles of biological systems. These characteristics make them well-suited to address the unique challenges and requirements of wastewater treatment facilities while enhancing their security and operational efficiency.

12.2 Related Work

AI researchers have made significant contributions in the field of applying AI for wastewater management systems and their revival. Their work has focused on various aspects of wastewater treatment, monitoring, and management, with the aim of improving efficiency, sustainability, and environmental impact. Here are some key areas of research and related work [12–15]:

1. Predictive Maintenance: AI researchers have developed predictive maintenance models using machine learning algorithms. These models analyse historical data and sensor readings to predict equipment failures in wastewater treatment plants. By identifying potential issues before they occur, maintenance can be scheduled more efficiently, reducing downtime and repair costs.
2. Real-time Monitoring: AI-based monitoring systems continuously collect and analyse data from various sensors within wastewater treatment facilities. Researchers have designed algorithms to detect anomalies, such as sudden spikes in pollutant levels or unusual equipment behavior. Early detection allows for rapid responses to prevent system failures and maintain the quality of treated water.
3. Process Optimization: AI plays a crucial role in optimizing treatment processes. Researchers have developed control systems that use AI algorithms to dynamically adjust water flow rates, chemical dosing, and treatment stages. This not only improves treatment efficiency but also reduces resource consumption and operational costs.
4. Energy Efficiency: AI research has led to the development of energy management systems that use AI to control pumps, aeration systems, and other energy-consuming components in wastewater treatment plants. These systems ensure that energy is used more efficiently, reducing energy costs and minimizing the carbon footprint.
5. Water Quality Prediction: AI models have been created to predict water quality in real-time. By analyzing data from multiple sources, including weather forecasts and upstream pollution sources, researchers can predict the quality of incoming wastewater. This information allows treatment plants to adjust their processes accordingly to meet regulatory standards and minimize the environmental impact.
6. Data Analytics: AI researchers have developed advanced data analytics techniques for handling the vast amounts of data generated in wastewater treatment. This includes the use of deep learning for image recognition in sewage treatment, as well as natural language processing for automated data reporting and anomaly detection.

7. Emerging Technologies: Researchers have explored the integration of emerging technologies, such as Internet of Things (IoT) sensors, Blockchain, and remote sensing, in combination with AI to enhance the resilience and efficiency of wastewater management systems. These technologies offer new opportunities for data collection, transparency, and decentralized management.
8. Optimization of Resource Recovery: Some AI research has been dedicated to optimizing the recovery of valuable resources from wastewater, such as phosphorus and nitrogen. AI-driven systems help identify the most efficient methods for resource extraction, contributing to sustainability and circular economy initiatives.
9. Policy and Decision Support: AI models have been used to support decision-making in wastewater management, helping authorities and plant operators make informed choices about infrastructure investments, treatment strategies, and environmental policies.

The work performed by AI researchers in the field of wastewater management systems has led to improved operational efficiency, cost reduction, and a reduced ecological footprint. Their contributions are vital for ensuring the sustainable management of wastewater and the protection of natural water resources. The on-going research in this area continues to drive innovation and pave the way for a more efficient and environmentally friendly approach to wastewater treatment and management.

12.3 Novel Security Framework

When designing a mathematical model for applying a bio-inspired variant of AI-based cyber threat detection for a wastewater treatment system, several assumptions must be made to simplify the modeling process and make it feasible. It's important to acknowledge these assumptions, as they can significantly impact the model's accuracy and applicability. Here are some common assumptions [16, 17]:

1. Homogeneous Sensor Data

Assumption: Sensor data is assumed to be homogenous and consistent in quality, with uniform time intervals between data points. In reality, sensor data may exhibit variations in quality, frequency, and precision.

2. Steady-State Processes

Assumption: The model may assume that the wastewater treatment processes are in a steady state, meaning they are not subject to rapid changes or disturbances. In practice, processes can exhibit dynamic behavior due to fluctuations in influent characteristics or operational changes.

3. Complete Sensor Coverage

Assumption: The model assumes that all critical aspects of the wastewater treatment system are monitored by sensors. In reality, some areas may lack sensor coverage, leaving blind spots in threat detection.

4. Known Threat Profiles

Assumption: The model assumes that all potential cyber threats and their signatures are known in advance. This assumption may not hold true, as new threats can emerge over time.

5. Uniform Cybersecurity Protocols

Assumption: The model may assume uniform cybersecurity protocols and security configurations across all components of the wastewater treatment system. In practice, there may be variations in security measures and practices.

6. Static Network Topology

Assumption: The model may consider a fixed and static network topology. In reality, network configurations may change due to maintenance, upgrades, or reconfigurations, affecting the cyber threat landscape.

7. No Insider Threats

Assumption: Insider threats from employees or contractors with access to the system are not considered. In practice, insider threats can be a significant concern.

8. No Hardware or Software Failures

Assumption: The model may disregard the possibility of hardware or software failures that could generate false alarms or mask real threats.

9. Ideal Data Transmission

Assumption: Data transmission from sensors to the AI agents is assumed to be ideal, without issues like data loss, latency, or interference. In reality, these issues can affect data quality.

10. Single Threat Detection Approach

Assumption: The model might focus on a single threat detection approach, such as bio-inspired AI, while ignoring the possibility of using multiple, complementary detection techniques.

11. No Human Error

Assumption: The model may not account for human error in system operation and threat response, assuming that all human actions are flawless.

12. Constant Attack Patterns

Assumption: The model might assume that cyber threats have constant attack patterns that do not change over time. In reality, cyber threats are constantly evolving and adapting.

It's crucial to be aware of these assumptions when designing the mathematical model, as they impact the model's realism and its ability to represent the complexities of real-world wastewater treatment systems and cybersecurity challenges [18–20]. The validity of these assumptions should be regularly evaluated and adjusted as needed to improve the model's accuracy and effectiveness in addressing cyber threats.

Designing a mathematical model for applying a bio-inspired variant of AI-based cyber threat detection for a wastewater treatment system involves formulating the underlying principles and mechanisms of the system. In this case, the bio-inspired AI approach is influenced by natural systems, such as the immune system or swarm intelligence. Below is a high-level mathematical model for this concept:

Variables and Parameters used:

1. n: Number of sensors or data sources.
2. m: Number of AI agents (inspired by biological entities, e.g., antibodies, immune cells).
3. t: Time index.

Constants used:

1. α: A parameter controlling the response of AI agents to threats.
2. β: A parameter determining the interaction strength among AI agents.
3. γ: A parameter governing the influence of environmental factors (e.g., temperature, humidity) on the system.
4. θ: A threshold for identifying anomalies or cyber threats.

State Variables:

1. X(t): A vector representing the state of the system at time t, including sensor readings and the health status of AI agents.

AI Agent Characteristics

1. Antigen Recognition Rate (ARR): A parameter representing how quickly AI agents recognize threats.
2. Response Strength (RS): A measure of how vigorously AI agents respond to detected threats.
3. Memory Capacity (MC): The number of past threats an AI agent can remember.

The characteristics of AI agents, namely Antigen Recognition Rate (ARR), Response Strength (RS), and Memory Capacity (MC), play a crucial role in their effectiveness in recognizing and responding to threats in various systems. Let's delve into these characteristics and their implications:

1. Antigen Recognition Rate (ARR):
   (a) Definition: ARR represents how quickly AI agents recognize threats. It measures the speed at which the agent processes incoming data to identify potential threats.

(b) Implications:
(c) High ARR: AI agents with a high ARR can quickly identify threats as they emerge, reducing response time and potentially minimizing the impact of threats.
(d) Low ARR: AI agents with a low ARR may take longer to recognize threats, which can lead to delayed response and potential risks.

2. Response Strength (RS):

(a) Definition: RS is a measure of how vigorously AI agents respond to detected threats. It reflects the intensity of the agent's actions when a threat is identified.
(b) Implications:
(c) High RS: AI agents with a high RS respond aggressively and proactively to threats, which can be effective in neutralizing threats but may also increase the risk of false positives and resource consumption.
(d) Low RS: AI agents with a low RS respond more conservatively, which may reduce the risk of false alarms but could also result in slower or insufficient threat mitigation.

3. Memory Capacity (MC):

(a) Definition: MC refers to the number of past threats an AI agent can remember. It determines the agent's ability to maintain a history of detected threats for reference.
(b) Implications:
(c) High MC: AI agents with a high MC can store and recall a significant number of past threats. This can be beneficial for pattern recognition, trend analysis, and forensic investigations.
(d) Low MC: AI agents with a low MC may only retain a limited history of threats, potentially limiting their ability to recognize recurring threats or assess long-term trends.

These characteristics are critical in tailoring AI agents for specific applications. The optimal values for ARR, RS, and MC will depend on the context, the nature of the threats, and the desired trade-offs between speed, accuracy, and resource utilization. Balancing these characteristics is a complex task that requires careful consideration to meet the requirements of the system or environment in which the AI agents operate. Additionally, these characteristics can be fine-tuned and adjusted over time to adapt to evolving threat landscapes and operational needs.

Mathematical Model

The mathematical model can be described using a system of differential equations:

1. Sensor Data Evolution:

$$\mathrm{dtdX(t)} = \mathrm{f(X(t))} + \alpha \ \Sigma \mathrm{imRSi(t)} \tag{12.1}$$

where f(X(t)) represents the natural evolution of the sensor data, and the second term represents the influence of AI agents' responses to detected threats.

The equation provided represents the evolution of sensor data in a system over time, taking into account both the natural evolution of the data and the influence of AI agents' responses to detected threats. Let's break down the components of the equation:

The first term ('f(X(t))') accounts for the inherent dynamics of the data, while the second term ('α·ΣimRSi(t)') reflects the impact of AI agents' responses to threats. The constant 'α' determines how much these responses influence the rate of change in the data. The cumulative effect of responses from multiple AI agents is considered by summing their individual response strengths ('RSi(t)').

This equation can be used to model and analyze the behavior of sensor data in a system where AI agents are actively responding to detected threats, potentially influencing the data's evolution. It provides a mathematical foundation for studying the interplay between natural data dynamics and AI-driven responses in threat detection and mitigation scenarios.

2. AI Agent Health Status:

$$\mathrm{dtdRSi(t)} = \beta\, \mathrm{ARRi(t)} \cdot \mathrm{X(t)} - \gamma\, \mathrm{RSi(t)} \tag{12.2}$$

This equation describes how the response strength of each AI agent changes over time. It increases when the AI agent detects a threat (based on antigen recognition rate) and decreases gradually if there are no threats.

The equation provided describes the evolution of the response strength (RS) of each AI agent over time. This equation takes into account both the influence of detecting threats and the natural decay or decrease in response strength when no threats are present. Let's break down the components of the equation:

This equation describes how the response strength of each AI agent changes over time. It increases when the AI agent detects a threat, as indicated by the ARR and the state of the sensor data. It decreases gradually when there are no threats, as determined by the decay constant 'γ'. The balance between detection and decay influences how quickly and strongly the AI agent responds to evolving conditions in the system.

This equation can be used to model and analyse the dynamic behavior of AI agents in a threat detection system, providing insights into how their response strengths evolve in response to changing conditions and the detection of threats.

3. Antigen Recognition Rate (ARR) Dynamics:

$$\mathrm{ARRi(t)} = \sum \mathrm{jnXj(t - MCi)} \tag{12.3}$$

The antigen recognition rate of an AI agent is based on the cumulative effect of past sensor data, where MC_i is the memory capacity of AI agent i.

The equation provided describes how the Antigen Recognition Rate (ARR) of an AI agent evolves over time. The ARR is calculated based on the cumulative effect of past sensor data, with consideration for the memory capacity of the AI agent. Let's break down the components of the equation:

In above Eq. 12.3 describes how the antigen recognition rate of an AI agent is calculated based on the cumulative effect of past sensor data. The agent's sensitivity to recognizing threats is influenced by the historical sensor data, and the memory capacity ('MCi') defines the time window over which past data is considered. This equation is essential for understanding how the AI agent's sensitivity to threats is shaped by its historical data. It allows the agent to adapt and learn from past experiences, helping it become more effective in recognizing threats or patterns of concern over time. The larger the memory capacity, the further back in time the agent considers, potentially capturing longer-term trends or recurring patterns in the data [21].

Cyber Threat Detection

Cyber threats are detected when the response strength of AI agents surpasses a certain threshold (θ):

If $RSi(t) > \theta$, then a threat is detected.

System Initialization:

- Initialize the sensor data X(t) and AI agents' health status RS_i(t) and ARR_i(t) at $t = 0$.

The mathematical model described is a comprehensive representation of a bio-inspired AI system for cyber threat detection in a dynamic environment, where AI agents emulate characteristics found in biological entities like antibodies or immune cells. This model integrates sensor data evolution, AI agent health, and the detection of cyber threats while considering both the dynamic nature of data and the adaptive behavior of AI agents. Below concepts are developed for a better understanding of the proposed method:

1. Sensor Data Evolution (dtdX(t)): The model accounts for the evolution of sensor data over time, influenced by both the natural dynamics of the data and the responses of AI agents to detected threats. This reflects the real-world scenario where sensor data is continually changing and can be influenced by AI-driven actions.
2. AI Agent Health Status (dtdRSi(t)): The model captures how the response strength (RS) of each AI agent changes over time. The response strength increases when a threat is detected (based on the Antigen Recognition Rate and the current state of the sensor data) and decreases gradually when no threats are present. This dynamic response allows AI agents to adapt to the evolving threat landscape.
3. Antigen Recognition Rate (ARR) Dynamics: The model considers how the antigen recognition rate of an AI agent is influenced by the cumulative effect of past sensor data. The memory capacity (MC) parameter determines how far

back in time the agent considers when calculating the ARR. This adaptive recognition rate enables AI agents to learn from historical data and become more effective over time.
4. Cyber Threat Detection: The model defines the criteria for detecting cyber threats based on the response strength of AI agents surpassing a certain threshold (θ). This mechanism ensures that threats are identified when the AI agents respond strongly to anomalies in the data, aligning with their role as threat detectors.
5. System Initialization: The model begins with the initialization of sensor data, AI agents' health status, and ARR at a starting point ($t = 0$). This initialization is a critical step to set the initial conditions for the simulation.
6. Environmental Factors (γ): The equation for AI agent health status incorporates an environmental factor (γ) that influences the rate at which the response strength decreases when no threats are detected. This environmental factor can model the impact of changing conditions or external factors on AI agent behavior.

Overall, this mathematical model provides a framework for simulating a bio-inspired AI system for cyber threat detection in a dynamic and evolving environment. It captures the adaptation, learning, and responsiveness of AI agents, similar to how the human immune system adapts to new threats over time [22, 23]. The model can be used for further analysis, simulation, and optimization of AI-based threat detection systems in various real-world applications.

As observable, the mathematical model described provides a solid foundation for simulating and potentially implementing a bio-inspired AI-based cyber threat detection system for wastewater treatment, drawing inspiration from the self-defense mechanisms found in nature. Here are some key advantages and potential applications of such a system in the context of wastewater treatment [24–27]:

1. Early Threat Detection: The bio-inspired AI system can effectively detect cyber threats in real-time or near real-time, enabling quick response and mitigation. This is crucial in safeguarding the critical infrastructure of wastewater treatment facilities.
2. Adaptation to Changing Conditions: The model's adaptive nature allows it to evolve with changing conditions and emerging threats. Wastewater treatment systems are subject to fluctuations, and an adaptable threat detection system can respond effectively to new challenges.
3. Reduced False Positives: By considering the historical context through the Antigen Recognition Rate (ARR) dynamics, the system can reduce false alarms and improve the accuracy of threat detection, reducing unnecessary operational disruptions.
4. Optimized Resource Allocation: The system can optimize resource allocation for threat response by varying the Response Strength (RS) based on the perceived threat level. This can result in more efficient resource utilization.
5. Customization and Integration: The model can be further customized to fit the specific requirements of a wastewater treatment facility. It can be integrated with the facility's existing cybersecurity infrastructure, enhancing overall security.

6. Continuous Learning: The model's ability to learn from past data enables it to become more effective over time. It can adapt to the unique characteristics and challenges of a particular facility.
7. Regulatory Compliance: The system can help wastewater treatment facilities meet regulatory requirements related to cybersecurity and data protection, ensuring that they operate in accordance with industry standards and government regulations.
8. Resilience and Sustainability: Implementing a robust threat detection system is essential for the resilience and sustainability of wastewater treatment facilities, which are critical for public health and environmental protection.

To implement this model, practical considerations would include data integration, sensor deployment, AI algorithm selection, and real-time monitoring. Collaboration with experts in AI, cybersecurity, and wastewater treatment engineering would be vital to ensure the successful application of this bio-inspired AI system in wastewater treatment facilities. It holds the potential to enhance the security and reliability of these essential infrastructures while leveraging the principles of nature's self-defense mechanisms.

12.4 Experimental Results

For conducting experiment for the proposed approach we have used full scale waste water treatment plant data from kaggle [28]. It includes data related to six years, i.e., from 2014 to 2019 of energy consumption, climate, and wastewater characteristics of Melbourne eastern wastewater treatment plant. It has 19 attributes and 1382 records in the dataset. The various attributes used in the dataset are Average Outflow, Average Inflow, Energy Consumption, Ammonia, Biological Oxygen Demand, Chemical Oxygen Demand, Total Nitrogen, Average Temperature, Maximum temperature, Minimum temperature, Atmospheric pressure, Average humidity, Total rainfall, Average visibility, Average wind speed, Maximum wind speed, Date. The proposed method was implemented in python 3.8 language in jupyter notebook and the code for generating the various plots were also implemented in python 3.8 language.

From Fig. 12.1 it can be observed that the convergence rate of the proposed variant of bio-inspired mechanism is directly proportional to the rate of accuracy. From the plot, it can be easily represented that the accuracy of cyber threat detection is nearly 94% with a convergence rate of about 50%.It should be discussed that a rate of convergence of 50% is difficult to achieve with limited number of runs and iterations in each run. Nevertheless this inherent challenge is handled by the complex and high powered GPUs deployed to implement the proposed algorithm

Figure 12.2 shows the graceful upgradation of the data utility rate wrt the increase in convergence rate. It can be observed that in a real time implementation and application of the proposed method there will be quite a trade-off between data utility and degree of security provided by the system.

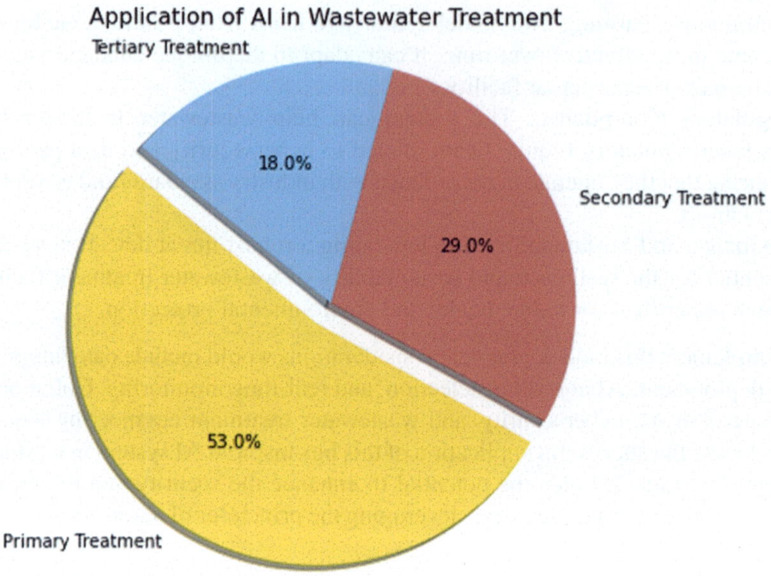

Fig. 12.1 Application of AI in broad spectrum of waste water treatment systems

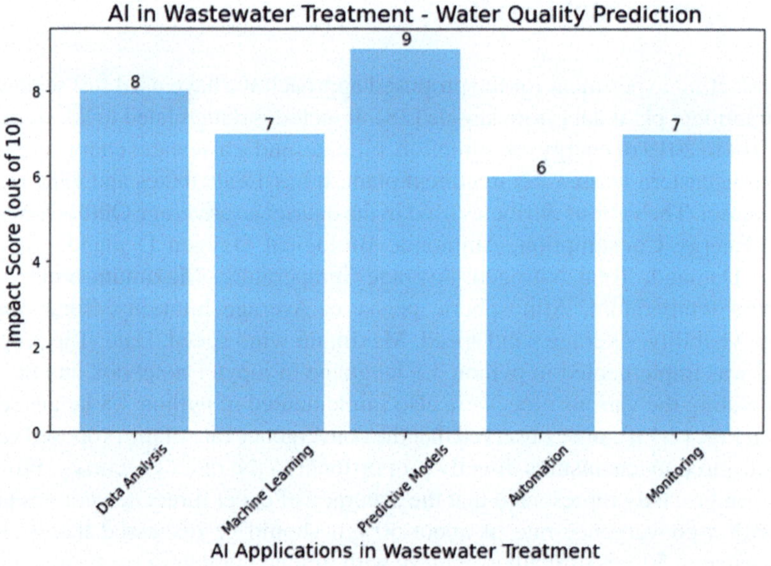

Fig. 12.2 Impact of AI Applications in Wastewater Treatment [12–15]

Figure 12.3 summarizes the performance of proposed method with current popular methods of cyber threat detection in waste water treatment systems. The proposed method outperforms heavily when compared to methods of firewalls, signature based detection and packet analysis methods. Only the honeypot methods give a tough competition to the proposed method.

It can be easily inferred from Fig. 12.4 that the number of iterations is directly proportional to the rate of antigen recognition as it increases with number of iterations.

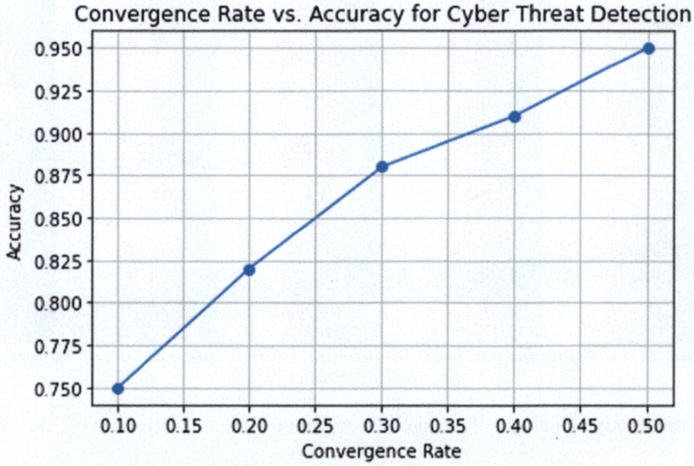

Fig. 12.3 Convergence rate of the algorithm Vs. Accuracy

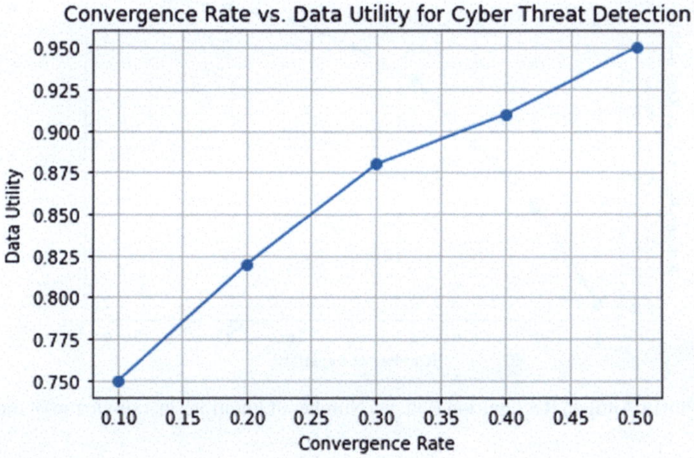

Fig. 12.4 Convergence Rate versus Data Utility

From Fig. 12.5 it can be observed that the response strength of the proposed mechanism increases as the number of AI agents involved in the process of cyber threat detection increases (Fig. 12.6).

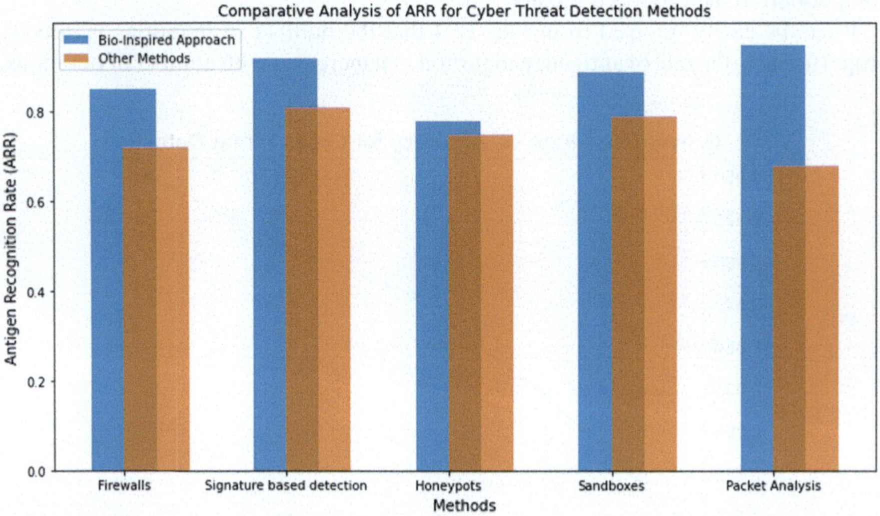

Fig. 12.5 Comparative Analysis of Antigen Recognition Rate for various cyber threat detection mechanisms

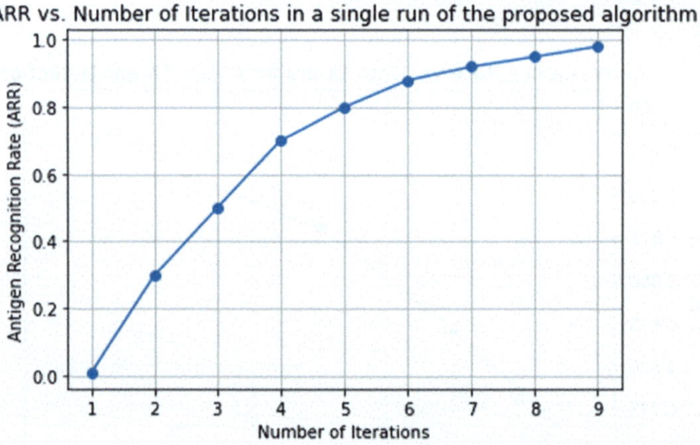

Fig. 12.6 Plot of Antigen Recognition Rate vs. Number of iterations in a single run of the proposed method

12.5 Future Directions

The application of bio-inspired AI for threat detection in wastewater treatment systems holds great potential for improving the efficiency, sustainability, and resilience of these critical infrastructure components. However, several research challenges need to be addressed to fully realize this potential. Here are some of the future research challenges and how they can be handled [29, 30]:

1. Data Collection and Integration:
   (a) Challenge: Wastewater treatment systems generate vast amounts of data from various sensors and sources. Integrating and managing this data can be a significant challenge.
   (b) Solution: Researchers can develop data integration frameworks and data preprocessing techniques to standardize data formats, enhance data quality, and enable real-time data analysis.
2. Complexity of Biological Systems:
   (a) Challenge: Modeling the intricate biological processes in wastewater treatment systems accurately is complex and computationally demanding.
   (b) Solution: AI researchers can work on developing advanced models that combine multiple bio-inspired algorithms and machine learning techniques to simulate and optimize the complex interactions in these systems.
3. Adaptation to Dynamic Conditions:
   (a) Challenge: Wastewater treatment systems are subject to varying input conditions and environmental factors, making it difficult to maintain a stable AI-based threat detection system.
   (b) Solution: Researchers can focus on developing adaptive AI algorithms that can continuously learn and adjust their models to changing conditions, similar to how biological systems adapt.
4. Interpretable AI:
   (a) Challenge: Making AI-driven threat detection systems interpretable is crucial for ensuring trust and regulatory compliance.
   (b) Solution: Research can focus on developing explainable AI (XAI) techniques that provide clear explanations of AI decision-making, which will be essential for operators and regulators to understand and trust the system.
5. Real-time Monitoring and Control:
   (a) Challenge: Real-time decision-making is essential for mitigating threats in wastewater treatment. Achieving low-latency AI models is a significant challenge.

   (b) Solution: Researchers can explore edge computing solutions, distributed AI, and the use of neuromorphic hardware to enable real-time decision-making at the edge of the wastewater treatment system.

6. Scalability and Generalization:
   (a) Challenge: AI models developed for a specific wastewater treatment plant may not generalize well to others. Achieving scalability and generalization is important.
   (b) Solution: Researchers can work on transfer learning and domain adaptation techniques to make AI models more easily transferable and adaptable to different treatment facilities.

7. Robustness and Security:
   (a) Challenge: Ensuring the security of AI-based threat detection systems and their resilience against adversarial attacks is crucial.
   (b) Solution: Research can focus on developing AI systems with robustness and security in mind, including methods for detecting and mitigating attacks and vulnerabilities.

8. Regulatory Compliance:
   (a) Challenge: Adhering to existing regulations and standards, and ensuring that AI systems meet regulatory requirements, can be complex and time-consuming.
   (b) Solution: Collaboration between AI researchers, environmental scientists, and regulatory bodies is essential to define and implement guidelines and standards for AI-based threat detection in wastewater treatment systems.

9. Cost and Resource Constraints:
   (a) Challenge: Many wastewater treatment facilities have limited budgets and resources, which can constrain the adoption of advanced AI technologies.
   (b) Solution: Research can focus on cost-effective AI solutions, leveraging open-source tools, and developing strategies for technology transfer to facilities with limited resources.

10. Ethical Considerations:
   (a) Challenge: As with any AI application, ethical concerns related to data privacy, bias, and transparency need to be addressed in the context of wastewater treatment systems.
   (b) Solution: Research can emphasize the development of ethical AI frameworks, guidelines, and policies that consider these concerns in the design and implementation of bio-inspired AI-based threat detection systems.

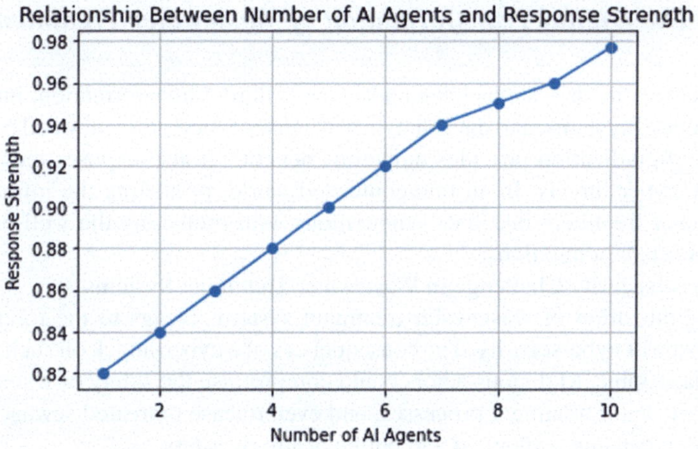

Fig. 12.7 Plot of Number of AI agents Vs. Response Strength

By addressing these research challenges will be crucial for the successful application of bio-inspired AI in wastewater treatment systems. Interdisciplinary collaboration between AI experts, environmental scientists, and industry professionals is essential to overcome these challenges and advance the field (Fig. 12.7).

12.6 Contributions of Proposed Research

The convergence of technology and urban infrastructure has given rise to the concept of smart cities, where innovative solutions are deployed to improve the quality of life and enhance resource management. Wastewater treatment systems play a pivotal role in the smart city landscape, ensuring the sustainable management of water resources and safeguarding public health. However, as these systems become increasingly digitized, they are exposed to new threats, particularly in the realm of cyber-attacks. To address this emerging challenge, the integration of hybrid AI-based bio-inspired methods for cyber threat detection is of paramount importance. This article explores the significance of these advanced techniques in protecting the critical infrastructure of wastewater treatment systems and ensuring the resilience of smart cities [31].

The Growing Significance of Smart Cities and Wastewater Management [32]

Smart cities, driven by technological innovation, are committed to creating sustainable, efficient, and secure urban environments. Wastewater treatment systems, a cornerstone of urban infrastructure, play a crucial role in achieving these goals. These systems not only ensure the environmentally responsible disposal of wastewater but also harness the potential to recycle and reuse water resources. As smart cities continue to evolve, wastewater treatment systems must be at the forefront of technological advancements.

12.6.1 *The Digital Transformation of Wastewater Treatment*

Wastewater treatment systems have undergone a digital transformation, integrating sensors, control systems, and data analytics to enhance their efficiency and reliability. While this digitalization provides numerous benefits, it also exposes these critical systems to cyber threats. In an interconnected world, protecting the infrastructure of wastewater treatment becomes synonymous with protecting the well-being and security of urban populations.

The Cybersecurity Challenge in Wastewater Treatment Systems

The digitalization of wastewater treatment systems brings to the forefront the pressing issue of cybersecurity. The consequences of a cyber-attack on such a system can be catastrophic. Malicious actors could compromise the safety of water quality, disrupt wastewater treatment processes, and even release untreated sewage into the environment, causing ecological and public health disasters.

12.6.2 *Emerging Threat Landscape*

The threat landscape in the realm of wastewater treatment is evolving rapidly. Threat actors are increasingly targeting critical infrastructure, seeking to exploit vulnerabilities for financial gain, political motives, or simply to create chaos. Attacks such as ransomware, distributed denial of service (DDoS), and insider threats can have devastating consequences if not proactively addressed.

12.6.3 *Complexity and Connectivity*

Wastewater treatment systems are complex networks of sensors, control systems, and data centers that communicate with each other and external entities. This interconnectedness offers efficiency and real-time monitoring but also creates multiple entry points for potential cyber-attacks. Securing these systems requires a holistic approach that accounts for the full spectrum of vulnerabilities.

12.6.4 *The Role of AI-Based Bio-Inspired Methods*

In addressing the cybersecurity challenge in wastewater treatment systems, the integration of hybrid AI-based bio-inspired methods stands out as a powerful solution. These methods leverage the principles of artificial intelligence, biomimicry, and advanced data analytics to detect and mitigate cyber threats effectively

12.6.5 Machine Learning and Anomaly Detection

Machine learning, a subset of AI, is a key component of cyber threat detection. Machine learning models can be trained on historical data to understand the normal behavior of a wastewater treatment system. Any deviation from this norm, which may indicate a cyber-threat, can be flagged as an anomaly. This proactive approach enables the system to detect attacks in real-time, minimizing potential damage.

12.6.6 Biomimicry and Biological Models

Bio-inspired methods draw inspiration from biological systems, which have evolved to adapt and respond to changing environmental conditions. Such methods are capable of enhancing the robustness and adaptability of cyber threat detection systems. For example, algorithms inspired by the immune system can identify and neutralize threats in a manner similar to how the human body responds to pathogens.

12.6.7 Advanced Data Analytics

The volume of data generated by wastewater treatment systems is immense. Advanced data analytics, driven by AI, can process this data in real-time to identify patterns and trends associated with cyber threats. This enables the system to predict and prevent attacks before they can disrupt the operation of the treatment plant.

12.6.8 The Crucial Role of Hybrid Approaches

While individual AI-based and bio-inspired methods offer substantial benefits, a hybrid approach that combines their strengths is especially effective. In the context of cybersecurity for wastewater treatment systems, hybrid methods offer several key advantages.

12.6.8.1 Enhanced Detection Accuracy

Hybrid methods leverage the strengths of both AI-based and bio-inspired approaches, leading to more accurate threat detection. By combining the learning capabilities of AI with the adaptive and self-healing properties of bio-inspired systems, hybrid approaches can identify even highly sophisticated and previously unseen cyber threats.

12.6.8.2 Increased Robustness

Wastewater treatment systems are exposed to various environmental and operational conditions that can challenge their cybersecurity measures. Hybrid AI-based bio-inspired methods enhance the robustness of the system, adapting to changing circumstances and providing a higher degree of protection against evolving threats.

12.6.8.3 Adaptation to New Threats

The cybersecurity landscape is dynamic, with threat actors constantly developing new attack methods. Hybrid approaches are well-suited to adapt to emerging threats by drawing on their flexible and learning-oriented components. This adaptability ensures that the wastewater treatment system remains resilient in the face of evolving risks.

12.7 Conclusion

In the age of smart cities, where technology and urban infrastructure converge to create more efficient, sustainable, and secure environments, wastewater treatment systems hold a pivotal role. The digital transformation of these systems brings unprecedented benefits but also exposes them to the ever-evolving threat landscape of cyber-attacks. To safeguard the critical infrastructure of wastewater treatment and ensure the resilience of smart cities, the integration of hybrid AI-based bio-inspired methods for cyber threat detection is of paramount importance. Hybrid approaches that leverage the strengths of artificial intelligence and bio-inspired methods offer enhanced detection accuracy, increased robustness, and the ability to adapt to new threats. These methods are instrumental in protecting the well-being of urban populations, the environment, and the economic stability of smart cities. As we move forward in our journey towards more intelligent and sustainable urban environments, it is clear that the role of hybrid AI-based bio-inspired methods for cyber threat detection in wastewater treatment systems will continue to grow in importance. Their integration is not just a cybersecurity measure; it is a fundamental step towards securing the future of our smart cities and ensuring the resilience of the critical infrastructure that sustains them. The intersection of artificial intelligence (AI) with a variant of bio-inspired approaches presents a novel and highly promising solution for bolstering the cybersecurity of wastewater treatment systems. This synergy not only enhances the robustness of these critical systems but also underscores the dynamic nature of innovation in the realm of cybersecurity and smart infrastructure. As we stand on the precipice of an increasingly digital and interconnected urban future, the security of our critical infrastructures takes center stage. The protection of wastewater treatment systems through the collaborative intelligence of AI and bio-inspired approaches is a testament to the ever-evolving nature of technology and innovation. It exemplifies the relentless pursuit of a safer, more resilient, and truly smart city landscape. With the fusion of technology and nature, the horizon for the protection of urban infrastructure has been broadened, offering a path to a more secure and sustainable future for smart cities and their residents.

References

1. Zhao L et al (2020) Application of artificial intelligence to wastewater treatment: a bibliometric analysis and systematic review of technology, economy, management, and wastewater reuse. Process Safety Environ Prot 133:169–182
2. Malviya A, Jaspal D (2021) Artificial intelligence as an upcoming technology in wastewater treatment: a comprehensive review. Environ Technol Rev 10(1):177–187
3. Krovvidy S et al (1991) An AI approach for wastewater treatment systems. Appl Intell 1:247–261
4. Altowayti, WAH et al (2022) The role of conventional methods and artificial intelligence in the wastewater treatment: a comprehensive review. Processes 10(9):1832
5. Nourani, V, Gozen E, Abba SI (2018) Wastewater treatment plant performance analysis using artificial intelligence–an ensemble approach. Water Sci Technol 78(10):2064–2076
6. Wen, CH, Vassiliadis CA (1998) Applying hybrid artificial intelligence techniques in wastewater treatment. Eng Appl Artif Intell 11(6):685–705
7. Nourani V, Asghari P, Sharghi E (2021) Artificial intelligence based ensemble modeling of wastewater treatment plant using jittered data. J Clean Prod 291:125772
8. Wang, Y et al (2023) A review on applications of artificial intelligence in wastewater treatment. Sustainability 15(18):13557
9. Zhang, S et al (2023) Artificial intelligence in wastewater treatment: a data-driven analysis of status and trends. Chemosphere, p 139163
10. Aghdam, E et al (2023) Predicting quality parameters of wastewater treatment plants using artificial intelligence techniques. J Cleaner Prod 405:137019
11. Sahu, S et al (2023) Harnessing the potential of microalgae-bacteria interaction for eco-friendly wastewater treatment: a review on new strategies involving machine learning and artificial intelligence. J Environ Manag 346:119004
12. Kshirsagar, PR et al (2022) IOT based smart wastewater treatment model for industry 4.0 using artificial intelligence. Sci Program 2022
13. Deepnarain, N et al (2020) Artificial intelligence and multivariate statistics for comprehensive assessment of filamentous bacteria in wastewater treatment plants experiencing sludge bulking. Environ Technol Innovation 19:100853
14. Alvi, M et al (2023) Deep learning in wastewater treatment: a critical review. Water Res, p 120518
15. Pisa, I et al (2018) A recurrent neural network for wastewater treatment plant effuents' prediction. Actas de las XXXIX Jornadas de Automática, Badajoz, 5–7 de Septiembre de 2018
16. Guo, Z et al (2020) Data-driven prediction and control of wastewater treatment process through the combination of convolutional neural network and recurrent neural network. RSC Adv 10(23):13410–13419
17. Baruch, IS et al (2005) Adaptive recurrent neural network control of biological wastewater treatment. Int J Intell Syst 20(2):173–193
18. Wongburi P, Park JK (2022) Prediction of sludge volume index in a wastewater treatment plant using recurrent neural network. Sustainability 14(10):6276
19. Qiao, J, Xiaoqi H, Honggui H (2012) Recurrent neural network-based control for wastewater treatment process. In: Advances in neural networks–ISNN 2012: 9th International symposium

on neural networks. Shenyang, China, July 11–14, 2012. Proceedings, Part II 9. Springer Berlin Heidelberg
20. Wongburi P, Park JK (2023) Prediction of wastewater treatment plant effluent water quality using recurrent neural network (RNN) models. Water 15(19):3325
21. Alvi, M et al (2022) Cost effective soft sensing for wastewater treatment facilities. IEEE Access 10:55694–55708
22. Qiao, J et al (2017) Wastewater treatment control method based on a rule adaptive recurrent fuzzy neural network. Int J Intell Comput Cybern 10(2):94–110
23. Alharbi M, Hong P-Y, Laleg-Kirati T-M (2022) Sliding window neural network based sensing of bacteria in wastewater treatment plants. J Process Control 110:35–44
24. Fang, X et al (2022) LSTM-based modelling for coagulant dosage prediction in wastewater treatment plant. In: Proceedings of the 2022 3rd international conference on artificial intelligence in electronics engineering
25. Zhang, D et al (2018) Manage sewer in-line storage control using hydraulic model and recurrent neural network. Water Res Manage 32:2079–2098
26. Mihály, NB, Vasile MC (2022) Optimization of the wastewater treatment plant aeration using artificial neural networks models. Comput Aided Chem Eng 51. Elsevier, 2022. 1375–1380
27. Heo, S et al (2020) Data-driven hybrid model for forecasting wastewater influent loads based on multimodal and ensemble deep learning. In: IEEE Transactions on industrial informatics 17(10):6925–6934
28. Xu, B et al (2023) A novel long short-term memory artificial neural network (LSTM)-based soft-sensor to monitor and forecast wastewater treatment performance. J Water Process Eng 54:104041
29. Chang P, Li ZeYu (2021) Over-complete deep recurrent neutral network based on wastewater treatment process soft sensor application. Appl Soft Comput 105:107227
30. Dairi, A et al (2019) Deep learning approach for sustainable WWTP operation: a case study on data-driven influent conditions monitoring. Sustain Cities Soc 50:101670
31. Djeddou, M et al. Sludge age predictive modeling in full-scale wastewater treatment plant using recurrent neural network
32. Chang, P, Shirao Z, Zichen W (2023) Soft sensor of the key effluent index in the municipal wastewater treatment process based on transformer. IEEE Trans Ind Inf

Chapter 13
Ethical Considerations in AI-Driven Wastewater Treatment

Seema Singh, Ramanpreet Kaur, and Nitin Kumar Agrawal

Abstract The escalating urbanization and industrialization worldwide have escalated the production of wastewater, necessitating efficient treatment methods to safeguard both human health and the environment. Traditional approaches to wastewater treatment are increasingly being complemented or replaced by innovative technologies driven by artificial intelligence (AI). This chapter explores the intricate intersection of AI and wastewater treatment, focusing not only on technical advancements but also on the ethical considerations inherent in this transformative journey. Wastewater treatment, critical for environmental management, involves removing pollutants before water is reintroduced into natural ecosystems. The integration of AI promises enhanced efficiency, precision, and adaptability in these processes, from predictive analytics optimizing plant operations to machine learning identifying emerging contaminants. The chapter navigates through ethical dimensions such as transparency, accountability, privacy, bias, inclusivity, and environmental impact assessment. It provides a comprehensive guide for stakeholders on responsible adoption of AI in wastewater treatment, fostering sustainability and efficiency while addressing societal and environmental concerns.

Keywords Artificial Intelligence (AI) · Wastewater treatment · Environmental management · Ethical considerations · Sustainability

---

S. Singh (✉)
Centre of Research Impact and Outcome, Chitkara University, Rajpura 140417, Punjab, India
e-mail: seema.1413@chitkara.edu.in

R. Kaur
Laboratory for Open Systems and Networks, Jozef Stefan Institute, 1000 Ljubljana, Slovenia
e-mail: ramanpreet.kaur@ijs.si

N. K. Agrawal
Department of Applied Sciences and Humanities, Moradabad Institute of Technology, Moradabad 244001, India

M. C. Garg (ed.), *The AI Cleanse: Transforming Wastewater Treatment Through Artificial Intelligence*, Springer Water, https://doi.org/10.1007/978-3-031-67237-8_13

13.1 Introduction

The exponential growth of urbanization and industrialization has led to an unprecedented increase in the generation of wastewater. As societies strive for economic development, the necessity of efficient wastewater treatment methods becomes paramount to safeguard human health and protect the environment. The traditional approaches to wastewater treatment are being increasingly supplemented, if not replaced, by innovative technologies driven by artificial intelligence (AI). This chapter delves into the intricate intersection of AI and wastewater treatment, aiming to explore not only the technical aspects but also the ethical considerations inherent in this transformative journey [1]. Wastewater treatment, a critical component of environmental management, involves the removal of pollutants and contaminants from water before it is discharged back into natural ecosystems [2]. The integration of AI into this domain has ushered in a new era of possibilities, promising enhanced efficiency, precision, and adaptability in treatment processes. From predictive analytics optimizing treatment plant operations to machine learning algorithms identifying emerging contaminants, AI is reshaping the landscape of wastewater management [3].

This chapter navigates through the ethical dimensions of AI applications in wastewater treatment, recognizing the need for responsible and sustainable practices. It goes beyond the technical intricacies of AI algorithms and explores the broader impact of AI on society, the environment, and the decision-making processes in wastewater treatment. The scope encompasses various dimensions, including transparency, accountability, privacy, bias, inclusivity, environmental impact assessment, and compliance with regulatory frameworks. By scrutinizing these ethical considerations, the chapter aims to provide a roadmap for stakeholders involved in the implementation of AI-driven wastewater treatment solutions [4]. As the nexus between technology and environmental stewardship evolves, it becomes imperative to strike a delicate balance between innovation and ethical responsibility. This chapter serves as a guide for researchers, practitioners, policymakers, and industry professionals, fostering a deeper understanding of the ethical challenges and opportunities that accompany the integration of AI in wastewater treatment. In doing so, it sets the stage for a thoughtful and responsible adoption of AI technologies in the pursuit of sustainable and efficient wastewater management practices [5].

Wastewater treatment stands as a linchpin in the global effort to maintain ecological balance and preserve human health. The process involves the removal of contaminants and pollutants from water before its return to the environment. Traditionally, this has been achieved through a series of physical, chemical, and biological processes. However, the increasing complexity of contaminants, coupled with the growing volume of wastewater, necessitates a more sophisticated and adaptable approach [6]. AI has emerged as a transformative force in enhancing the efficiency and efficacy of wastewater treatment processes. By leveraging advanced algorithms, machine learning, and data analytics, AI systems can optimize the operation of treatment plants, predict equipment failures, and identify patterns in pollutant concentrations. This chapter explores how AI, with its capacity for data analysis and pattern recognition, is revolutionizing wastewater treatment methodologies, making them more adaptive and responsive to the dynamic challenges posed by urbanization and industrialization.

The evolution of AI in environmental sciences marks a paradigm shift in our ability to monitor, analyze, and mitigate environmental issues. Early applications of AI in this domain focused on data analysis, enabling researchers to process vast datasets and extract meaningful insights. As computing power increased, machine learning algorithms gained prominence, allowing systems to learn from data patterns and make predictions. In wastewater treatment, this evolution has translated into more accurate predictions of pollutant levels, proactive maintenance of treatment facilities, and precise decision-making [7]. The integration of AI in environmental sciences also brings forth the concept of smart environmental monitoring. AI-driven sensors and devices can continuously collect and analyze real-time data, providing a comprehensive understanding of the environmental dynamics. This capability is particularly crucial in wastewater treatment, where the composition of inflowing water can vary significantly. AI allows for real-time adjustments in treatment processes, ensuring optimal performance even in the face of changing conditions.

13.2 AI Integration and Wastewater Treatment

The integration of AI in wastewater treatment processes is multifaceted, addressing various challenges faced by traditional methods as shown in Fig. 13.1. One key application lies in predictive analytics, where AI algorithms analyse historical data to forecast potential issues in treatment plant operations. This proactive approach enables preventive maintenance, reducing downtime and ensuring the continuous functionality of treatment facilities. Machine learning algorithms play a pivotal role in optimizing treatment processes [8]. These algorithms learn from historical data patterns and adjust operational parameters to achieve optimal efficiency. For instance, they can dynamically control aeration rates, chemical dosages, and sedimentation processes, responding in real-time to variations in wastewater characteristics. This adaptability enhances treatment performance and minimizes resource usage, contributing to both economic and environmental sustainability. AI's capability to identify emerging contaminants is another critical aspect of its integration into wastewater treatment. Traditional methods may struggle to detect novel pollutants, but AI, through pattern recognition, can identify aberrations in water quality. This is particularly valuable in addressing the challenges posed by emerging pollutants such as pharmaceuticals and personal care products [9]. AI's ability to quickly adapt to new information ensures that treatment processes remain effective against evolving threats.

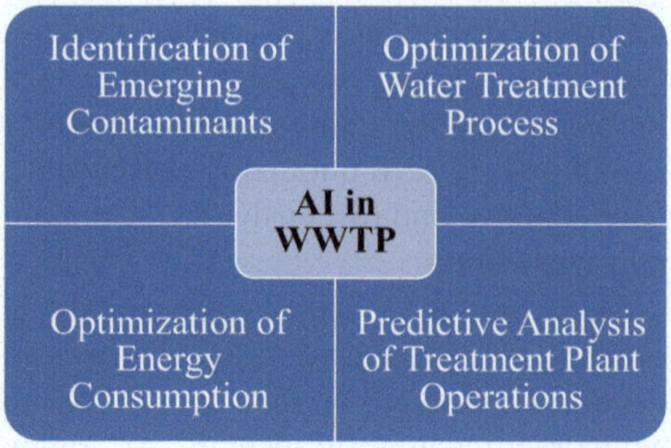

Fig. 13.1 Integration of AI in WWTP

Furthermore, AI facilitates the optimization of energy consumption in wastewater treatment plants. By analysing energy usage patterns and incorporating predictive models, AI systems can recommend energy-efficient strategies, reducing operational costs and the environmental footprint of treatment facilities [10]. This not only aligns with sustainability goals but also enhances the economic viability of wastewater treatment operations. In summary, the foundations of AI in wastewater treatment represent a transformative leap in the field of environmental management. The amalgamation of data-driven insights, machine learning capabilities, and adaptive decision-making positions AI as a key ally in addressing the challenges posed by modern wastewater treatment [11]. As the integration of AI continues to evolve, it promises not only increased efficiency in treatment processes but also a more sustainable and resilient approach to safeguarding our water resources.

13.3 Ethical Understanding and Challenges of AI Frameworks

As AI becomes an integral part of wastewater treatment processes, it brings forth a myriad of ethical considerations that demand careful examination as shown in Fig. 13.2. Ethical considerations in AI extend beyond the immediate technical applications and delve into the broader societal implications of these technologies. In the context of wastewater treatment, understanding the ethical dimensions involves recognizing the potential impacts on human well-being, environmental integrity, and equitable access to resources. One fundamental ethical consideration revolves around the transparency of AI algorithms. The opacity of complex AI models can pose challenges in understanding how decisions are made. In the realm of wastewater treatment, where the consequences of algorithmic decisions directly impact water quality and environmental health, ensuring transparency becomes paramount. Stakeholders, including policymakers, environmentalists, and the general public, should have access to understandable and interpretable information about AI-driven processes to foster trust and accountability.

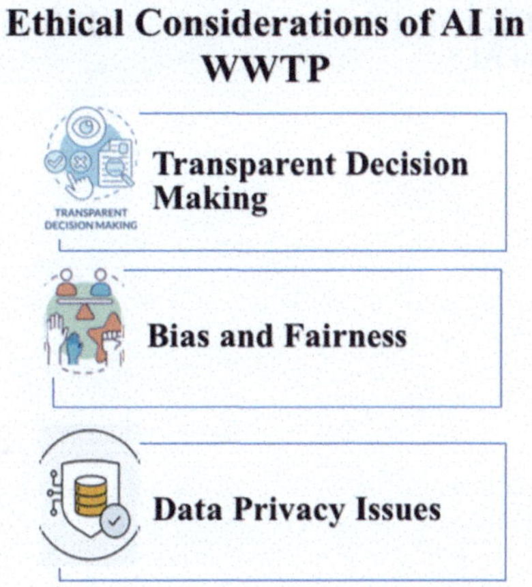

Fig. 13.2 Ethical considerations of AI

Additionally, the ethical use of data is a crucial aspect. Wastewater treatment involves handling sensitive information related to water quality, environmental conditions, and infrastructure. Ethical considerations include safeguarding data privacy, ensuring informed consent for data collection, and preventing the misuse of data. Striking a balance between the utility of data for optimizing treatment processes and respecting individual privacy rights becomes a central ethical challenge [12]. Bias and fairness are ethical dimensions that permeate AI applications in various domains, and wastewater treatment is no exception [13]. If AI algorithms are trained on biased datasets, they can perpetuate or exacerbate existing disparities. In wastewater treatment, this could manifest as unequal distribution of resources or inadequate treatment in certain communities. Recognizing and addressing bias is fundamental to ensuring fair and equitable outcomes in the implementation of AI-driven solutions.

The integration of AI in wastewater treatment, while promising numerous benefits, also introduces ethical challenges (see Fig. 13.3) that necessitate thoughtful consideration. One such challenge is the potential for job displacement. As AI automates certain tasks in treatment plants, there may be a reduction in the need for manual labour. This raises ethical questions about the responsibility of stakeholders to provide retraining and alternative employment opportunities for those affected. Another ethical challenge is the issue of accountability. Traditional wastewater treatment processes often involve clear lines of responsibility, where human operators make decisions and can be held accountable for the outcomes. With the introduction of AI, decision-making becomes decentralized, and accountability may become diffuse [14]. Determining who is responsible for AI-driven decisions, especially in the event of system failures or errors, poses a significant ethical dilemma [15]. The use of AI in decision-making processes also raises concerns about the potential for loss of human control. While AI systems are designed to optimize processes and enhance efficiency, relinquishing too much control to automated systems can lead to unintended consequences. Ethical considerations revolve around the need for human

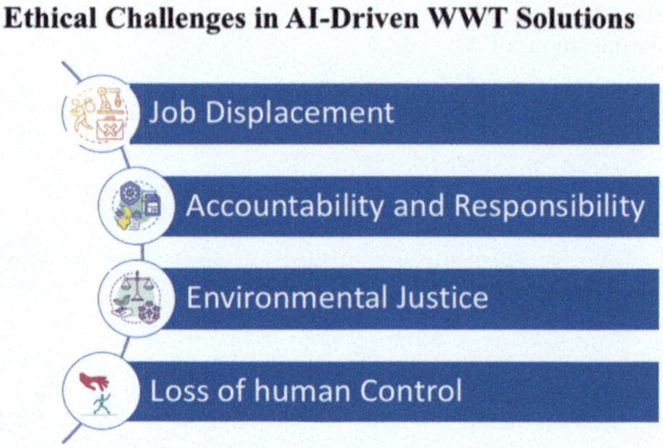

Fig. 13.3 Ethical challenges of AI

oversight, intervention capabilities, and fail-safes to prevent catastrophic outcomes in wastewater treatment.

Additionally, the ethical challenges in AI-driven solutions extend to issues of environmental justice. If AI systems are implemented without considering the socio-economic and cultural context of the communities they serve, there is a risk of exacerbating existing inequalities [16]. For example, if a wastewater treatment plant in an economically disadvantaged area relies heavily on AI, the community may face challenges in understanding, trusting, or influencing the decisions made by these systems.

Recognizing the ethical challenges inherent in AI-driven wastewater treatment solutions underscores the critical importance of establishing and adhering to ethical guidelines. Ethical guidelines serve as a compass, guiding the responsible development, deployment, and management of AI technologies in the context of wastewater treatment. One key aspect of ethical guidelines is the promotion of transparency, privacy regulations is imperative [17]. Implementing robust security measures, obtaining informed consent, and establishing data anonymization protocols contribute to ethical data management practices.

To address the challenge of bias and fairness, ethical guidelines should emphasize the importance of fairness audits and ongoing monitoring of AI systems. Regular assessments of algorithms for potential biases, especially in the context of demographic and geographic variations, help mitigate the risk of discriminatory outcomes in wastewater treatment processes. Human accountability should be a central tenet of ethical guidelines in AI-driven wastewater treatment [18]. Establishing mechanisms for clear lines of responsibility, defining the limits of automated decision-making, and ensuring human oversight in critical processes are essential components. These guidelines should also incorporate strategies for upskilling and reskilling workers to navigate the evolving landscape of AI in wastewater treatment [19].

Environmental justice considerations should be explicitly addressed in ethical guidelines. Incorporating community engagement practices, involving stakeholders in decision-making processes, and conducting impact assessments on vulnerable communities help mitigate the risk of unintended consequences and ensure that the benefits of AI-driven wastewater treatment are equitably distributed [20]. Therefore, the establishment and adherence to ethical guidelines are pivotal in navigating the ethical complexities associated with AI in wastewater treatment. As technology continues to advance, these guidelines serve as a compass, guiding the responsible development, deployment, and management of AI systems to ensure not only technical efficacy but also ethical integrity in safeguarding our water resources and environmental well-being [21].

13.4 Transparency and Accountability of AI Algorithms

Transparency is a cornerstone of ethical AI implementation in wastewater treatment. Ensuring transparency in AI algorithms involves making the decision-making processes understandable and accessible to stakeholders. In the context of wastewater treatment, where the consequences of algorithmic decisions directly impact water quality and environmental health, transparency becomes paramount [22]. To achieve transparency, it is essential to provide clear documentation of the AI algorithms used in wastewater treatment processes. This documentation should outline the data inputs, processing steps, and decision outputs. Additionally, efforts should be made to communicate these processes in a comprehensible manner, avoiding overly technical language that may alienate non-expert stakeholders. Visualization tools, interactive interfaces, and educational outreach programs can be employed to enhance transparency and promote a better understanding of the AI-driven decision-making processes. Open-source initiatives also contribute to transparency by allowing external scrutiny of the AI algorithms. Making the source code and model architectures publicly accessible enables independent researchers, environmental organizations, and concerned citizens to review and validate the algorithms [11]. This open approach builds trust and confidence in the AI systems deployed for wastewater treatment.

Accountability in AI-driven wastewater treatment processes is a complex and critical dimension. As decision-making becomes more automated and decentralized, establishing clear lines of responsibility is essential to mitigate the risks associated with unintended consequences or system failures. Defining accountability involves delineating the roles and responsibilities of various stakeholders in the AI-driven wastewater treatment ecosystem. This includes specifying the obligations of AI developers, operators, regulators, and the broader community. Decision-making authority, intervention protocols, and mechanisms for addressing errors or discrepancies should be clearly outlined to ensure that accountability is not diffuse but rather firmly rooted in the decision-making processes [23]. Human oversight plays a crucial role in ensuring accountability. While AI systems can optimize processes, humans must retain the ability to intervene, question decisions, and address ethical concerns. Establishing mechanisms for human review, intervention, and control safeguards against the abdication of responsibility to automated systems. Regular training and upskilling programs for human operators are also vital to enable them to navigate the evolving landscape of AI in wastewater treatment.

Legal frameworks and industry standards can further contribute to accountability. Clear regulations outlining the responsibilities of AI developers, operators, and users, as well as consequences for non-compliance, provide a foundation for accountability [24]. Aligning AI practices with established industry standards ensures a common understanding of ethical norms and expectations, fostering a culture of accountability within the wastewater treatment sector.

The lack of transparency and accountability in AI-driven wastewater treatment processes can have far-reaching implications, both environmentally and societally. One immediate consequence is the erosion of public trust. If stakeholders, including communities, regulatory bodies, and environmental organizations, perceive a lack of transparency in how AI algorithms operate and make decisions, trust in the technology diminishes. This erosion of trust can lead to resistance, skepticism, and reluctance to accept AI-driven solutions, hindering their successful implementation [25]. Environmental consequences may arise when AI algorithms lack transparency and accountability. Inaccurate or biased decision-making can result in suboptimal treatment processes, leading to the release of inadequately treated water into natural ecosystems. This can have detrimental effects on aquatic life, biodiversity, and overall ecosystem health. In the absence of accountability, identifying and rectifying such issues becomes challenging, exacerbating the environmental impact. Social implications include the potential for unequal distribution of benefits and burdens [26]. If the decision-making processes of AI systems are not transparent, vulnerable communities may be disproportionately affected by suboptimal wastewater treatment. Lack of accountability may result in the neglect of environmental justice considerations, leading to the exacerbation of existing socio-economic disparities.

Moreover, the absence of transparency and accountability can impede the learning and improvement of AI systems. Without clear mechanisms for feedback, error correction, and continuous refinement, AI algorithms may perpetuate mistakes, biases, or inefficiencies. This stagnation hinders the evolution of AI-driven wastewater treatment technologies and their potential to adapt to changing environmental conditions and emerging contaminants. So, ensuring transparency and accountability in AI-driven wastewater treatment processes is imperative for building trust, minimizing environmental impact, and fostering equitable outcomes [27]. Transparency enhances understanding, promotes public engagement, and facilitates scrutiny, while accountability establishes clear responsibility and oversight. The implications of lacking transparency and accountability extend beyond technical challenges, impacting the socio-environmental fabric of communities. Ethical AI implementation in wastewater treatment demands a commitment to openness, responsibility, and continuous improvement.

13.5 Data Collection and Privacy Issues

The integration of AI in wastewater treatment brings forth a host of privacy concerns, particularly concerning the collection and use of data. Wastewater treatment processes rely on the continuous monitoring of water quality, flow rates, and other environmental parameters. As AI-enhanced systems leverage extensive datasets to optimize operations, there is an inherent risk of infringing on individuals' privacy rights. Data collection in wastewater treatment involves the gathering of information from various sources, including sensors, monitoring devices, and even public records. The challenge arises when personally identifiable information (PII) is inadvertently or intentionally included in these datasets [28]. For instance, data related to specific industrial discharges, residential water usage, or health-related parameters may be linked to individuals or businesses, posing a potential threat to privacy. In the context of AI-driven wastewater treatment, the privacy issues extend beyond the mere collection of data. The analysis and processing of this data to derive actionable insights and make informed decisions can inadvertently reveal sensitive information. Balancing the need for detailed data to optimize treatment processes with the imperative to protect individuals' privacy becomes a critical ethical consideration.

Balancing the utilization of data for optimizing wastewater treatment processes with the imperative to protect privacy requires a nuanced approach. An essential step is the implementation of data anonymization and aggregation techniques. By removing or encrypting personally identifiable information and aggregating data at a group level, the utility of the information for optimizing treatment processes is maintained while mitigating privacy risks. Privacy-preserving technologies, such as homomorphic encryption and federated learning, can further enhance the protection of sensitive data [2]. Homomorphic encryption allows computations to be performed on encrypted data without decrypting it, ensuring that even during analysis, sensitive information remains confidential. Federated learning enables model training across decentralized datasets, reducing the need for centralized storage of sensitive information. Informed consent is a foundational principle in addressing privacy concerns. Clearly communicating the purpose of data collection, the types of data involved, and the intended use of the data to stakeholders is essential [29]. Obtaining explicit consent from individuals or businesses whose data is being utilized in AI-enhanced wastewater treatment processes establishes a transparent and ethical foundation for data utilization.

Educating the public about the benefits and risks of AI applications in wastewater treatment is also crucial. Fostering awareness about the measures in place to protect privacy, the ethical guidelines governing data use, and the overall societal benefits of AI-driven wastewater treatment helps build trust among stakeholders. Public engagement and feedback mechanisms further empower communities to actively participate in decisions regarding the balance between data utilization and privacy protection.

13.6 Bias and Fairness in AI Algorithms

13.6.1 Recognizing and Addressing Bias

As AI becomes increasingly integrated into wastewater treatment processes, the ethical consideration of bias and fairness takes center stage. Bias in AI algorithms refers to the presence of systematic and unfair preferences or prejudices in decision-making processes. In the context of wastewater treatment, where equitable access to clean water is paramount, recognizing and addressing bias is essential to prevent disparities in treatment outcomes. Recognizing bias begins with an awareness of the data used to train and test AI algorithms [30]. If historical data used to train the model reflects existing biases or inequalities, the AI system may perpetuate and even amplify these biases in its decisions. For example, if historical wastewater treatment data reflects a bias in favour of certain communities or industries, the AI system may inadvertently favour those groups, leading to unequal treatment.

Addressing bias involves implementing strategies to mitigate its impact on decision-making processes. One approach is to conduct a thorough bias audit during the development and training phases of AI algorithms. This involves systematically evaluating the dataset for potential biases and adjusting the algorithms accordingly. Techniques such as re-sampling, re-weighting, and introducing fairness constraints during model training can help rectify imbalances in the data and promote fair outcomes [31]. Moreover, transparency in AI algorithms is crucial for recognizing and addressing bias. By making the decision-making processes transparent and understandable, stakeholders can actively identify and rectify instances of bias. Open-source initiatives and collaboration with external experts can provide additional perspectives and contribute to unbiased AI development.

13.6.2 Ensuring Fairness in Wastewater Treatment Processes

Ensuring fairness in wastewater treatment processes powered by AI is imperative to prevent environmental injustice and uphold ethical standards. Fairness involves treating all stakeholders, communities, and industries equitably, irrespective of demographic, geographic, or socioeconomic factors. Achieving fairness requires a proactive and ongoing commitment to identify and rectify any biases that may emerge in AI-driven decision-making. One key aspect of ensuring fairness is the careful selection and curation of training datasets. Diverse and representative datasets, encompassing a broad spectrum of water sources, industrial discharges, and community profiles, contribute to unbiased AI models [32]. Striving for inclusivity in data collection and avoiding underrepresentation or overrepresentation of certain groups help in preventing inadvertent biases in wastewater treatment algorithms. Implementing fairness-aware algorithms during the development phase is essential. These algorithms incorporate fairness constraints that explicitly address and rectify potential biases. Fairness-aware models aim to distribute the benefits and burdens of wastewater treatment equitably, preventing the perpetuation of historical inequalities.

Another strategy for ensuring fairness is to involve diverse stakeholders in the decision-making processes related to AI-driven wastewater treatment. Including representatives from different communities, environmental organizations, and industries fosters a collective understanding of fairness considerations. Participatory approaches ensure that the perspectives of marginalized or vulnerable groups are taken into account, mitigating the risk of biased decision-making. Continuous monitoring and auditing of AI algorithms for fairness are integral components of ensuring ongoing equity in wastewater treatment. Regular assessments help identify emerging biases or unintended consequences that may arise as environmental conditions, industrial practices, or societal dynamics evolve [33]. This adaptive approach ensures that fairness is not just a one-time consideration but a dynamic and responsive aspect of AI-driven wastewater treatment. Examining case studies on bias mitigation in AI-driven wastewater treatment provides valuable insights into practical strategies and lessons learned in addressing fairness concerns. One illustrative case involves a wastewater treatment plant in an urban area that implemented an AI system for optimizing chemical dosages based on historical data. The initial analysis revealed a bias in favour of certain industrial discharges, leading to suboptimal treatment for residential wastewater. Recognizing this bias, the plant adjusted the algorithm to prioritize equitable treatment outcomes, ensuring that all sources received fair consideration [34]. In another case, a municipality deployed an AI system for predicting pollutant concentrations in wastewater. The initial models exhibited biases that disproportionately affected communities located near industrial zones. Through a collaborative effort with environmental justice organizations and community representatives, the municipality conducted a thorough bias audit and revised the algorithms to account for the specific challenges faced by these communities [35]. The revised AI models demonstrated improved fairness and more accurately predicted pollutant levels for all stakeholders.

These case studies highlight the importance of a proactive approach to bias mitigation and fairness in AI-driven wastewater treatment. Recognizing bias, involving diverse stakeholders, and adapting algorithms based on real-world feedback contribute to the successful implementation of fair and equitable AI systems. So, bias and fairness in AI algorithms for wastewater treatment are critical ethical considerations. Recognizing and addressing bias involves transparency, thorough audits, and proactive measures during development [36]. Ensuring fairness in wastewater treatment processes requires diverse and representative datasets, fairness-aware algorithms, stakeholder involvement, and ongoing monitoring. Case studies provide practical insights into the challenges and successful strategies for mitigating bias and promoting fairness in AI-driven wastewater treatment, ultimately contributing to more equitable and ethical water management practices.

13.7 Community Engagement and Inclusivity

13.7.1 Involving Communities in Decision-Making

Community engagement and inclusivity play a pivotal role in the ethical implementation of AI in wastewater treatment. The involvement of communities in decision-making processes ensures that the diverse needs, concerns, and perspectives of stakeholders are considered, fostering transparency and building trust in the use of advanced technologies. In the context of wastewater treatment, where the impacts directly affect local communities and the environment, community engagement becomes a cornerstone of responsible AI deployment. Involving communities in decision-making begins with establishing clear communication channels [37]. Transparent communication about the goals, benefits, and potential risks of AI-driven wastewater treatment processes is essential. This communication should be accessible, avoiding technical jargon and providing information in multiple languages to cater to diverse communities. Town hall meetings, community forums, and digital platforms can be utilized to facilitate open and ongoing dialogue. Participatory decision-making processes empower communities to contribute to the design, development, and deployment of AI systems in wastewater treatment. Seeking input on issues such as data privacy, algorithmic decision-making, and the overall goals of the technology ensures that community values are taken into account [38]. Community representatives can be involved in advisory committees or collaborate directly with AI developers and wastewater treatment experts to shape policies and practices.

Moreover, community engagement involves addressing concerns and gathering feedback. Conducting impact assessments and risk analyses in collaboration with communities helps identify potential challenges and opportunities associated with AI implementation. Responding to feedback, adapting algorithms based on community input, and providing mechanisms for grievance resolution demonstrate a commitment to inclusive decision-making.

13.7.2 Considering Diverse Perspectives in AI Implementation

Considering diverse perspectives in the implementation of AI in wastewater treatment acknowledges the unique challenges and opportunities faced by different communities. Recognizing the socio-economic, cultural, and environmental diversity of stakeholders ensures that AI systems are designed to be adaptable, responsive, and respectful of the specific needs of each community. One key aspect of considering diverse perspectives is addressing potential biases in AI algorithms [34]. Biases can emerge from historical data that may disproportionately represent certain groups or industries. By actively involving diverse communities in the development and testing of AI systems, developers can identify and rectify biases early in the process. Diverse perspectives contribute to more comprehensive datasets, leading to fairer and more equitable wastewater treatment outcomes.

Cultural considerations also play a role in AI implementation. Some communities may have unique practices, beliefs, or values that impact their relationship with water resources. Understanding and respecting these cultural nuances is essential to avoid inadvertently imposing technology that may conflict with local customs or traditions. Inclusive AI development involves incorporating cultural sensitivity into the design and operation of wastewater treatment systems [39]. In addition to cultural considerations, addressing socio-economic disparities is crucial. Low-income communities or those facing environmental justice challenges may have different access to resources, information, and decision-making power. Inclusivity in AI implementation involves actively seeking out the perspectives of marginalized or vulnerable groups, ensuring that technology benefits all stakeholders and does not exacerbate existing inequalities.

13.7.3 Benefits of Inclusive Practices in Wastewater Treatment

The benefits of inclusive practices in wastewater treatment extend beyond ethical considerations to encompass improved efficiency, effectiveness, and long-term sustainability. Inclusive decision-making processes lead to a more comprehensive understanding of the local context, enabling the customization of AI algorithms to address specific challenges faced by different communities. By involving communities in the decision-making and implementation of AI in wastewater treatment, there is an increased likelihood of successful adoption and acceptance. Communities that feel heard and respected are more likely to support and cooperate with the deployment of AI technologies [40]. This positive collaboration enhances the overall effectiveness of wastewater treatment processes, as community members become active partners in environmental stewardship. Inclusivity also contributes to the development of robust and adaptive AI systems [41]. Diverse perspectives provide a wealth of insights that can uncover potential pitfalls, unintended consequences, or overlooked opportunities. Integrating this knowledge into the design and refinement of AI algorithms enhances the technology's resilience to changing environmental conditions, emerging contaminants, and evolving community needs.

Moreover, inclusive practices foster a sense of ownership and responsibility among communities. When stakeholders actively participate in decision-making processes related to AI-driven wastewater treatment, they are more likely to be invested in the success of these technologies. This sense of ownership can lead to improved maintenance of treatment infrastructure, adherence to guidelines, and overall environmental awareness. In the long term, inclusive practices contribute to the sustainability of AI-driven wastewater treatment solutions. Sustainable technologies are those that not only address immediate challenges but also adapt to changing conditions and promote equitable access to resources [42]. Inclusive decision-making ensures that AI systems evolve in tandem with the needs and aspirations of communities, creating a resilient and socially responsible framework for wastewater management.

So, community engagement and inclusivity are foundational principles for the ethical implementation of AI in wastewater treatment. Involving communities in decision-making, considering diverse perspectives, and embracing inclusive practices lead to more transparent, responsive, and sustainable AI-driven solutions. The benefits extend beyond ethical considerations to encompass improved efficiency, effectiveness, and the empowerment of communities in safeguarding their water resources and environment.

13.8 Environmental Impact Assessment

13.8.1 Evaluating the Environmental Consequences of AI

As the integration of AI in wastewater treatment advances, conducting comprehensive Environmental Impact Assessments (EIAs) becomes imperative. EIAs are crucial tools for evaluating the potential environmental consequences of AI-driven wastewater treatment processes. They provide a systematic framework to identify, assess, and mitigate any adverse effects on the environment, ensuring that technological advancements align with principles of environmental stewardship [43]. One primary consideration in the EIA process is the energy consumption associated with AI algorithms and computing infrastructure. AI models often require significant computational power for training and operation. Evaluating the carbon footprint and energy consumption of AI-driven wastewater treatment systems is essential to understand their environmental impact. This assessment allows for the identification of opportunities to optimize energy efficiency, explore renewable energy sources, and minimize the overall environmental footprint.

Another aspect of the EIA involves assessing the impacts of data centers and server farms that support AI applications. These facilities, critical for processing and storing large datasets used in wastewater treatment, can have substantial environmental consequences, including land use, water consumption, and greenhouse gas emissions. Identifying environmentally sustainable hosting solutions and data storage practices contributes to mitigating the environmental footprint of AI technologies. Furthermore, evaluating the life cycle of AI technologies is essential [42]. This includes assessing the environmental impacts associated with the production, usage, and disposal of AI hardware components. Understanding the full life cycle allows for the implementation of strategies to reduce electronic waste, promote recycling, and minimize the environmental impact of manufacturing processes.

13.8.2 Sustainable Practices in AI-Driven Wastewater Treatment

The integration of sustainable practices in AI-driven wastewater treatment is a key objective for minimizing environmental impact. Sustainable practices go beyond the evaluation of potential harm and actively seek ways to enhance the positive contributions of AI technologies to environmental stewardship. One sustainable practice involves optimizing algorithms and models to reduce resource consumption. Through ongoing research and development, AI algorithms can be fine-tuned to achieve the same or improved treatment outcomes with reduced computational demands [42]. This optimization not only lowers energy consumption but also aligns with broader sustainability goals by minimizing the overall environmental impact of AI-driven wastewater treatment.

Implementing predictive maintenance strategies is another sustainable practice. By leveraging AI to predict equipment failures and optimize maintenance schedules, wastewater treatment plants can minimize downtime and extend the lifespan of machinery. This not only reduces the need for frequent replacements but also contributes to efficient resource utilization and a decrease in the environmental footprint associated with manufacturing and disposing of equipment. Incorporating real-time monitoring and adaptive control in wastewater treatment processes enhances sustainability [40]. AI-driven systems can continuously adjust operational parameters based on changing environmental conditions, ensuring optimal treatment efficiency. This adaptability not only improves treatment outcomes but also minimizes resource usage, contributing to both economic and environmental sustainability.

Additionally, embracing decentralized and edge computing approaches can enhance the sustainability of AI applications in wastewater treatment. Rather than relying solely on centralized data centers, distributing computing power to local facilities or edge devices can reduce the need for large-scale infrastructure. This approach minimizes energy transmission losses, enhances system resilience, and aligns with sustainable practices by promoting localized and efficient computing.

13.8.3 Balancing Technological Advancements with Environmental Stewardship

Balancing technological advancements with environmental stewardship is a central theme in the ethical deployment of AI in wastewater treatment. Striking this balance requires a nuanced approach that prioritizes innovation while proactively addressing and mitigating potential environmental impacts. One strategy for balancing technological advancements is to adopt a circular economy approach. This involves designing AI systems with the end of life in mind, emphasizing recycling, reusing components, and minimizing waste. Integrating circular economy principles into the development and deployment of AI technologies aligns with broader environmental sustainability goals and ensures responsible resource management. Collaboration between AI developers, environmental scientists, and policymakers is crucial for achieving this balance [42]. Engaging in interdisciplinary dialogue allows for a holistic understanding of the environmental implications of AI-driven wastewater treatment. Researchers can work collaboratively to develop technologies that not only advance treatment capabilities but also adhere to rigorous environmental standards and contribute positively to overall sustainability.

Regulatory frameworks also play a key role in balancing technological advancements with environmental stewardship. Governments and regulatory bodies can establish guidelines and standards that incentivize the development and adoption of environmentally sustainable AI technologies. This may include offering incentives for energy-efficient AI models, promoting eco-friendly data storage practices, or imposing regulations to ensure responsible disposal of electronic components. Education and awareness initiatives contribute to achieving a balanced approach [34]. By fostering a culture of environmental consciousness within the AI community, developers and practitioners can actively seek solutions that prioritize sustainability. Training programs, workshops, and awareness campaigns can promote responsible practices and encourage the integration of environmental considerations into the ethos of AI development. Therefore, conducting Environmental Impact Assessments, incorporating sustainable practices, and balancing technological advancements with environmental stewardship are integral components of the ethical deployment of AI in wastewater treatment. By proactively addressing potential environmental consequences, optimizing resource utilization, and collaborating across disciplines, stakeholders can harness the benefits of AI technologies while upholding principles of environmental responsibility. This balanced approach ensures that AI-driven wastewater treatment not only advances technological capabilities but also contributes positively to environmental sustainability and long-term ecological well-being.

13.9 Regulatory Frameworks and Compliance

The integration of AI into wastewater treatment processes necessitates a robust regulatory framework to ensure ethical and responsible deployment. The current regulatory landscape for AI in wastewater treatment is evolving, with a focus on addressing potential risks, ensuring transparency, and safeguarding environmental and public health. Regulatory bodies, both at national and international levels, are gradually recognizing the need to establish guidelines specific to AI applications in critical sectors like wastewater treatment. Existing environmental regulations, such as those governing water quality standards and discharge permits, provide a foundational framework. However, there is a growing recognition that AI introduces unique challenges that require dedicated attention [44].

In some regions, general AI principles and ethics guidelines are being adapted to the environmental context. For instance, principles related to transparency, accountability, and fairness are increasingly considered essential for AI-driven wastewater treatment processes. As technology continues to advance, regulatory agencies are actively engaging with AI developers, environmental scientists, and stakeholders to create standards that ensure the responsible use of AI technologies.

Navigating compliance challenges in the realm of AI-driven wastewater treatment involves addressing a range of complex issues. One significant challenge is the lack of standardized regulations specifically tailored to AI applications in environmental management. Wastewater treatment plants often find themselves in uncharted territory, having to interpret and adapt existing regulations to the unique considerations introduced by AI technologies. Data privacy and security are critical compliance challenges in AI-driven wastewater treatment. As these systems handle sensitive information related to water quality, industrial discharges, and infrastructure, ensuring compliance with data protection regulations is paramount. Wastewater treatment plants must navigate the complexities of data anonymization, consent management, and secure data storage to meet regulatory requirements [45].

Another compliance challenge revolves around transparency and explainability of AI algorithms. Regulatory agencies are increasingly emphasizing the importance of understanding and interpreting AI-driven decision-making processes. This requirement poses a challenge for wastewater treatment plants to implement algorithms that are not only effective but also transparent and interpretable. Meeting this challenge involves adopting algorithms that can provide clear explanations of their decisions and establishing protocols for transparent communication with regulatory bodies and the public [45]. Furthermore, compliance challenges include issues related to bias and fairness in AI algorithms. Ensuring that AI systems do not perpetuate or exacerbate existing environmental inequalities is a complex task [24, 26, 27]. Wastewater treatment plants must actively work to identify and mitigate biases in their algorithms, conduct fairness assessments, and incorporate ethical considerations into their compliance strategies.

To address the compliance challenges and promote responsible AI deployment in wastewater treatment, proposing ethical standards specific to this domain is crucial. These standards should encompass a set of principles that guide the development, implementation, and management of AI technologies while ensuring alignment with environmental goals and regulatory requirements.

Transparency and Accountability: Ethical standards should prioritize transparency in AI algorithms used in wastewater treatment processes. This involves making the decision-making processes understandable and accessible to regulatory bodies, stakeholders, and the general public. Accountability mechanisms should be established to clearly delineate responsibilities for the outcomes of AI-driven decisions, ensuring that human operators can intervene when necessary.

Fairness and Non-Discrimination: Ethical standards should emphasize fairness in AI applications, particularly in the context of wastewater treatment. AI algorithms should be designed to avoid discriminatory outcomes and address biases in data.

Regular fairness audits and assessments should be conducted to ensure equitable treatment for all communities, industries, and stakeholders.

Data Privacy and Security: Ethical standards should establish stringent protocols for data privacy and security in AI-driven wastewater treatment. Compliance with data protection regulations, informed consent for data collection, and secure data storage practices should be prioritized. These standards should also include guidelines for anonymizing and protecting sensitive information to prevent unauthorized access or misuse.

Environmental Impact Assessment: Ethical standards should mandate the conduct of comprehensive Environmental Impact Assessments for AI-driven wastewater treatment systems. This involves evaluating the environmental consequences, energy consumption, and life cycle impacts associated with AI technologies. Compliance should require ongoing monitoring and mitigation measures to minimize the environmental footprint.

Community Engagement and Inclusivity: Ethical standards should encourage active community engagement and inclusivity in decision-making processes related to AI in wastewater treatment. This involves seeking input from diverse stakeholders, incorporating community perspectives into algorithmic decision-making, and fostering collaboration between AI developers and local communities (Table 13.1).

Compliance Reporting and Auditing: Ethical standards should establish clear reporting mechanisms and auditing processes for AI-driven wastewater treatment. Compliance reports should be submitted to regulatory bodies, detailing the adherence to ethical standards, transparency in decision-making, and mitigation of potential biases. Regular audits can ensure ongoing compliance and identify areas for improvement. So, proposing and adhering to ethical standards for AI in wastewater treatment is essential for navigating compliance challenges and ensuring responsible deployment. These standards should align with existing environmental regulations while addressing the unique considerations introduced by AI technologies. By prioritizing transparency, fairness, data privacy, and community engagement, ethical standards contribute to the development of AI-driven wastewater treatment systems that not only meet regulatory requirements but also foster environmental sustainability and public trust.

13.10 Conclusion

In conclusion, the ethical considerations in AI-driven wastewater treatment are multifaceted, requiring a holistic approach to ensure responsible deployment. The exploration of ethical dimensions in this context has encompassed diverse facets, including transparency, accountability, fairness, privacy, community engagement, and environmental impact assessment in the following point.

Table 13.1 Ethical standards summary for AI in wastewater treatment

Ethical standards	Key principles	Implementation guidelines
Transparency and Accountability	– Clear communication about AI algorithms and decision-making processes – Establishing accountability mechanisms for AI-driven decisions	– Develop algorithms with understandable decision-making processes – Clearly define responsibilities and intervention protocols
Fairness and non-discrimination	– Identifying and mitigating biases in data and algorithms – Actively involving diverse perspectives in decision-making	– Conduct regular fairness audits and assessments – Foster collaboration between AI developers and local communities
Data privacy and security	– Compliance with data protection regulations – Informed consent for data collection and secure data storage	– Anonymize and protect sensitive information – Adhere to legal and regulatory frameworks for data privacy
Environmental impact assessment	– Evaluating the environmental consequences and life cycle impacts of AI technologies	– Conduct comprehensive Environmental Impact Assessments – Implement ongoing monitoring and mitigation measures
Community engagement and inclusivity	– Actively involving communities in decision-making processes – Seeking input from diverse stakeholders	– Establish clear communication channels with communities – Incorporate community perspectives into algorithmic decision-making
Compliance reporting and auditing	– Establishing reporting mechanisms for adherence to ethical standards – Conducting regular audits of AI-driven wastewater treatment systems	– Submit compliance reports to regulatory bodies detailing adherence to ethical standards – Implement auditing processes for ongoing compliance and improvement

a. Transparency emerged as a foundational principle, emphasizing the need for clear and understandable communication about AI algorithms and decision-making processes. This transparency not only builds trust among stakeholders but also allows for scrutiny, accountability, and informed decision-making. Accountability is paramount in the decentralized and automated decision-making landscape introduced by AI. Establishing clear lines of responsibility, intervention protocols, and mechanisms for human oversight is crucial to mitigate risks and ensure that AI-driven wastewater treatment remains aligned with ethical standards.

b. Fairness considerations recognize the importance of avoiding biases and discriminatory outcomes in AI algorithms. By actively addressing biases in data and algorithms and involving diverse perspectives in the decision-making process, fairness becomes an integral aspect of ethical AI deployment.
c. Privacy concerns highlight the need to balance data utilization for optimization with the imperative to protect individuals' privacy rights. Strategies such as data anonymization, informed consent, and adherence to legal and regulatory frameworks contribute to addressing these privacy challenges.
d. Community engagement and inclusivity emerged as ethical imperatives, recognizing the importance of involving communities in decision-making processes. By actively seeking input, considering diverse perspectives, and fostering collaboration, AI-driven wastewater treatment systems can be developed and implemented with the support and understanding of local communities.
e. Environmental impact assessment is crucial to evaluate the consequences of AI technologies on the environment. Assessing energy consumption, life cycle impacts, and potential risks associated with AI-driven wastewater treatment contributes to minimizing the environmental footprint and promoting sustainability.
f. The future of AI-driven wastewater treatment holds great promise for advancing environmental sustainability and improving water management practices. As technology continues to evolve, the integration of AI is expected to enhance the efficiency, accuracy, and adaptability of wastewater treatment processes.
g. Predictive analytics and real-time monitoring powered by AI can revolutionize how wastewater treatment plants respond to changing environmental conditions and emerging contaminants. This dynamic approach allows for proactive decision-making, optimized resource utilization, and the prevention of environmental harm.
h. Advancements in machine learning algorithms, particularly those related to deep learning and neural networks, are anticipated to further improve the accuracy of predictive models. These sophisticated algorithms can uncover complex patterns and relationships within wastewater data, leading to more precise and effective treatment strategies.
i. Decentralized and edge computing models are likely to gain prominence, offering localized and efficient processing of data. This shift can contribute to reduced energy consumption, enhanced system resilience, and increased accessibility to AI-driven wastewater treatment technologies, especially in remote or underserved areas.
j. Collaboration and knowledge-sharing among researchers, environmental scientists, policymakers, and industry stakeholders will be pivotal for steering the future of AI-driven wastewater treatment ethically. Open-source initiatives, interdisciplinary research, and international cooperation can foster a global community committed to ethical AI implementation in water management.

k. A call to action resonates for all stakeholders involved in AI-driven wastewater treatment—from developers and operators to regulators and community members. Ethical implementation requires a collective commitment to principles that prioritize transparency, fairness, privacy, community engagement, and environmental sustainability.
l. For developers and operators, the call to action involves incorporating ethical considerations into every stage of AI system development. This includes prioritizing transparency in algorithmic decision-making, conducting regular bias assessments, implementing robust privacy measures, and actively engaging with communities to address their concerns.
m. Regulatory bodies play a crucial role in setting the stage for ethical implementation. The call to action for regulators is to establish clear and adaptive frameworks that account for the evolving landscape of AI-driven wastewater treatment. These frameworks should incentivize ethical practices, provide guidelines for compliance, and facilitate ongoing monitoring of AI technologies.
n. Communities are integral stakeholders with a vested interest in the responsible deployment of AI technologies. The call to action for communities is to actively participate in decision-making processes, seek information about AI applications, and contribute local knowledge to enhance the ethical considerations embedded in wastewater treatment systems.
o. Environmental organizations and advocacy groups have a role in holding stakeholders accountable for ethical practices. The call to action for these groups involves fostering awareness, conducting independent assessments, and advocating for policies that prioritize environmental sustainability and social responsibility.

In conclusion, the ethical implementation of AI in wastewater treatment demands a collective commitment to principles that safeguard transparency, fairness, privacy, community engagement, and environmental stewardship. By heeding this call to action, stakeholders can contribute to the development of AI-driven wastewater treatment systems that not only optimize efficiency but also prioritize ethical considerations, ensuring a sustainable and responsible future for water management.

References

1. Arnon TA, Ezra S, Fishbain B (2019) Water characterization and early contamination detection in highly varying stochastic background water, based on machine learning methodology for processing real-time UV-spectrophotometry. Water Res 155:333–342
2. Basco-Carrera L, Warren A, van Beek E, Jonoski A, Giardino A (2017) Collaborative modelling or participatory modelling? A framework for water resources management. Environ Model Softw 91:95–110
3. Cath C (2018) Governing artificial intelligence: ethical, legal and technical opportunities and challenges introduction. Philos Trans R Soc A Math Phys Eng Sci 376:8
4. Char DS, Shah NH, Magnus D (2018) Implementing machine learning in health care addressing ethical challenges. N Engl J Med 378:981–983

5. Chatterjee S, Sarkar S, Dey N, Sen S, Goto T, Debnath NG, Ieee (2017) Water quality prediction: multi objective genetic algorithm coupled artificial neural network-based approach. In: IEEE 15th International conference on industrial informatics
6. Dawson CW, Wilby RL (2001) Hydrological modelling using artificial neural networks. Progr Phys Geogr: Earth Environ 25:80–108
7. Dignum V (2018) Ethics in artificial intelligence: introduction to the special issue. Ethics Inf Technol 20:1–3
8. Jobin A, Ienca M, Vayena E (2019) The global landscape of AI ethics guide lines. Nat Mach Intell 1:389–399
9. Jordan MI, Mitchell TM (2015) Machine learning: trends, perspectives, and prospects. Science 349:255–260
10. Kasprzyk JR, Nataraj S, Reed PM, Lempert RJ (2013) Many objective robust decisions making for complex environmental systems undergoing change. Environ Model Softw 42:55–71
11. Seo Y, Kim S, Kisi O, Singh VP (2015) Daily water level forecasting using wavelet de composition and artificial intelligence techniques. J Hydrol 520:224–243
12. Van Thienen P (2019) Responsible AI forth ewater sector? https://www.kwrwater.nl/actueel/responsible-ai-for-the-water-sector/: KWR.
13. Russell SJ (2015) Ethics of artificial intelligence. Nature 521:415–416
14. Sun AY, Scanlon BR (2019) How can big data and machine learning benefit environment and water management: a survey of methods, applications, and future directions. Environ Res Lett 14:28
15. Russell SJ (2019) Human compatible: artificial intelligence and the problem of control. London, Viking- Penguin Random House
16. Russell SJ (2016) Should we fear supersmart robots. Sci Am 314:58–59
17. Strobl RO, Robillard PD (2006) Artificial intelligence technologies in surface water quality monitoring. Water Int 31:198–209
18. Zechman EM (2013) Integrating evolution strategies and genetic algorithms with agent-based modeling for flushing a contaminated water distribution system. J Hydroinf 15:798–812
19. Winfield AFT, Jirotka M (2018) Ethical governance is essential to building trust in robotics and artificial intelligence systems. Philos Trans R Soc A Math Phys Eng Sci 376:13
20. Singh KP, Gupta S (2012) Artificial intelligence-based modeling for predicting the disinfection by-products in water. Chemom Intell Lab Syst 114:122–131
21. Lorenzo M, Picó Y (2019) Wastewater-based epidemiology: current status and future prospects. Curr Opin Environ Sci Health 9:77–84
22. Park SH, Kim YH, Lee JY, Yoo S, Kim CJ (2019) Ethical challenges regarding artificial intelligence in medicine from the perspective of scientific editing and peer review. Sci Editing 6:91–98
23. Yu CX, Yin XN, Yang ZF, Dang Z (2019) Sustainable water resource management of regulated rivers under uncertain inflow conditions using a noisy genetic algorithm. Int J Environ Res Public Health 16
24. Koehler J (2018) Exploring policy perceptions and responsibility of devolved decision-making for water service delivery in Kenya's 47 county governments. Geoforum 92:68–80
25. Labadie JW (2014) Advances in water resources systems engineering: applications of machine learning
26. Smith R, Kasprzyk J, Dilling L (2019) Testing the potential of multiobjective evolutionary algorithms (MOEAs) with Colorado water managers. Environ Model Softw 117:149–163
27. Bagriacik A, Davidson RA, Hughes MW, Bradley BA, Cubrinovski M (2018) Comparison of statistical and machine learning approaches to modeling earthquake damage to water pipelines. Soil Dyn Earthq Eng 112:76–88
28. Neto RCJ, Mérindol P, Theoleyre F (2023) Enabling privacy by anonymization in the collection of similar data in multi-domain IoT. Computer Commun 203:60–76
29. Buchanan BG (2005) A (very) brief history of artificial intelligence. AI Mag 26:53–60
30. Ferrara E (2024) The Butterfly Effect in artificial intelligence systems: implications for AI bias and fairness. Mach Learn Appl 15:100525

31. Akter S, Sultana S, Mariani M, Wamb SF, Spanaki K, Dwivedi YK (2023) Advancing algorithmic bias management capabilities in AI-driven marketing analytics research. Ind Mark Manag 114:243–261
32. Xie Q, Xu Q, Zhu D, Rao K, Dai Q (2020) Fair allocation of wastewater discharge permits based on satisfaction criteria using data envelopment analysis. Util Policy 66:101078
33. Alvi M, Batstone D, Mbamba CK, Keymer P, French T, Ward A, Dwyer J, Cardell-Oliver R (2023) Deep learning in wastewater treatment: a critical review. Water Res 245:120518
34. Kazadi Mbamba C, Batstone D (2023) Optimization of deep learning models with genetic algorithms for forecasting performance in water industry. Comput Chem Eng:108276
35. Wang D, Li X, Xin P, Liu A, Qiao J (2024) Supplementary heuristic dynamic programming for wastewater treatment process control. Expert Syst Appl 247:123280
36. Ba-Alawi AH, Loy-Benitez J, Kim S, Yoo C (2022) Missing data imputation and sensor self-validation towards a sustainable operation of wastewater treatment plants via deep variational residual autoencoders. Chemosphere 288:132647
37. Townley C, Koop C (2024) Exploring the potential and limits of digital tools for inclusive regulatory engagement with citizens. G Info Q 41:101901
38. Burton F, Tchobanoglous G, Tsuchihashi R, Stensel H, Metcalf and Eddy Inc. (2013) Wastewater engineering: treatment and resource recovery. McGraw-Hill Education
39. Jawad J, Hawari AH, Zaidi SJ (2021) Artificial neural network modeling of wastewater treatment and desalination using membrane processes: a review. Chem Eng J 419:129540
40. Solomatine DP, Ostfeld A (2008) Data-driven modelling: some past experiences and new approaches. J Hydroinform 10:3–22
41. Alhaffa MBA, Joury E, Eriksson A (2023) Community engagement and crowdsourcing for effective disaster response and rescue operations during the earthquake in Syria. Int J Disaster Risk Red 98:104096
42. Li K, Duan H, Liu L, Qiu R, Akker B van den, Ni BJ, Chen T, Yin H, Yuan Z, Ye L (2022) An integrated first principal and deep learning approach for modeling nitrous oxide emissions from wastewater treatment plants. Environ Sci Technol 56:2816–2826
43. Newhart KB, Holloway RW, Hering AS, Cath TY (2019) Data-driven performance analyses of wastewater treatment plants: a review. Water Res 157:498–513
44. Buterin V, Illum J, Nadler M, Schär F, Soleimani A (2024) Blockchain privacy and regulatory compliance: towards a practical equilibrium. Blockchain: Res Appl 5(1):100176
45. Meade MJ, Ju X, Hunter D, Jamieson L (2023) Compliance of orthodontic practice websites with ethical, legal and regulatory advertising obligations. Int Orthodontics 21(1):100727

Chapter 14
Economic Implications and Cost–Benefit Analysis of AI in Wastewater Treatment

Priyanka Ray Choudhury, Preeti Karki, Alakto Choudhury, Priyanka Narad, and Abhishek Sengupta

Abstract In recent times, Artificial Intelligence (AI) has revolutionized several sectors by offering innovative solutions to complex challenges. In wastewater treatment (WWT), artificial intelligence (AI) models have been widely used to predict efficiency, evaluate performance, and optimize processes in search of more cost-effective solutions. This is because AI models have fast learning and precise prediction capabilities. The current chapter delves into the different AI models that can and have been implemented for the purpose of water management and treatment, improving the accuracy of the pollutant removal and offering insights into potential future directions for innovative research in the field. It also showcases caste studies exhibiting the successful implementation of AI-based WWT solutions. There is also a deep emphasis on the economic implications of integrating AI technology into WWT systems and an extensive review of the cost–benefit analysis of such implementations. The intersection of artificial intelligence and WWT is becoming an increasingly important area of focus for both researchers, policymakers and the common people as stakeholders due to the environmental concerns and the increasing need for sustainable solutions. It also seeks to give significant insights into the possible advantages and problems that may arise due to the implementation of new AI technologies for the purpose of environmental management by assessing not only the costs and benefits associated with AI adoption in this area but also investigating the economic repercussions of this technology.

Keywords Artificial intelligence · WWT · Economic implications · Cost benefit · Environmental impact

---

P. R. Choudhury · P. Karki · A. Choudhury · A. Sengupta (✉)
Amity Institute of Biotechnology, Amity University, Noida, Uttar Pradesh 201313, India
e-mail: asengupta@amity.edu

P. Narad
Indian Council of Medical Research, Ansari Nagar, New Delhi 110029, India

M. C. Garg (ed.), *The AI Cleanse: Transforming Wastewater Treatment Through Artificial Intelligence*, Springer Water, https://doi.org/10.1007/978-3-031-67237-8_14

14.1 Introduction

AI has led to some of the greatest innovations leading enabling transformative advancements across various industries such as healthcare [70], manufacturing [7], finance [13], urban development [40] and also towards environmental sustainability [82]. One of the key motivators for implementation of AI is to reduce costs [29] and improve the pre-existing water treatment systems by enhancing the efficiency and improving the decision-making process. Artificial neural networks are especially helpful in observing data patterns and trends which might be missed by the human eye. These trends can give helpful insights illustrating the importance of AI in various applications.

One of the greatest and ever-increasing challenges of the global economy is maintaining the freshwater supply for various agricultural, industrial and household needs although these are also the primary sources of water pollution as well [71]. The disposal and treatment of such wastewater is also a big issue due to its high impact on health and environment [79, 80]. There exist a wide variety of pollutants such as heavy metals, oils, detergents, algae, chemical wastes and other animal and artificial wastes which are to be accounted for in any WWT plan. In addition to that, a wastewater plant also needs to account for the water pH, temperature, aeration, mixing rate, carbon dioxide and light intensity among other parameters [71]. However, by collecting data for all these parameters, it is possible to create a data-driven model, mainly AI, for a lot of waste-water related issues and to optimize the functioning of pre-existing wastewater treatment plants (WWTPs).

The integration of AI into WWT has a potential to significantly increase economic output, considering both opportunities as well as the challenges associated with it. Globally, the AI approach to circular economy and waste management have drawn increasing attention especially among the pertinent stakeholders. The increased problems caused by the growing volume of industrial and municipal wastewaters make it difficult to meet the objectives for both economic growth and environmental sustainability. Due to the linear economic link between the approach and the present challenges, the strategic treatment is intended for industries and agriculture [42]. Polluted water and wastewater streams have severely affected human health and ecosystem for decades as a whole since they could contain harmful xenobiotic compounds which are released by various growing industrial, domestic and agricultural sectors.

Furthermore, because AI provides reliable answers to real-world issues such as WWT and machine defect diagnosis, it has been widely applied in many engineering fields [4]. Application of AI or Machine learning (ML) is dependable and creative when it comes to several existing technologies for WWT, such as the membrane filtration, adsorption process, and advanced oxidation systems [55]. One of the most significant economic benefits of AI in WWT is the possibility for cost reductions in operations.

Real-time optimization of treatment processes using ML algorithms can result in reduced energy consumption, lesser chemical usage, and an improvement in the system efficiency. AI driven resource recovery finds valuable materials in wastewater, resulting in increased income production from recycling and reuse, which lowers costs over time with continuous usage. Moreover, this chapter aims to provide a complete and comprehensive analysis of the economic implications of AI in WWT, highlighting the diverse ways in which advancements in technology influences costs, benefits, and overall financial considerations.

14.2 AI Applications in WWT

14.2.1 Smart WWT Methods

The incorporation of AI models in conjunction to frameworks such as Internet of Things (IoT) along with conventional methods, has emerged as an essential component in the development of smart WWT (WWT) systems, specifically for the reuse of sewage water [33]. AI models are effective instruments for modeling, prediction as well as optimization of different tasks including WWT processes. These models may be used to remove various pollutants from water, including dyes, organics, sediments, heavy metals, nutrients, microbiological contamination, pharmaceuticals and pesticides (see Fig. 14.1) [5].

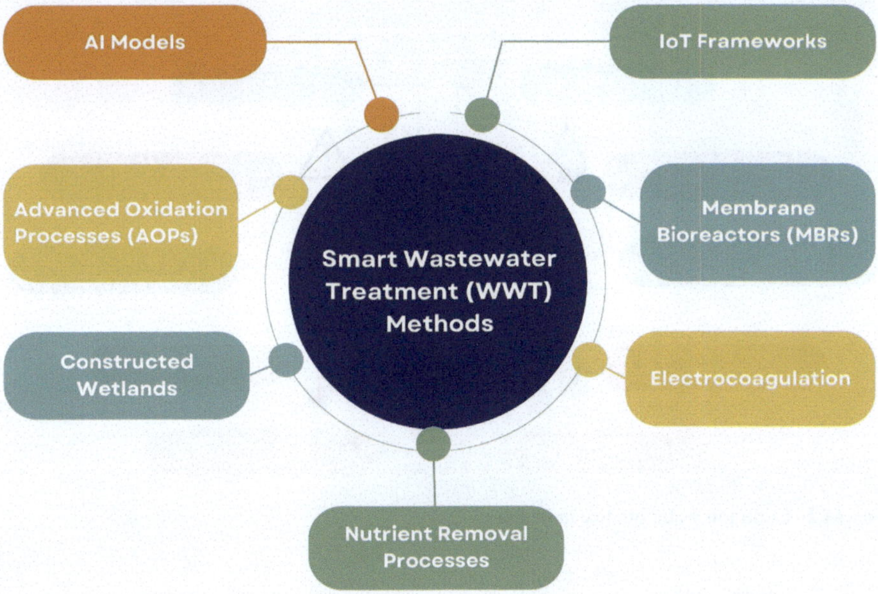

Fig. 14.1 Smart wastewater treatment methods

14.2.2 Prediction of BOD and COD of Wastewater

Prediction of water quality and monitoring for data collection is one of the primary applications of AI in WWT. It can be employed in predicting the changes in quality indicators of water such as pH, turbidity, Chemical Oxygen Demand (COD), and Biological Oxygen Demand (BOD). For monitoring of influent water, parameters such as flow rate, initial pollutant concentration, BOD, COD, pH, and temperature can be easily obtained and used for creating AI models. When monitoring effluent water, the parameters that are typically employed to assess the efficacy of WWTPs are the aforementioned quality indicators for water i.e. BOD, COD and pH, etc. [81].

14.2.3 Continuous Monitoring of Water Quality in a WWT Plant

Sensors are used to continually check influent and effluent water quality yielding large volumes of data [74]. In an attempt to facilitate fast and accurate real-time monitoring and the automation of the WWT process through online controls, data analysis, and real-time sensing, several researchers have designed a variety of sensors to help in the process [14, 21, 87]. Common water quality indicators are given in Fig. 14.2.

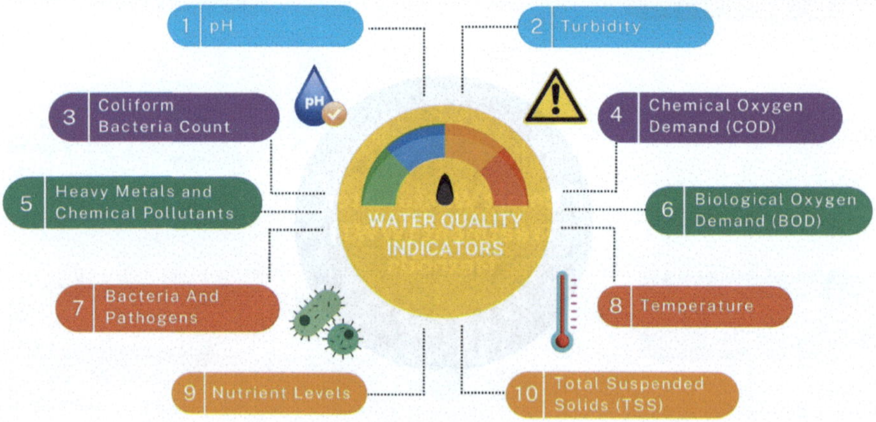

Fig. 14.2 Common water quality indicators

14.2.4 AI Implementation in Membrane Processes

The application of AI models in membrane processes and membrane bioreactors (MBRs) optimization and design has shown encouraging outcomes in laboratory size research. MBRs, by combining membrane filtration and biological treatment, provide benefits such as biomass concentration, reduced sludge generation, and increased quality of effluents. Although membrane fouling is a concern with MBRs, researchers are investigating AI-driven solutions in response to these challenges. AI models, such as radial basis function artificial neural network (RBF-ANN) and hybrid artificial neural network (ANN) response surface methodology (RSM) have been successfully applied to a variety of membrane types, including forward osmosis, reverse osmosis, ultrafiltration, nanofiltration, and microfiltration in order to estimate permeate flux and eliminate membrane fouling [63].

14.2.5 AI in Water Disinfection and Pollutant Removal

AI has also been implemented for disinfection and chlorination by-product management, ANNs have been employed for disinfection and disinfection by-product prediction with additional applications including Support Vector Machines (SVMs), genetic algorithms (GA) and fuzzy inference systems [6].

AI models are used in process parameter optimisation and process performance prediction, illustrating its many applications in enhancing treatment efficiency and in minimizing the total expenses involved in the WWT process. Research indicates that AI models may accurately predict the best parameters for increasing biomass, improving decontamination efficacy, treating sulphate wastewaters, and removing pollutants at high rates. Deep learning models improve real-time monitoring and forecasting of water quality in industrial WWT systems, whereas image-based AI models work well in conjunction with spectroscopic techniques to detect pollutants [81]. By monitoring sensor data and past maintenance records or historical data, AI plays a critical role in predictive maintenance by assisting in anticipating equipment problems. This preventative strategy optimizes equipment longevity, lowers maintenance costs, and reduces production downtime. By deploying artificial intelligence (AI) for anomaly detection, operational data anomalies may be found, reducing treatment process interruptions and guaranteeing regulatory compliance. Furthermore, through data-driven modifications and consumption trends, AI-driven energy optimisation lowers environmental impact and operating costs while promoting sustainable WWT.

In a study led by Messikh et al., the emulsion liquid membrane process employed the RBF-ANN model to estimate and evaluate copper removal efficiency, demonstrating a quicker training capability compared to other neural networks. The predicted values were closely aligned with experimental data, achieving an coefficient of detemination (R2) value equal to 0.997 [58]. In another research, Adaptive

Neuro-Fuzzy Inference System (ANFIS) was employed to assess the efficiency of Cd^{2+} ions adsorption from aqueous solutions in the case of *Typha domingensis*, revealing that the adsorption was primarily influenced by pH [22].

ANNs have been employed for WWT extensively to study the efficacy of AI implementation and the potential benefits associated with it. One such study by Mandal et al., aimed to predict the efficiency of algal remediation of As^{3+} and As^{5+} from wastewater, achieving maximum phyco-remediation of 85% (for As^{3+}) and 88% (for As^{5+}). The established ANN model demonstrated high accuracy (R2 0.9998) in determining the elimination process of the ions under different conditions [54]. In another AI application, Zaferani et al., found that simulating the effect of thiosemicarbazide modified chitosan for removal of Pb^{2+}, Multilayer Perceptron-Artificial Neural Network (MLP-ANN) exhibited better prediction accuracy (R2 0.990) as compared to RSM [86]. ANN models have been used to accurately predict the degradation of photo-induced polycyclic aromatic hydrocarbons in seawater [41]. MLP-ANN and RBF-ANN models were employed for simulation and optimization of the methylene blue and malachite green removal efficiency. They found that MLP-ANN model provided superior predictions [9, 67]. Multiple Linear Regression (MLR) and ANN-GA have also been used to simulate the adsorption tendencies of triamide on carbon nanotubes, with the ANN model demonstrating more accurate prediction in adsorption efficiency than the MLR model [27]. ANN optimization has also extended to the removal of micropollutants such as—bisphenol A, carbamazepine, ketoprofen, tonalide, etc. in water with the aid of chitosan/zeolite fixed-bed column [78]. While observing the solar degradation of 2-nitrophenol in water, ANN performed better than RSM in prediction accuracy by achieving a better R2 value (equal to 0.9877) [11]. Convoluted Neural Network (CNN) was applied for classifying microbeads in wastewater to determine microplastic contaminant presence and concentration, achieving a classification performance of 89% based on microscopic images [85]. The simulation of a sequencing batch reactor (SBR) based on MLP-ANN and RBF-ANN resulted in very high removal efficiency for total phosphate (TP), COD and NH_4^+–N (86%, 79%, and 93%, respectively) [10]. The R2 values for all MLP-ANN and RBF-ANN models ranged from 0.90 to 0.99, with the MLP-ANN consistently exhibiting higher R2 and lower Root Mean Square Error (RMSE) values than the RBF-ANN model. The SDAE deep learning network predictions outperformed five other tested models for COD, NH_4–N, and TN removal [73]. BP-ANN models effectively predicted COD and NH_4–N removal during SBR, achieving a 3.3% experimental error and an R2 value of 0.95 [48]. BP-ANN GA and MLP-ANN models have been effectively simulated and optimized for the removal of Pb^{2+} and malachite green from aqueous solutions, it achieved removal rates of at least 98.7% and R2 values greater than 0.99 [20]. Random forest (RF) model demonstrated good accuracy and prediction performance (R2 0.973) in modeling the adsorption of heavy metals such as Zn^{2+}, Cu^{2+}, Pb^{2+}, Cd^{2+}, Ni^{2+}, As^{3+} in biochar [90]. Therefore, ANN and its combinations with different models are being widely used for the removal of heavy metals from wastewater, physical and chemical treatment processes are commonly employed, with.

14.2.6 AI in Wastewater Reusability

Efforts in sustainable wastewater management can be improved by considering waste-water reuse as a viable strategy. Wastewater reuse along with AI can help in achieving recovering purified water, energy and materials typically involved in the treatment process, minimizing wastage of materials and improving cost efficiency. Reusing wastewater can also improve environmental quality and generate benefits economically and increase water savings [12]. Previously, Neural networks have been employed to evaluate potential for wastewater reusability by means of contact aeration for groundwater recharge [16]. Parameters such as rainfall index is an important input to enhance the cost effectiveness of wastewater reuse, depending on weather conditions, it can provide useful insights in the water reuse potential. A study conducted by Akhoundi and Nazif prioritized WWT and reuse applications with the aid of evidence-based reasoning methods; this provided a coordinated and comprehensive approach to assessing water reuse sustainability. Reusability of water can be feasible in sectors such as agriculture, groundwater recharging by artificial means and industrial applications [3].

In addition, ML techniques and cutting-edge imaging technologies are becoming increasingly significant in the evaluation and management of water quality. Real-time internet-based monitoring systems that utilize ANN are important for overcoming issues pertaining to maintenance and operation at WWTPs. It is used to optimize WWT systems to protect land and enhance tertiary treatment processes such as reverse osmosis (RO) and nanofiltration. ML techniques such as SVM, RF, and decision trees are also extensively used in drinking water management and groundwater quality analysis [2] wherein they have shown success in detection of impurities, source water quality prediction and control management. Through the use of imaging techniques, contaminants and pollutants in water can be visually identified and categorized, promoting more focused and effective cleaning techniques. These clustering imaging methods use complex algorithms to examine image data and identify related elements, helping to identify contaminants and cluster them according to classes based on similarities. This ground-breaking method makes it easier to understand the current status of the water quality and facilitates the creation of personalized purification processes that are meant to specifically target identified pollutants. This kind of visual characterization of certain types of water contamination allows for more effective and targeted water purification [6]. Various AI models used in wastewater treatment are summarized below in Table 14.1.

Table 14.1 AI models that have been applied in wastewater treatment

Model	Used in	References
ANN	BOD, COD, heavy metal and pollutant estimation, real-time monitoring, maintenance and operation, reducing energy consumption and membrane filtration	[47, 61, 63]
SVM	Disinfection and by-product management, groundwater analysis	[2, 6]
GA	Disinfection and by-product management	[6]
ANFIS	Decreasing energy consumption, increasing adsorption efficiency, increasing performance and efficiency, dye-removal in textile water	[22, 47, 64]
MLR	Organic pollutant removal	[26]

14.3 Economic Implications of AI in WWT

The integration of artificial intelligence (AI) into waste water management possesses significant economic potential. The extent to which AI technology supports and improves general efficiency will determine its economic implications. The performance of the water and sewer treatment infrastructure can be optimized using AI-powered analytics tools [46]. It is not only responsible for improving the environment and public health, but also saves money for municipalities and facility operators. Artificial intelligence (AI) facilitates increased efficiency, compliance security, and cost savings by optimizing water treatment for specific reuse cases and by continually monitoring effluent quality to anticipate any potential anomalies. It opens up new market possibilities by offering data-driven services and innovative solutions that meet the expanding need for effective and sustainable wastewater management. This contributes to fresh opportunities for market expansion and revenue generation. AI software may also identify abnormalities and faults early, allowing quick solutions and decision-making. Utilization patterns are also analyzed to effectively schedule network and plant maintenance before problems worsen. Furthermore, with the help of large volume of relevant data, AI algorithms optimize treatment processes for chemical and energy savings [69].

AI enables the sustainable reuse of purified and treated wastewater. Predictive analytics matches use cases for reuse, including agricultural or industrial irrigation, based on the amount and quality of treated output [46]. Thus, in drought-prone areas, there is less of a demand for freshwater intake and less water stress. Wastewater reuse also opens up new sources of income for treatment facilities. Technical plant management, engineering, data analytics, and other related fields provide new opportunities even as AI replaces manual monitoring professions. The transfer of displaced operators may be aided by workforce retraining programs [77]. Ultimately, AI's increased operational capacity and reliability result in decreased expenses for public health.

AI also possesses the capability to reduce and optimize energy costs and consumption. There have been various studies conducted towards the goal of cost and energy reduction for WWT. Han et al. introduced an enhanced MOOC approach and devised a flexible kernel function model capable of depicting intricate dynamic processes related to quality of water and total energy consumption. In comparison to the adaptive multi-objective differential evolution algorithm as well as the Proportional Integral (PI) controller strategy, their MOOC strategy resulted in decreases of 1.6%, 1.2%, and 2.2% in energy consumption values during different weather conditions i.e. dry, rainy, and stormy respectively [34]. In another study, Kuisak et al., created a model employing a Data Mining (DM) method. Their method worked on optimization of the Activated Sludge Process (ASP) and focused on controlling dissolved oxygen (DO) in the wastewater that was being treated. Had they prioritized energy savings over effluent quality, there was a possibility to reduce airflow by 15% [49].

Asadi et al. utilized a DM-based model for optimizing the aeration process, leading to a 31.4% reduction in energy consumption due to decreased aeration oxygen, while maintaining effluent quality higher than the standard [8]. The pump system was a crucial part in WWT energy consumption and was modeled using ANN model to reduce energy consumption and fluid flow rate, achieving a model error of less than 3% [89]. Innovative predictive control with the combination of statistical learning and deep reinforcement learning, was implemented to decrease electricity consumption, resulting in a 16.7% decrease as compared to normal working conditions [23].

Data-driven neural networks have also been created to enhance sewage pumping system performance, ensuring energy conservation with an average 10% energy savings. The most significant energy-saving case depicted a reduction in energy usage up to 1/4th (25%) [88]. Fuzzy logic (FL) control was employed to decrease the energy requirements of WWTP and demonstrated that aeration fuzzy control could get energy savings of more than 10% while maintaining effective removal [24]. The application of fuzzy neural networks demonstrated the effectiveness of ANFIS controllers in attaining the economic goals through computer simulations. This powerful and effective DO control tool showed the potential to save nearly 33% of operating costs [59].

Santis et al., deployed a two-tier hierarchical control strategy, encompassing model predictive control (MPC) and feedforward (FF). The model was devised to meet operating cost goals while also meeting quality and energy consumption used in WWT. This strategy resulted in an overall cost reduction of 0.8% along with a 6% decrease in aeration energy requirement [72].

There are other miscellaneous costs involved in WWTPs which can be minimized with the help of AI model integration. Utilizing a model-free RL approach proved effective in minimizing operating costs while maintaining the quality of water at acceptable levels, thereby enhancing the performance of WWTPs [35]. In comparison to benchmark simulation, model-based operating costs demonstrated a 6% reduction in costs. A BP-ANN has also been constructed for prediction of the adsorption of As^{3+} on cerium hydrochloride, contributing to a decrease in the adsorbent material costs [54].

Another example showcases cost reduction in the operational costs by focusing on aeration systems. In an attempt to simulate aeration in aerated submerged biofilm WWT, a fuzzy system was integrated with a BP-ANN. The AI system, addressing the varying inflow rates along with organic load requirements, controlled the airflow rate of the WWT process, resulting in approximately 38% savings in operational costs employing fuzzy control [15]. A hybrid Neuro-Fuzzy (NF) control system based on rough sets, including ANN, GA, and FL, were developed to enhance the accuracy of the control system. It was able to efficiently achieve real-time control objectives, addressing the uncertainties that are associated with the WWT process. The operational cost of the RsFLC system, on average, was 1.2% lower than the FL control system [18]. In an attempt to reduce greenhouse emissions, Sweetapple et al., devised a system employing the non-dominant sorting GA II. Their investigation into the efficiency of an economical method for reduction of greenhouse gas emissions during WWT revealed a maximum reduction of emission up to 18.5%, potentially reducing operational costs by 4.1% [76].

The examples given above demonstrate that AI-driven technology has the ability to lower the risk of wastewater leaks, malfunctioning of equipment, or inefficient treatment by using water more wisely and analyzing data. In this way, contaminated public areas and nearby biological ecosystems are kept safe away from hazardous effects. AI in wastewater subsequently assists countries in achieving reduced waste levels and avoiding fines and in meeting the global environmental initiatives and goals. Along with environmental effects and financial benefits, these innovations draw in private impact investors. Long-term economic growth is supported by AI's ability to essentially cut costs, create new revenue sources, promote job realignment, and generate major environmental advantages in wastewater management.

14.4 Cost Benefit Analysis Framework of Wastewater Treatment Projects

Cost–benefit analysis (CBA) is a widely popular economic tool due to its rational and systematic way of decision making. It ensures the economic feasibility of proposal implementation by assuming that a project can only be feasible and undertaken if the benefits associated with it are greater than the total costs involved. The net profit is calculated as a difference between the total benefits and total costs involved (see Eq. 14.1). These analyses can be carried out for different proposals and projects and can further be benchmarked analytically using Data Envelopment Analysis (DEA) [36]

$$\mathrm{NP} = \sum B_i - \sum C_i \tag{14.1}$$

Here, B_i and C_i are the value of the benefit and cost of item i respectively. A project is economically viable if the net profit (NP) > 0 and unviable if NP < 0. Usually, the highest value of NP which is > 0 is considered the best option [17].

While it is relatively easier to estimate the costs, in the case of WWT and management, there are numerous environmental benefits which are either neglected or omitted altogether due to not having any "market" value or due to being difficult to calculate. This issue was addressed by Molinos-Senante et al., in their study where they employed a concept of shadow-pricing to quantify the environmental benefits derived from wastewater management to estimate the CBA [60]. Other methods employed in the estimation of environmental benefits and costs in CBA include the widely popular contingent valuation method (CVM) [19, 28].

The application of AI in this paradigm enables process optimization along with predictive modeling, error detection, and results in higher efficacy. AI implementation in CBA could allow for ease in estimation of costs pertaining to installation, maintenance and evaluation of the newly emerging technologies in WWT. Additionally, owing to its dynamic nature, AI technology can enable optimization in resource management and can be helpful in re-assessment of CBA at different time points depending on any changes that may occur in the situation or emergence of newer methods for environmental benefit estimation. By leveraging CBA in AI led wastewater management, we can ensure that investments can yield financial returns and be environmentally responsible simultaneously.

AI has been previously implemented for cost–benefit analysis in sectors such as agriculture, in the field of cattle management, in travel, tourism and hospitality [31, 38] and with decent success. Gao et al., undertook a comprehensive analysis involving cost–benefit and data envelopment methods on 35 large-scale MBRs, each possessing a treatment capacity exceeding 10,000 m^3/d. Through an examination of investment and operational data, they assessed marginal costs and environmental benefits, factoring in shadow prices of pollutants. As a consequence of their study, they derived a net profit averaging around 35 CNY/m^3 or approximately 4.9 USD/m^3. They employed a non-radial directional distance function (NDDF) within their DEA framework after which, the average cost efficiency was found to be 0.77 and energy efficiency of the MBRs were found to be 0.66. Nonparametric tests revealed significant variations in the distributions of net profit and cost/energy efficiencies based on geographical location, effluent standards, and operational years of the MBRs. Economic panel modeling further demonstrated the impact of external factors, such as regional economic levels and population density, on cost/energy efficiencies. Their study suggested that MBRs exhibit noteworthy techno-economic advantages in instances of stringent effluent standards and underground locations. Furthermore, improvements in technological advancements and policy incentives could enhance the overall techno-economy of MBRs and also demonstrated a workable framework of CBA for WWTP project implementation and planning [25].

However, there are also some counter arguments for usage of CBA and AI in WWT. These arguments stem from the fact that certain concepts such as life and nature cannot be denoted a specific monetary value since these evaluations depend on nature and context. Also, there are risks which cannot be computed even by the AI if not previously trained to do so and even if, these are computed somehow and assigned a value, any existing regulations would be erroneously imputed as barrier making it seem like the benefits are trivial in comparison to the investments [75]. This could thwart WWT projects. Another argument against the usage of CBA is given by Ackerman and Heinzerling in their book where they contend that assignment of monetary values by CBA to things such as human-life and environment can lead to highly misleading outputs and end up justifying unethical actions an example of which they state being conclusions like—"toxic waste being preferably be dumped on to the developing nations since, life is already short in such places, along with low wages and poor productivity meaning that the dumping will be inexpensive in both action and consequence" [1].

Despite the arguments and counterarguments provided for and against employment of CBA and AI in WWT, it is arguable that these methods have time and again been employed to improve efficacy and save costs on infrastructure and maintenance. It is up to the user to decide whether the implementation of AI or usage of CBA for assessing projects feasibility will be desirable for their WWT outcomes.

14.5 Environmental Impact

AI in WWT can lead to better environmental control if used properly. There have been multiple studies on sustainable waste-water management [52, 53]. It can be employed in assessing the reusability of wastewater [16] reusing wastewater could lead to increase in water savings and increase economic benefits and also improve environment quality [12]. Wastewater reuse can be directed to agricultural and industrial applications, evidence-based reasoning methods have been employed to assess the sustainability of wastewater reuse [3].

It can also be used to make quick responses in cases of potential pollution incidents. With the aid of sensors, AI can detect anomalies and react accordingly to minimize the harmful effects at the source. This proactive approach helps prevent environmental degradation and protects aquatic ecosystems. Minimization of energy requirements is yet another positive impact of AI implementation in WWT. It has been successful in describing complex dynamic processes involved in water quality and energy consumption [34], aeration optimization for energy consumption reduction [8], predictive control to minimize power consumption [23] and in creating data-driven networks prioritizing pumping performance while conserving energy [88].

14.6 Challenges and Limitations

There are a variety of AI models that can be implemented for WWT plants however, any model is not a one size fits all and needs to be built and optimized according to need and data available. An ideal model needs to be carefully built depending on its advantages and the characteristics of the problem to achieve the best results [81]. Models such as RNN, CNN, KNN and FNN are often computationally expensive and are difficult to train due to their complex model structure [32, 68, 90]. Similar problems are encountered in DNNs and SVMs with the additional issue of overfitting tendencies in DNN and lack of applicability in larger datasets in case of SVMs [68]. Although PCA is widely used for its ease of use, it is also highly sensitive to noise in data and the dimensionality reduction achieved by it can also filter out important information [50]. Models built on ANFIS architecture have been highly successful in WWT [64] however, even in such models, the high computational costs and difficulty to define appropriate member functions can be a challenge in development [90].

Another issue that might arise is the black box issue of AI models, there is still little understanding as to how AI systems make decisions and why they arrive at their respective answers or predictions. Their internal logic leading to the decision taken or prediction made is still inscrutable and can lead to loss of confidence in the system altogether. This discrepancy, along with biased data or poor data reporting could lead to biased models which may not perform ideally [66]. Since AI heavily relies on data in its learning process to ensure optimum prediction capabilities, poor data management and availability would mean that it will not be possible to accurately build a model for the problem. Any model built on data for a specific WWT plant would not be implementable on another WWT plant having other variables [81]. The complete use cases, advantages and disadvantages of various AI models are summarized in Table 14.2.

14.7 Case Studies

14.7.1 Efficiency of AI in WWT: "WWT Plant Performance Analysis Using Artificial Intelligence—An Ensemble Approach"

A typical WWT plant is a highly complicated and dynamic system and there are a lot of parameters which impact on its operation as well as performance [2]. These parameters are inclusive of but not limited to aeration control, effluent concentration prediction, BOD estimation, etc. [44] and most of the traditional approaches for such predictions functioned on linear approximations. Since the traditional linear models were insufficient for predicting and monitoring of these parameters pertaining to WWT and management, Nourani et al., employed three different methods—Feed Forward Neural Network (FFNN), ANFIS, SVM along with a classical MLR. They employed this ensemble model to evaluate the performance of Nicosia WWT plant.

For the development of the model, they took daily data from The New WWT Plant of Nicosia for over a period of 2015–2016. The features they selected for the model building covered various sets of input and output parameters taking all seasonal variations into account. The dataset thus obtained consisted of influent values such as pH_{inf}, BOD_{inf}, COD_{inf}, Conductivity ($Cond_{inf}$), Total Nitrogen (TN_{inf}) as its inputs for the models, and three effluent values as targets—BOD_{eff}, COD_{eff} and TN_{eff}. This dataset was then split into a ratio of 75% and 25% for testing and validation. The individual models when compared showed ANFIS as the best predictive model in this case however, it was also observed that the ensemble model using simple averaging, weighted and neural networks improved the prediction. Among ensemble techniques,

Table 14.2 Uses, advantages and disadvantages of various AI models

Model name	Uses	Advantages	Limitations	References
CNN	Used in classification, regression and prediction, sometimes used for segmentation as well	Can extract useful features from images and therefore useful for image-related modelling	It is computationally expensive and difficult to train	[68]
RNN		Highly useful for working with time-series data and doesn't have a limit to the length of inputs	It is also computationally expensive and difficult to train	[32]
FNN		It is suitable for use in complex non-linear problems and is easy to implement and interpret	The model architecture is complex thus making the model computationally expensive	[90]
DNN		It is also suitable for use in complex non-linear problems and enables fast and accurate prediction	It is difficult to train, easy to overfit and is computationally expensive as well	[68]
Multilayer perceptron neural network (MLPNN)	Used in supervised learning; Back-propagation algorithm is widely used	It is very consistent and highly accurate even during changes	It is possible to get a local minimum instead of global one, is also prone to overfitting and the number of hidden layers or neurons have to be based on trial and error	[84]
RBF-ANN	Basis function in this model can be wavelets or Gaussian	Good for generalization, trains fast and has a high tolerance for noise	Requires large volumes of training data and hidden layers	[84]
KNN	Used in classification and regression	It is suitable for non-linear classification and is easy to implement	Consumes lot of memory and is computationally expensive	[68]

(continued)

Table 14.2 (continued)

Model name	Uses	Advantages	Limitations	References
SVM	Used in regression, prediction and classification, based on the principle of minimization and structured risk	Needs lesser volume of data and can handle data with high dimensionality	Its is not suitable for datasets that are too vast and can become computationally expensive since the memory demands and CPU time requirement increases in batch-mode training	[84, 90]
PCA	Mainly used for clustering and dimensionality reduction	It is very easy to implement	It is sensitive to data noise and may lose some important information	[50]
GA	Used in optimization, classification and regression. Works by simulating natural selection and genetic mechanisms derived from evolutionary theory	It is flexible and can be combined with other models and provides good optimization. It also supports multi-objective optimization	Unsuitable for data with higher dimensions, is difficult to trains and its local search ability is poor. It is also computationally expensive	[81, 84]
ANFIS	Used in classification, regression and prediction. It integrates gradient descent method and least square method for training	It combines the advantages of ANN and FIS and reaches optimum results quickly. It also has a shorter learning period and prevents settling on local minimum	It is difficult to define an appropriate membership function, the total number of hidden layers and is computationally expensive	[81, 84, 90]

NNE was reported as the most effective method of combination and improved the performance of the model up to 24% [64].

14.7.2 Cost Benefit of AI in WWT: "ML Approaches for the Treatment of Textile Wastewater Using Sugarcane Bagasse (Saccharum officinarum) Biochar"

Increasing agricultural, food, beverages, cosmetic, plastic, rubber, printing, paper and pulp production has led to an increase in the quantity of dyes in effluents polluting the water supplies. Stable dyes contain complexes which are not completely biodegradable and are very difficult to handle [83]. These dyes can be carcinogenic and can lead to birth defects, skin irritation, and allergic dermatitis [30, 51] and lead to detrimental environmental effects. In the study conducted by Kumari et al., the primary focus was on application of ML approaches for the treatment of textile wastewater with the aid of *Saccharum officinarum* biochar which is obtained as a byproduct of sugarcane industry [47].

The entire workflow involved baggase procurement, followed by application of statistical tests such as ANOVA for data standardization. A total of 30 randomized experiments were conducted taking numerical variables—adsorbent dose, dye concentration, pH and temperature. SEM analysis was employed to investigate the mechanism of crystal violet adsorption onto the adsorbent. In the project, they employed multiple models—Response Surface Methodology (RSM), KNN, ANN, and ANFIS to investigate the numerical variables. The models were evaluated on the basis of statistical error functions such as root mean square error, hybrid fractional error function, correlation coefficient, mean square error and mean relative percent deviation [47].

In their analysis, they found that the ANN model was the most suitable, demonstrating an R^2 value of 0.968. This optimisation in the strategy to find ideal solution for dye removal from wastewater was highly cost effective due to reduction in the number of trials for assessment of biosorption and helped to assess the influence of non-essential constituents on the process as well. The repurposing of agricultural waste materials through AI to serve as biosorbents is a highly cost-effective strategy for growing demands of WWT solutions [47].

14.7.3 Predictive Capability of AI in WWT: "Predicting Effluent Quality in Full-Scale WWT Plants Using Shallow and Deep Artificial Neural Networks"

WWT strategies are complex and nonlinear systems. They need accurate and reliable models for the purpose of monitoring and optimization of their performance. Traditional mechanistic models for prediction have drawbacks inclusive of but not limited to—high computational costs, poor adaptability, lack of generalization and poor scalability [65]. Therefore, AI models, possessing the ability to learn from existing data and capture its nonlinear and hidden complexities and relationships, have been increasingly applied to WWTP modeling, assessment and prediction. The aim of this project was the development and comparison of different ANN models, based on performance. The models, including shallow neural networks (SNNs) such as FFNNs and RF, and deep neural networks (DNNs) such as CNN, recurrent neural network (RNN), and pre-train stacked auto-encoder (SAE), were employed for estimating the quality of the effluent on a full-scale waste water treatment plant based in Syria. This was done to prevent the shortcomings of conventional mechanistic methods. The effluent quality indicators in the study included COD, BOD and total suspended solids (TSS) [39].

The data was collected from a WWTP serving a population of about 500,000 people and treating an average flow rate of 80,000 m^3/day. This data contained 11 input variables (influent COD, BOD, TSS, PO4, NO3, T, pH, EC, TDS, flow rate, and retention time) and 3 output variables (effluent COD, BOD, and TSS), It was collected over a period of three years from 2017 to 2019 with a sampling frequency of once per week, therefore resulting in a total of 156 samples. The dataset thus obtained was then divided into two sets of 70% and 30% for training and testing respectively. The AI models were built with the aid of Python libraries such as TensorFlow, Keras, and Scikit-learn. The performance of the created models was evaluated using the following criteria—R2, RMSE, mean absolute error (MAE), and Nash–Sutcliffe efficiency (NSE). The optimal model for each output variable was selected on the basis of the highest R2 and NSE values as well as the lowest RMSE and MAE values [39].

The results obtained demonstrated that the DNN models outperformed the SNN models in predicting the effluent quality of the WWTPs. The best models for COD, BOD, and TSS were found to be CNN, SAE, and RNN, respectively. The R2 values of the best models ranged from 0.93 to 0.98, indicating a high degree of fit and accuracy. The RMSE and MAE values of the best models ranged from 3.67 to 9.81 mg/L and from 2.64 to 7.32 mg/L, respectively, indicating a low degree of error and bias. The NSE values of the best models ranged from 0.92 to 0.98, indicating a high degree of efficiency and reliability. This study demonstrated the feasibility and effectiveness of AI model applications, in this case, especially DNNs, for the prediction the effluent quality of WWT plants. The AI models can therefore provide valuable information and insights for the WWT plant operators and managers to improve the process control and optimization, reduce the operational costs, improve the environmental impacts, and ultimately, ensure the compliance with the discharge standards. These AI models can also serve as a basis for developing decision support systems and early warning systems for WWTPs [39].

14.8 Policy Implication

Policy implications can be derived from two aspects—from putting emphasis on the importance of improvement of the WWT's energy efficiency and from regional integrated development. These goals can be achieved by investing in energy self-recovery technologies and proper planning, designing and coordination of such projects [37]. In addition to these motivations, policies should be implemented such that projects which invest heavily in technology for increasing the efficiency of the process and generate economies of scale are encouraged more and should be required to use renewable sources of energy to ensure that the process is environmentally sustainable. Due to lack of infrastructure in developing nations, lower GDP and lack of trust in newer technologies, such nations would need a stronger policy for encouragement along with dynamization to increase the GDP and willingness to participate in the sustainability endeavors [43].

Other policies that have been implemented previously include recycling and reuse of treated water for irrigation and other miscellaneous purposes [45]. Privatization efforts have also been employed to ensure 100% water reusability and recyclability in Saudi Arabia [62]. While the feasibility and success of such policies will only be certain with time. However, there are some crucial issues that might deter such policy implementation. One such issue is the public trust and readiness to adopt the technology as one of the major stakeholders [56].

14.9 Future Directions

The application of AI in WWT has the potential to revolutionize the industry by improving efficiency, sustainability, and environmental protection. AI models can help optimize the process, predict the efficiency, and evaluate the performance of WWTPs leading to more cost-effective solutions and better decision-making processes. The future implications of AI in WWT are vast and varied, ranging from real-time monitoring and control to predictive maintenance and risk management.

One of the most significant benefits of AI in WWT is real-time monitoring and control. AI models can analyze large amounts of data from various sources, such as sensors, cameras, and weather forecasts, to detect anomalies, predict failures, and optimize the WWT processes. For example, AI models can predict the influent flow rate, pollutant concentration, and effluent quality based on historical data and real-time measurements, and adjust the treatment parameters accordingly to meet the discharge standards and minimize the environmental impact. AI models can also detect leaks, blockages, and other malfunctions in the WWTP infrastructure, and alert the operators to take corrective actions before the situation worsens. Real-time monitoring and control can improve the reliability, safety, and efficiency of WWTPs, and reduce the operational costs and risks.

Another future implication of AI in WWT is predictive maintenance. AI models can analyze the performance data of the WWTP equipment, such as pumps, valves, and membranes, and predict their remaining useful life, failure probability, and maintenance needs. Predictive maintenance can help prevent unexpected breakdowns, reduce downtime, and extend the equipment lifespan, resulting in significant cost savings and environmental benefits. AI models can also optimize the maintenance schedule, prioritize the critical components, and recommend the best maintenance strategy based on the cost–benefit analysis and risk assessment. Predictive maintenance can improve the asset management, reliability, and sustainability of WWTPs, and enhance the customer satisfaction and public trust.

Another future implication of AI in WWT is in the field of risk management and compliance to standards. AI models can assess the environmental risks associated with the WWTP operations, such as the discharge of pollutants, the emission of greenhouse gases, and the consumption of energy and water. AI models can also evaluate the social and economic risks, such as the public health impacts, the regulatory compliance, and the stakeholder engagement. Risk management can help WWTPs identify the potential hazards, quantify the likelihood and consequence of the risks, and develop the mitigation and contingency plans. AI models can also simulate the scenarios of the WWTP operations under different conditions, such as climate change, population growth, and technological innovation, and evaluate the resilience and adaptability of the WWTPs. Risk management can improve the sustainability, transparency, and accountability of WWTPs, and foster the innovation and collaboration among the stakeholders.

14.10 Conclusion

AI has seen applicability in various fields with great success. In the future, with better data collection protocols and implementation of standardization policies, it will be possible for large scale implementation of AI models for various aspects of WWTs as well. Such data-driven approaches can also help to reduce cost of water treatment by optimizing resource usage and minimizing by-product and also prevent potential future losses by accurately predicting material needs and troubleshooting without needing continuous human supervision. These optimizing strategies, predictive maintenance, resource optimization and enhanced efficiency as a result of integrating AI in WWT strategies could therefore contribute to higher cost–benefit ratio and encourage sustainable projects for WWT plants.

As we have already summarized in this chapter, AI implementation and integration has already been successful in various research and projects leading to good returns in terms of value in a sustainable manner. It is therefore advisable for policymakers to further encourage the adaptation of such new technologies and gain public trust to ensure a seamless transition toward AI-integrated WWT systems. Public awareness campaigns, educational initiatives, and transparent communication about the benefits and safety measures associated with AI in water treatment can play a crucial role in building confidence among the community and stakeholders.

To conclude, the future implications of AI in WWT are promising and revolutionizing. Real-time monitoring and control, predictive maintenance, and risk management are just scratch the surface of the plethora of examples of how AI can transform the WWTP industry. However, there are also challenges and limitations that need to be addressed, such as the data quality, the model interpretability, the ethical and legal issues, human–machine interaction and current lack of policies for standardization and benchmarking. Therefore, it is essential to develop a holistic and integrated approach to AI in WWT, which involves not only the technical and financial aspects but also the social, economic, and environmental dimensions. By doing so, we can ensure that AI in WWT is not only smart but also sustainable, equitable, and resilient.

References

1. Ackerman F, Heinzerling L, Massey R (2005) Applying cost-benefit to past decisions: was environmental protection ever a good idea. Adm Law Rev 57
2. Ahmadi MM, Mahdavirad H, Bakhtiari B (2017) Multi-criteria analysis of site selection for groundwater recharge with treated municipal wastewater. Water Sci Technol 76(4):909–919. https://doi.org/10.2166/WST.2017.273
3. Akhoundi A, Nazif S (2018) Sustainability assessment of wastewater reuse alternatives using the evidential reasoning approach. J Clean Prod 195:1350–1376. https://doi.org/10.1016/j.jclepro.2018.05.220
4. Al Aani S, Bonny T, Hasan SW, Hilal N (2019) Can machine language and artificial intelligence revolutionize process automation for water treatment and desalination? Desalination 458:84–96. https://doi.org/10.1016/j.desal.2019.02.005
5. Alam G, Ihsanullah I, Naushad M, Sillanpää M (2022) Applications of artificial intelligence in water treatment for optimization and automation of adsorption processes: recent advances and prospects. Chem Eng J 427:130011. https://doi.org/10.1016/J.CEJ.2021.130011
6. Alprol AE;, Mansour AT;, Ibrahim ME, Ashour E-D;, Kantamaneni K, Rathnayake U, Alprol AE, Tageldein Mansour A, Ezz M, Ibrahim E-D, Ashour M (2024) Artificial intelligence technologies revolutionizing wastewater treatment: current trends and future prospective. Water 16(2):314. https://doi.org/10.3390/W16020314
7. Arinez JF, Chang Q, Gao RX, Xu C, Zhang J (2020) Artificial intelligence in advanced manufacturing: current status and future outlook. J Manuf Sci Eng Trans ASME 142(11). https://doi.org/10.1115/1.4047855/1085487
8. Asadi A, Verma A, Yang K, Mejabi B (2017) Wastewater treatment aeration process optimization: a data mining approach. J Environ Manage 203:630–639. https://doi.org/10.1016/j.jenvman.2016.07.047
9. Asfaram A, Ghaedi M, Hajati S, Goudarzi A (2016) Synthesis of magnetic γ-Fe_2O_3-based nanomaterial for ultrasonic assisted dyes adsorption: modeling and optimization. Ultrason Sonochem 32:418–431. https://doi.org/10.1016/j.ultsonch.2016.04.011
10. Bagheri M, Mirbagheri SA, Ehteshami M, Bagheri Z (2015) Modeling of a sequencing batch reactor treating municipal wastewater using multi-layer perceptron and radial basis function artificial neural networks. Process Saf Environ Prot 93:111–123. https://doi.org/10.1016/j.psep.2014.04.006
11. Bararpour ST, Feylizadeh MR, Delparish A, Qanbarzadeh M, Raeiszadeh M, Feilizadeh M (2018) Investigation of 2-nitrophenol solar degradation in the simultaneous presence of $K_2S_2O_8$

and H_2O_2: using experimental design and artificial neural network. J Clean Prod 176:1154–1162. https://doi.org/10.1016/J.JCLEPRO.2017.11.191
12. Bozkurt H, van Loosdrecht MCM, Gernaey KV, Sin G (2016) Optimal WWTP process selection for treatment of domestic wastewater—a realistic full-scale retrofitting study. Chem Eng J 286:447–458. https://doi.org/10.1016/j.cej.2015.10.088
13. Cao L (2020) AI in finance: a review. SSRN Electron J. https://doi.org/10.2139/SSRN.3647625
14. Carreres-Prieto D, García JT, Carrillo JM, Vigueras-Rodríguez A (2023) Towards highly economical and accurate wastewater sensors by reduced parts of the LED-visible spectrum. Sci Total Environ 871:162082. https://doi.org/10.1016/J.SCITOTENV.2023.162082
15. Chen JC, Bin CN (2007) Mining the fuzzy control rules of aeration in a submerged biofilm wastewater treatment process. Eng Appl Artif Intell 20(7):959–969. https://doi.org/10.1016/j.engappai.2006.11.012
16. Chen JC, Chang NB, Shieh WK (2003) Assessing wastewater reclamation potential by neural network model. Eng Appl Artif Intell 16(2 SPEC):149–157. https://doi.org/10.1016/S0952-1976(03)00056-3
17. Chen R, Wang XC (2009) Cost-benefit evaluation of a decentralized water system for wastewater reuse and environmental protection. Water Sci Technol 59(8):1515–1522. https://doi.org/10.2166/WST.2009.156
18. Chen WC, Bin CN, Chen JC (2003) Rough set-based hybrid fuzzy-neural controller design for industrial wastewater treatment. Water Res 37(1):95–107. https://doi.org/10.1016/S0043-1354(02)00255-5
19. Del Saz-Salazar S, Hernández-Sancho F, Sala-Garrido R (2009) The social benefits of restoring water quality in the context of the water framework directive: a comparison of willingness to pay and willingness to accept. Sci Total Environ 407(16):4574–4583. https://doi.org/10.1016/J.SCITOTENV.2009.05.010
20. Dil EA, Ghaedi M, Asfaram A, Hajati S, Mehrabi F, Goudarzi A (2017) Preparation of nanomaterials for the ultrasound-enhanced removal of Pb2+ ions and malachite green dye: chemometric optimization and modeling. Ultrason Sonochem 34:677–691. https://doi.org/10.1016/j.ultsonch.2016.07.001
21. Emaminejad SA, Morgan VL, Kumar K, Kavathekar A, Ragush C, Shuai W, Jia Z, Huffaker R, Wells G, Cusick RD (2022) Statistical and microbial analysis of bio-electrochemical sensors used for carbon monitoring at water resource recovery facilities. Environ Sci (Camb) 8(10):2052–2064. https://doi.org/10.1039/D1EW00653C
22. Fawzy M, Nasr M, Adel S, Nagy H, Helmi S (2016) Environmental approach and artificial intelligence for Ni(II) and Cd(II) biosorption from aqueous solution using Typha domingensis biomass. Ecol Eng 95:743–752. https://doi.org/10.1016/j.ecoleng.2016.07.007
23. Filipe J, Bessa RJ, Reis M, Alves R, Póvoa P (2019) Data-driven predictive energy optimization in a wastewater pumping station. Appl Energy 252. https://doi.org/10.1016/j.apenergy.2019.113423
24. Fiter M, Güell D, Comas J, Colprim J, Poch M, Rodríguez-Roda I (2005) Energy saving in a wastewater treatment process: an application of fuzzy logic control. Environ Technol 26(11):1263–1270. https://doi.org/10.1080/09593332608618596
25. Gao T, Xiao K, Zhang J, Zhang X, Wang X, Liang S, Sun J, Meng F, Huang X (2021) Cost-benefit analysis and technical efficiency evaluation of full-scale membrane bioreactors for wastewater treatment using economic approaches. J Clean Prod 301:126984. https://doi.org/10.1016/J.JCLEPRO.2021.126984
26. Ghaedi AM, Ghaedi M, Pouranfard AR, Ansari A, Avazzadeh Z, Vafaei A, Tyagi I, Agarwal S, Gupta VK (2016) Adsorption of Triamterene on multi-walled and single-walled carbon nanotubes: artificial neural network modeling and genetic algorithm optimization. J Mol Liq 216:654–665. https://doi.org/10.1016/j.molliq.2016.01.068
27. Ghaedi M, Shojaeipour E, Ghaedi AM, Sahraei R (2015) Isotherm and kinetics study of malachite green adsorption onto copper nanowires loaded on activated carbon: artificial neural network modeling and genetic algorithm optimization. Spectrochim Acta A Mol Biomol Spectrosc 142:135–149. https://doi.org/10.1016/j.saa.2015.01.086

28. Godfrey S, Labhasetwar P, Wate S (2009) Greywater reuse in residential schools in Madhya Pradesh, India—a case study of cost–benefit analysis. Resour Conserv Recycl 53(5):287–293. https://doi.org/10.1016/J.RESCONREC.2009.01.001
29. Golding LP, Nicola GN (2019) A business case for artificial intelligence tools: the currency of improved quality and reduced cost. J Am Coll Radiol 16(9):1357–1361. https://doi.org/10.1016/J.JACR.2019.05.004
30. Grassi P, Drumm FC, Spannemberg SS, Georgin J, Tonato D, Mazutti MA, Gonçalves JO, Oliveira MLS, Dotto GL, Jahn SL (2020) Solid wastes from the enzyme production as a potential biosorbent to treat colored effluents containing crystal violet dye. Environ Sci Pollut Res 27(10):10484–10494. https://doi.org/10.1007/S11356-020-07664-0/METRICS
31. Griffith AP, Boyer CN, Rhinehart JD, Carter C (2020) Cost-benefit analysis of timed A.I. and natural service in beef cattle. https://doi.org/10.22004/AG.ECON.303636
32. Guo Z, Du B, Wang J, Shen Y, Li Q, Feng D, Gao X, Wang H (2020) Data-driven prediction and control of wastewater treatment process through the combination of convolutional neural network and recurrent neural network. RSC Adv 10(23):13410–13419. https://doi.org/10.1039/D0RA00736F
33. Haimi H, Mulas M, Corona F, Vahala R (2013) Data-derived soft-sensors for biological wastewater treatment plants: an overview. Environ Model Softw 47:88–107. https://doi.org/10.1016/J.ENVSOFT.2013.05.009
34. Han HG, Zhang L, Liu HX, Qiao JF (2018) Multiobjective design of fuzzy neural network controller for wastewater treatment process. Applied Soft Comput J 67:467–478. https://doi.org/10.1016/j.asoc.2018.03.020
35. Hernández-Del-Olmo F, Llanes FH, Gaudioso E (2012) An emergent approach for the control of wastewater treatment plants by means of reinforcement learning techniques. Expert Syst Appl 39(3):2355–2360. https://doi.org/10.1016/j.eswa.2011.08.062
36. Hernández-Sancho F, Sala-Garrido R (2009) Technical efficiency and cost analysis in wastewater treatment processes: a DEA approach. Desalination 249(1):230–234. https://doi.org/10.1016/J.DESAL.2009.01.029
37. Huang R, Shen Z, Wang H, Xu J, Ai Z, Zheng H, Liu R (2021) Evaluating the energy efficiency of wastewater treatment plants in the Yangtze River Delta: perspectives on regional discrepancies. Appl Energy 297:117087. https://doi.org/10.1016/J.APENERGY.2021.117087
38. Ivanov S, Webster C (2017) Adoption of robots, artificial intelligence and service automation by travel, tourism and hospitality companies—a cost-benefit analysis. Tour Econ 26(7):1065–1085. https://doi.org/10.1177/1354816619879583
39. Jafar R, Awad A, Jafar K, Shahrour I (2022) Predicting effluent quality in full-scale wastewater treatment plants using shallow and deep artificial neural networks. Sustainability 14(23):15598. https://doi.org/10.3390/SU142315598
40. Jha AK, Ghimire A, Thapa S, Jha AM, Raj R (2021) A review of AI for urban planning: towards building sustainable smart cities. In: Proceedings of the 6th international conference on inventive computation technologies, ICICT 2021, pp 937–944. https://doi.org/10.1109/ICICT50816.2021.9358548
41. Jing L, Chen B, Zhang B, Zheng J, Liu B (2014) Naphthalene degradation in seawater by UV irradiation: the effects of fluence rate, salinity, temperature and initial concentration. Mar Pollut Bull 81(1):149–156. https://doi.org/10.1016/j.marpolbul.2014.02.003
42. Kerdlap P, Low JSC, Ramakrishna S (2019) Zero waste manufacturing: a framework and review of technology, research, and implementation barriers for enabling a circular economy transition in Singapore. Resour Conserv Recycl 151:104438. https://doi.org/10.1016/J.RESCONREC.2019.104438
43. Khan SAR, Ponce P, Yu Z, Golpîra H, Mathew M (2022) Environmental technology and wastewater treatment: strategies to achieve environmental sustainability. Chemosphere 286:131532. https://doi.org/10.1016/J.CHEMOSPHERE.2021.131532
44. Kim M, Park HS, Jin GJ, Cho WH, Lee DK, Hyun MS, Choi CH, Kim HJ (2006) A novel combined biomonitoring system for BOD measurement and toxicity detection using microbial fuel cells. In: Proceedings of IEEE sensors, pp 1247–1248. https://doi.org/10.1109/ICSENS.2007.355855

45. Kreutzberger W, Robinson J, Bkayrat R, Bacon E (2012) Comprehensive study of water reuse role in integrated water resources management—Kingdom of Saudi Arabia. In: World environmental and water resources congress 2012: crossing boundaries, proceedings of the 2012 congress, pp 3366–3377. https://doi.org/10.1061/9780784412312.339
46. Krishnan SR, Nallakaruppan MK, Chengoden R, Koppu S, Iyapparaja M, Sadhasivam J, Sethuraman S (2022) Smart water resource management using artificial intelligence—a review. Sustainability 14(20):13384. https://doi.org/10.3390/SU142013384
47. Kumari S, Chowdhry J, Chandra Garg M (2024) AI-enhanced adsorption modeling: challenges, applications, and bibliographic analysis. J Environ Manage 351. https://doi.org/10.1016/j.jenvman.2023.119968
48. Kundu P, Debsarkar A, Mukherjee S (2013) Artificial neural network modeling for biological removal of organic carbon and nitrogen from slaughterhouse wastewater in a sequencing batch reactor. Adv Artif Neural Syst 2013:1–15. https://doi.org/10.1155/2013/268064
49. Kusiak A, Wei X (2013) Optimization of the activated sludge process. J Energy Eng-ASCE 139(1):12–17. https://doi.org/10.1061/(ASCE)EY.1943-7897.0000092
50. Lee MW, Hong SH, Choi H, Kim JH, Lee DS, Park JM (2008) Real-time remote monitoring of small-scaled biological wastewater treatment plants by a multivariate statistical process control and neural network-based software sensors. Process Biochem 43(10):1107–1113. https://doi.org/10.1016/J.PROCBIO.2008.06.002
51. Liu L, Gao ZY, Su XP, Chen X, Jiang L, Yao JM (2015) Adsorption removal of dyes from single and binary solutions using a cellulose-based bioadsorbent. ACS Sustain Chem Eng 3(3):432–442. https://doi.org/10.1021/SC500848M/SUPPL_FILE/SC500848M_SI_001.PDF
52. López-Morales CA, Rodríguez-Tapia L (2019) On the economic analysis of wastewater treatment and reuse for designing strategies for water sustainability: lessons from the Mexico Valley Basin. Resour Conserv Recycl 140:1–12. https://doi.org/10.1016/j.resconrec.2018.09.001
53. Man Y, Hu Y, Ren J (2019) Forecasting COD load in municipal sewage based on ARMA and VAR algorithms. Resour Conserv Recycl 144:56–64. https://doi.org/10.1016/j.resconrec.2019.01.030
54. Mandal S, Mahapatra SS, Sahu MK, Patel RK (2015) Artificial neural network modelling of As(III) removal from water by novel hybrid material. Process Saf Environ Prot 93:249–264. https://doi.org/10.1016/j.psep.2014.02.016
55. Martini S, Roni KA (2021) The existing technology and the application of digital artificial intelligent in the wastewater treatment area: a review paper. J Phys Conf Ser 1858(1). https://doi.org/10.1088/1742-6596/1858/1/012013
56. Massoud MA, Kazarian A, Alameddine I, Al-Hindi M (2018) Factors influencing the reuse of reclaimed water as a management option to augment water supplies. Environ Monit Assess 190(9). https://doi.org/10.1007/S10661-018-6905-Y
57. Mazaheri H, Ghaedi M, Ahmadi Azqhandi MH, Asfaram A (2017) Application of machine/statistical learning, artificial intelligence and statistical experimental design for the modeling and optimization of methylene blue and Cd(ii) removal from a binary aqueous solution by natural walnut carbon. Phys Chem Chem Phys 19(18):11299–11317. https://doi.org/10.1039/C6CP08437K
58. Messikh N, Chiha M, Ahmedchekkat F, Al Bsoul A (2015) Application of radial basis function neural network for removal of copper using an emulsion liquid membrane process assisted by ultrasound. Desalination Water Treat 56(2):399–408. https://doi.org/10.1080/19443994.2014.936513
59. Mingzhi H, Jinquan W, Yongwen M, Yan W, Weijiang L, Xiaofei S (2009) Control rules of aeration in a submerged biofilm wastewater treatment process using fuzzy neural networks. Expert Syst Appl 36(7):10428–10437. https://doi.org/10.1016/J.ESWA.2009.01.035
60. Molinos-Senante M, Hernández-Sancho F, Sala-Garrido R (2010) Economic feasibility study for wastewater treatment: a cost–benefit analysis. Sci Total Environ 408(20):4396–4402. https://doi.org/10.1016/J.SCITOTENV.2010.07.014
61. Moral H, Aksoy A, Gokcay CF (2008) Modeling of the activated sludge process by using artificial neural networks with automated architecture screening. Comput Chem Eng 32(10):2471–2478. https://doi.org/10.1016/J.COMPCHEMENG.2008.01.008

62. Mu'azu ND, Abubakar IR, Blaisi NI (2020) Public acceptability of treated wastewater reuse in Saudi Arabia: implications for water management policy. Sci Tot Environ 721. https://doi.org/10.1016/J.SCITOTENV.2020.137659
63. Niu C, Li X, Dai R, Wang Z (2022) Artificial intelligence-incorporated membrane fouling prediction for membrane-based processes in the past 20 years: a critical review. Water Res 216. https://doi.org/10.1016/J.WATRES.2022.118299
64. Nourani V, Elkiran G, Abba SI (2018) Wastewater treatment plant performance analysis using artificial intelligence—an ensemble approach. Water Sci Technol 78(10):2064–2076. https://doi.org/10.2166/WST.2018.477
65. Ongsulee P (2018) Artificial intelligence, machine learning and deep learning. In: International conference on ICT and knowledge engineering, pp 1–6. https://doi.org/10.1109/ICTKE.2017.8259629
66. Rai A (2020) Explainable AI: from black box to glass box. J Acad Mark Sci 48(1):137–141. https://doi.org/10.1007/S11747-019-00710-5/TABLES/1
67. Ranjbari S, Khatibi T, Vosough Dizaji A, Sajadi H, Totonchi M, Ghaffari F (2021) CNFE-SE: a novel approach combining complex network-based feature engineering and stacked ensemble to predict the success of intrauterine insemination and ranking the features. BMC Med Inform Decis Mak 21(1):1. https://doi.org/10.1186/s12911-020-01362-0
68. Rodríguez-Rangel H, Arias DM, Morales-Rosales LA, Gonzalez-Huitron V, Partida MV, García J (2022) Machine learning methods modeling carbohydrate-enriched cyanobacteria biomass production in wastewater treatment systems. Energies 15(7):2500. https://doi.org/10.3390/EN15072500
69. Rojek I, Jasiulewicz-Kaczmarek M, Piechowski M, Mikołajewski D (2023) An artificial intelligence approach for improving maintenance to supervise machine failures and support their repair. Appl Sci 13(8):4971. https://doi.org/10.3390/APP13084971
70. Rong G, Mendez A, Bou Assi E, Zhao B, Sawan M (2020) Artificial intelligence in healthcare: review and prediction case studies. Engineering 6(3):291–301. https://doi.org/10.1016/J.ENG.2019.08.015
71. Sahu S, Kaur A, Singh G, Kumar Arya S (2023) Harnessing the potential of microalgae-bacteria interaction for eco-friendly wastewater treatment: a review on new strategies involving machine learning and artificial intelligence. J Environ Manage 346:119004. https://doi.org/10.1016/J.JENVMAN.2023.119004
72. Santín I, Pedret C, Vilanova R (2015) Applying variable dissolved oxygen set point in a two level hierarchical control structure to a wastewater treatment process. J Process Control 28:40–55. https://doi.org/10.1016/J.JPROCONT.2015.02.005
73. Shi S, Xu G (2018) Novel performance prediction model of a biofilm system treating domestic wastewater based on stacked denoising auto-encoders deep learning network. Chem Eng J 347:280–290. https://doi.org/10.1016/J.CEJ.2018.04.087
74. Singh NK, Yadav M, Singh V, Padhiyar H, Kumar V, Bhatia SK, Show PL (2023) Artificial intelligence and machine learning-based monitoring and design of biological wastewater treatment systems. Bioresour Technol 369:128486. https://doi.org/10.1016/J.BIORTECH.2022.128486
75. Sunstein CR (2005) Cost-benefit analysis and the environment. 115(2):351–385. https://doi.org/10.1086/426308
76. Sweetapple C, Fu G, Butler D (2014) Multi-objective optimisation of wastewater treatment plant control to reduce greenhouse gas emissions. Water Res 55:52–62. https://doi.org/10.1016/J.WATRES.2014.02.018
77. Ungureanu N, Vlăduț V, Voicu G (2020) Water scarcity and wastewater reuse in crop irrigation. Sustainability (21):9055. https://doi.org/10.3390/SU12219055
78. Vakili M, Mojiri A, Kindaichi T, Cagnetta G, Yuan J, Wang B, Giwa AS (2019) Cross-linked chitosan/zeolite as a fixed-bed column for organic micropollutants removal from aqueous solution, optimization with RSM and artificial neural network. J Environ Manage 250. https://doi.org/10.1016/J.JENVMAN.2019.109434
79. Vanrolleghem PA, Verstraete W (1993) Simultaneous biokinetic characterization of heterotrophic and nitrifying populations of activated sludge with an on-line respirographic biosensor. Water Sci Technol 28(11–12):377–387. https://doi.org/10.2166/WST.1993.0678

80. Vassos TD (1993) Future directions in instrumentation, control and automation in the water and wastewater industry. Water Sci Technol 28(11–12):9–14. https://doi.org/10.2166/WST.1993.0640
81. Wang Y, Cheng Y, Liu H, Guo Q, Dai C, Zhao M, Liu DA, Wang Y, Cheng Y, Liu H, Guo Q, Dai C, Zhao M, Liu D (2023) A review on applications of artificial intelligence in wastewater treatment. Sustainability 15(18):13557. https://doi.org/10.3390/SU151813557
82. Wu C-J, Raghavendra R, Gupta U, Acun B, Ardalani N, Maeng K, Chang G, Behram FA, Huang J, Bai C, Gschwind M, Gupta A, Ott M, Melnikov A, Candido S, Brooks D, Chauhan G, Lee B, Lee H-HS, Akyildiz B, Balandat M, Spisak J, Jain R, Rabbat M, Hazelwood K (2022) Sustainable AI: environmental implications, challenges and opportunities. Proc Mach Learn Syst 4:795–813
83. Yagub MT, Sen TK, Afroze S, Ang HM (2014) Dye and its removal from aqueous solution by adsorption: a review. Adv Colloid Interface Sci 209:172–184. https://doi.org/10.1016/J.CIS.2014.04.002
84. Ye Z, Yang J, Zhong N, Tu X, Jia J, Wang J (2020) Tackling environmental challenges in pollution controls using artificial intelligence: a review. Sci Tot Environ 699. https://doi.org/10.1016/j.scitotenv.2019.134279
85. Yurtsever M, Yurtsever U (2019) Use of a convolutional neural network for the classification of microbeads in urban wastewater. Chemosphere 216:271–280. https://doi.org/10.1016/J.CHEMOSPHERE.2018.10.084
86. Zaferani SPG, Emami MRS, Amiri MK, Binaeian E (2019) Optimization of the removal Pb (II) and its Gibbs free energy by thiosemicarbazide modified chitosan using RSM and ANN modeling. Int J Biol Macromol 139:307–319. https://doi.org/10.1016/J.IJBIOMAC.2019.07.208
87. Zhang W, Tooker NB, Mueller AV (2020) Enabling wastewater treatment process automation: leveraging innovations in real-time sensing, data analysis, and online controls. Environ Sci (Camb) 6(11):2973–2992. https://doi.org/10.1039/D0EW00394H
88. Zhang Z, Kusiak A, Zeng Y, Wei X (2016) Modeling and optimization of a wastewater pumping system with data-mining methods. Appl Energy 164:303–311. https://doi.org/10.1016/J.APENERGY.2015.11.061
89. Zhang Z, Zeng Y, Kusiak A (2012) Minimizing pump energy in a wastewater processing plant. Energy 47(1):505–514. https://doi.org/10.1016/J.ENERGY.2012.08.048
90. Zhu J, Jiang Z, Feng L (2022) Improved neural network with least square support vector machine for wastewater treatment process. Chemosphere 308:136116. https://doi.org/10.1016/J.CHEMOSPHERE.2022.136116

Chapter 15
Regulatory Frameworks and Policy Recommendations for AI Enabled Wastewater Treatment

Ashwani Pant

Abstract Water, sanitation, and hygiene access is the most fundamental human need for health and well-being. The UN Sustainable Development Goal 6 is about "Ensure availability and sustainable management of water and sanitation for all" and now recognized as human right. There are currently over 4.2 billion people not having safe sanitation access and over 2.2 billion people without access to safe drinking water. As a result of Climate Change natural disasters like droughts and floods has become more frequent. Eighty percent of the world's wastewater is simply released back into nature. Wastewater management encompasses many tasks with the overarching goal of fostering efficient and ethical water usage and its treatment. AI has shown potential use in the field of vehicular production, banking, finance, agriculture, exploration of space, and AI creativity. Due to its obvious benefits in terms of efficiency, speed, and autonomy—all of which minimize the need for human intervention—artificial intelligence (AI) is increasingly being incorporated into the wastewater treatment sector. For ensuring sustainable management of wastewater UNEP is determined to focus on global action which requires collaboration amongst UN and Non-UN countries. Both developed and developing countries are to adopt and effectively implement strict regulations, Infrastructure and finance for wastewater management. The present study highlights the issues and challenges to ensure the safe and effective deployment of artificial intelligence in this critical sector. AI also raises legal, ethical, environmental, and privacy concerns which are required to be addressed globally.

Keywords Water sanitation · Hygiene access · Sustainable Development Goal 6 · Climate change · Wastewater management · Artificial intelligence (AI) · Sustainable development · Global action · Regulations · Ethical concerns

---

A. Pant (✉)
School of Law, Dr. Vishwanath Karad MIT World Peace University, Pune, Maharashtra, India
e-mail: ashwani.alm@gmail.com; ashwani.pant@mitwpu.edu.in

M. C. Garg (ed.), *The AI Cleanse: Transforming Wastewater Treatment Through Artificial Intelligence*, Springer Water, https://doi.org/10.1007/978-3-031-67237-8_15

15.1 Introduction

Water has a crucial role to play in sustaining ecosystems, serving as a fundamental resource for promoting human health and well-being and as a prerequisite for fostering economic growth. Hence, it has acquired a central position within the agenda framework of Sustainable Development 2030. The interconnections between Sustainable Development Goal 6 (SDG 6), which pertains to ensuring universal access to clean water and sanitation, and the remaining Sustainable Development Goals (SDGs) are significant [1]. Water sustainability is one of the fundamental towards attaining of the 'sustainable Development Goals (SDGs), 2015'. These goals are governed by three overarching principles: "the implementation of a human rights-based approach, ensuring that no one is excluded, and promoting gender equality and women's empowerment" [2] (Fig. 15.1).

Overuse, pollution, and insufficient recycling have all contributed to the world's water supply drying up during the past few decades. Life, biodiversity, ecological health, agriculture, and industry all rely heavily on a steady supply of water. The situation is more worsened due to the effects of global warming, increased population, and increased urbanization. Wastewater management is an essential part of a sustainable approach to water use. Over the past decade, wastewater treatment and reuse have improved, but an estimated 50% of the world's wastewater is still released

Fig. 15.1 The relationship of SDG 6 with other SDG's. *Source* GRID-Arendal/Studio Atlantis, 2020. https://www.grida.no/resources/13734, United Nations Economic and Social Commission for Asia and the Pacific (UN ESCAP), 2017 (Developed by the Author)

untreated and becoming a global issue. In 2013, municipal wastewater was expected to reach 330 billion m^3. Up to 360–380 billion m^3/year by 2015, according to calculations. This quantity of wastewater is expected to be five times the annual volume of water pouring over Niagara Falls by the end of 2023 [3]. World population is expected to grow by 2 billion to 10 billion by 2050, driving up wastewater volumes. UNEP predicts that by 2030, the world would be producing 470–497 billion m^3 of wastewater annually from home and municipal sources, an increase of 24–38% from the volume of wastewater produced in 2015 [3].

The potential co-benefits of implementing safe and appropriate wastewater management for resource recovery and reuse extend beyond the attainment of water security. These include enhancing overall health of environment and well-being, promoting health of human being, biodiversity conservation and preservation, decreasing use of fertilizers, and fostering diversification of production of energy through different resources and opening of economic development. Women in particular benefit from an inclusive water and wastewater management strategy since it improves their access to clean water and frees them up to work. Many countries have successfully reused wastewater, but scaled water reuse still challenging [3].

AI has displayed potential in different fields like agriculture, automobiles, banking, finance, exploration of space, creativity and now set towards revolutionizing wastewater treatment due to its efficiency, speed, and independence from humans. Artificial Intelligence in wastewater treatment is used to monitor efficiency of plant, determination of oxygen demand both biologically and chemically, predict hardness and turbidity, eliminates the dissolved nitrogen and sulphur. Commonly used AI models in wastewater treatment includes Artificial Neural Networks (ANN), Fuzzy Logic Algorithms (FL), and Genetic Algorithms (GA) [4].

This chapter delves into regulation of wastewater treatment which is one of the complex issues in the era of sustainable development. There is requirement for the development of a cross-sectoral global approach amongst states to have a robust and effective wastewater management system. In Sect. 2 of the chapter discuss the how AI ingresses into wastewater treatment and its different models used. Section 3 is the exploration of different challenges posed by AI like legal, ethical, environmental and privacy. Section 4 will discuss the various regulatory framework International as well as domestic dealing with wastewater treatment. Section 5 is the conclusion with suggestions to improve regulatory framework in the era of AI.

15.2 Ingression of AI in Wastewater Treatment

Wastewater which has various nutrients, suspended particles, microorganism is being released every day without treatment or reuse into environment [5]. The major sources of wastewater are domestic and municipal waste, agriculture, nuclear power plant, industries, hospitals. Amongst these municipal and domestic wastewaters are the predominant sources of sewage, thereby warranting concentrated research efforts in the field of wastewater treatment (WWT) [6]. The assessment of the characteristics of wastewaters from various sources is crucial in determining the suitable treatment technique, as these wastewaters might exhibit substantial variations and possess distinct physical and chemical properties. In light of the need to safeguard the finite water resources, preserve the environment, and promote human well-being, it is crucial to investigate the treatment of wastewater and its subsequent utilization as a valuable resource. The process of wastewater treatment (WWT) involves the utilization of several physical, chemical, and biological processes to effectively eliminate contaminants from sewage and release of treated water back to the nature [7]. WWT's main challenges include reducing water contamination to a safe level while reducing energy use and environmental effect. The clean water results from wastewater treatment (WWT) is widely used in agriculture and industry protecting public health and environment ultimately leading to sustainable development [8]. It is absolutely necessary to use effective and creative technologies to increase WWT's effectiveness, lower costs, and use less energy [4].

Human intellect is being impersonated by AI in learning, reasoning, making judgements and decisions through computer programs. AI through machine learning (ML) uses algorithms, data processing, statistical methods and optimization. Due to supercomputing capabilities and use of Big Data, ML now has been efficaciously being utilized in various fields [9]. WWT has prioritized monitoring of water quality, designing of process through laboratory research. To optimize and forecast WWT through adoption of AI and ML models is done [10] (Fig. 15.2).

15.2.1 AI Models in Wastewater Sector

Figure 15.3 illustrates the AI models that are frequently deployed in Wastewater Treatment (WWT) in the existing research. Prediction and optimization models are classified as "Artificial Neural Network (ANN), Machine Learning (ML), and Search Algorithm".

1. Artificial Neural Network which is used to stimulates the functioning of Biological Neural networks for data analysis. ANN models simulate neurons with unit nodes and process information by altering the weights of connections among several neurons. [11].
2. Without programming models developed in ML allows systems to learn from data [12]. After learning from data, ML seeks to develop accurate predicted models or decisions [13].
3. Besides ANN and ML, Search Algorithms (SA) like Genetic Algorithms (GA) and Genetic Programming (GP) are also used as AI models [9].
4. By combining two or more AI models, hybrid AI models can maximize model performance in prediction and optimization. Single AI model has disadvantages in wastewater treatment. Researchers are increasingly interested in hybrid AI

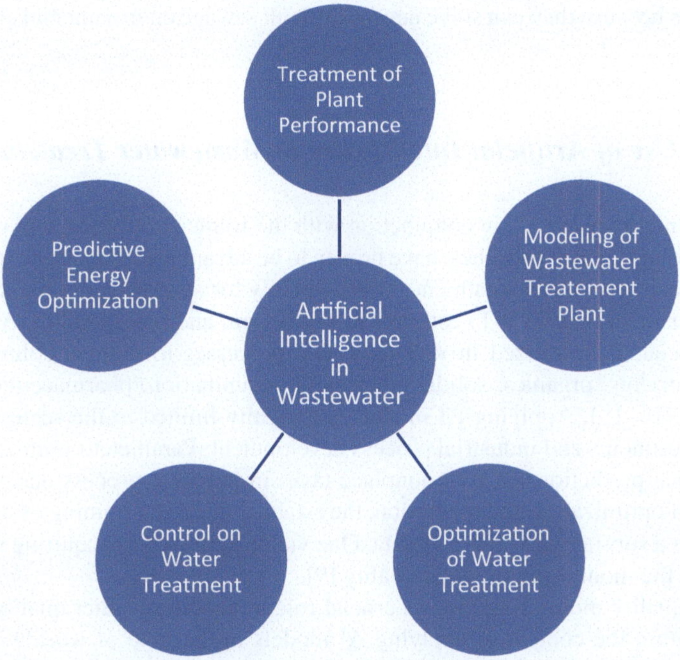

Fig. 15.2 Application of AI in wastewater. *Source* Malviya and Jaspal [4]

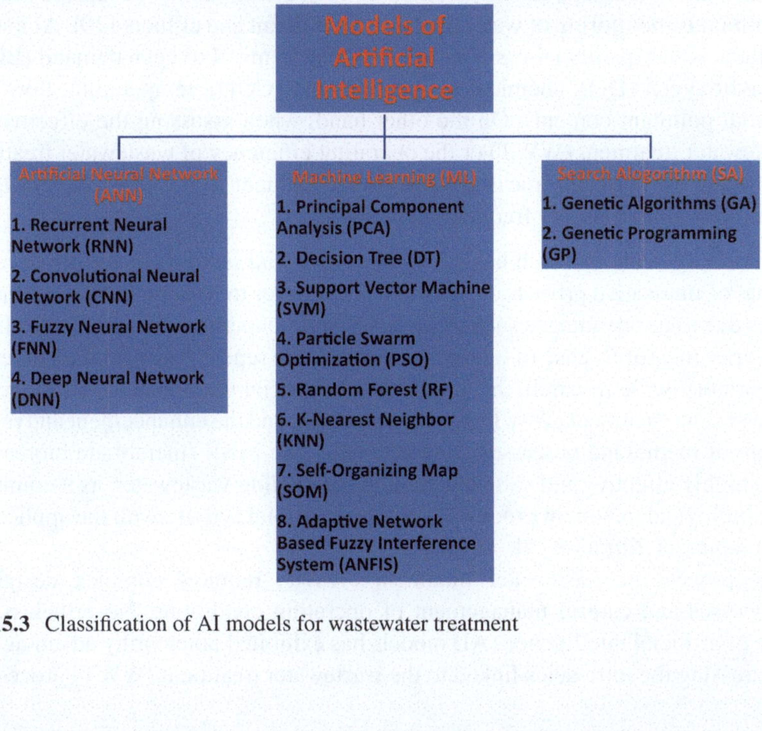

Fig. 15.3 Classification of AI models for wastewater treatment

models because they can solve new or difficult sewage environmental challenges [14].

15.2.2 Use of Artificial Intelligence in Wastewater Treatemnt

Artificial intelligence (AI) in conjunction with the framework of Internet of Things (IoT) and traditional approaches, have proven to be advantageous in the development of intelligent wastewater treatment (WWT) plants for sewage water reuse [15]. In wastewater treatment (WWT) estimation, prediction, and optimization, AI models are beneficial. It is utilized in various WWT processes to remove colours, toxic metals, nutrients, organics, solids, microbial contamination, pharmaceuticals, and pesticides [16–19]. Applying AI models is typically limited to the scales of laboratory experiments and industrial process development. Parameter optimization and performance prediction are commonplace tasks in practical process design. In the domains of optimization and prediction, the establishment and training of AI models necessitate a substantial volume of data. One viable approach to acquiring such data is through the monitoring of water quality [9].

Water quality monitoring plays a crucial role in acquiring water quality metrics or data within the context of applying AI models in the field of worldwide water treatment (WWT). Significant quantities of data can be gathered through the use of sensors used for the ongoing surveillance of influent and effluent water quality. A substantial amount of data can be obtained through the utilization of sensors used for the continuous monitoring of water quality both influent and effluent [20]. AI models for influent water quality assessment use 'pH, biochemical oxygen demand (BOD), dissolved oxygen (DO), chemical oxygen demand (COD), temperature, flow rate, and initial pollutant content'. On the other hand, when assessing the effectiveness of wastewater treatment (WWT) or the operating efficiency of wastewater treatment plants, many water quality measures, including 'effluent BOD, COD, pH, DO, and pollutant concentration' are frequently employed [9].

1. Laboratory-scale research has emerged as a crucial instrument for the development of innovative process designs and wastewater treatment (WWT) technologies due to its advantages such as low cost, short experiment length, controllable security measures, ease of operation, and reliable repeatability. In the context of laboratory-scale research, the utilization of artificial intelligence (AI) models in wastewater treatment (WWT) mostly centres around the enhancement and refinement of membrane processes or bioreactors. The MBR (membrane bioreactor) is a highly effective and valuable method for treating wastewater, as it combines the biological treatment process involving microbial activity with the application of membrane filtration [9].
2. The process of wastewater treatment (WWT) requires complex design of processes and careful management of operating conditions. Nevertheless, the use of artificial intelligence (AI) models has exhibited noteworthy advantages in addressing the intricacies linked to the wastewater treatment (WWT) procedure.

AI models effectively integrate time, temperature, pH, initial pollutant concentration, and influent water quality parameters [21]. AI models are mostly used to optimize process parameters and predict process performance in wastewater treatment process design [22].

15.3 Challenges and Issues in Using Ai in Wastewater Treatment

AI models have multiple benefits over traditional models and are successful in 'wastewater treatment (WWT)' plants, but they also have drawbacks and constraints that prevent their widespread use. AI applications in wastewater treatment tracking and monitoring (WWT) face the following challenges and prospects:

1. The Data Quality and Quantity: One of the challenges faced by AI systems is the necessity for substantial quantities of meticulously curated data in order to facilitate effective training and provide precise predictions. The availability and quality of data provide potential challenges in the context of wastewater treatment [23].
2. A significant challenge is having an accurate AI model in wastewater treatment that can predict precision in in complex processes [24].
3. Adherence to regulatory compliances in wastewater treatment is an essential requirement with water quality standards by the treatment plants. There is an absence of uniform regulation and legal framework to control the activities through AI.
4. In the digital era, use of AI in wastewater treatment plants are susceptible to cyber-attacks and crimes which ultimately can disrupt operations and can compromise the water quality.
5. AI applications in WWT also raises ethical considerations related to privacy, data use and decision-making in the available critical infrastructure.
6. Unavailability of trained and specialized operators who can work with AI systems effectively is also a major challenge worldwide.
7. In the era of technological transformation and rapid changes in the technology, ensuring long-term maintenance and reliability on AI systems can be a challenging issue (Table 15.1; Fig. 15.4).

15.4 Regulatory Framework and Policy

The issue of wastewater as a global concern has become prominent in the aftermath of World War II, with the majority of rules pertaining to water quality, management, and sanitation are developed subsequently. Wastewater management may handle issues related to water supply, pollution, waste recycling, water-borne diseases, and environment. However, it is crucial to emphasize the importance of regulating wastewater

Table 15.1 Perspective challenges of wastewater management

Challenges	Enablers	Specific issues	Scope/Opportunities
Water scarcity and pollution	Advanced treatment technologies	Textile dyeing processes releases toxic chemicals, Mining operations contaminates groundwater Agriculture runoff pollutes water bodies	Membrane filtration for effective color and pollutant removal in textile Bioremediation techniques for complex organic compounds Closed-loop water recycling in agriculture
Regulatory compliances and infrastructure	Digitalization and monitoring	Complex industrial processes require tailored treatment method Leather Industry releases toxic substances Inadequate wastewater treatment infrastructure	Real-time monitoring systems for timely intervention IoT sensors in textile industry for dyeing optimization Closed-loop water recycling in mining for reduced water intake and contamination risks
Deficits	Circular Economy Approaches	High water consumption in mining operations Agriculture wastewater remains untreated	Agricultural wastewater reuse for irrigation and nutrient provisions
Collaborations and knowledge sharing	Collaboration and Knowledge Sharing Regulatory Frameworks and Incentives	Hazardous chemical use in textiles	ZeroDischargeof Hazardous Chemicals (ZDHC) framework Economic incentives for adopting cleaner production processes Regulatory frameworks with fines for non-compliance
Data privacy and security	Deep Learning Methods	Sensitive data exposure and breaches Privacy concerns hinder data sharing	Implementation of secure data encryption protocols Secure data sharing platforms with controlled access
Multi-agent systems	Integration of AI in Decision Making Collaborative Platformsand Tools	Lack of coordination among agents	Development of AI-powered multi- agent frameworks

(continued)

Table 15.1 (continued)

Challenges	Enablers	Specific issues	Scope/Opportunities
Energy efficiency and sustainability	Innovative Technologies Policy Support and Funding	High energy consumption in treatment processes Carbon emissions from energy-intensive industries	Integration of renewable energy sources wastewater Government Incentives for energy- efficient technologies
Emerging contaminants	Research and development Analytical methods and technology	Presence of pharmaceuticals in wastewater Limited analytical methods for detection	Identification of effective treatment methods for emerging contaminants Development of advanced sensors and analytical techniques

Source Singh et al. [25]

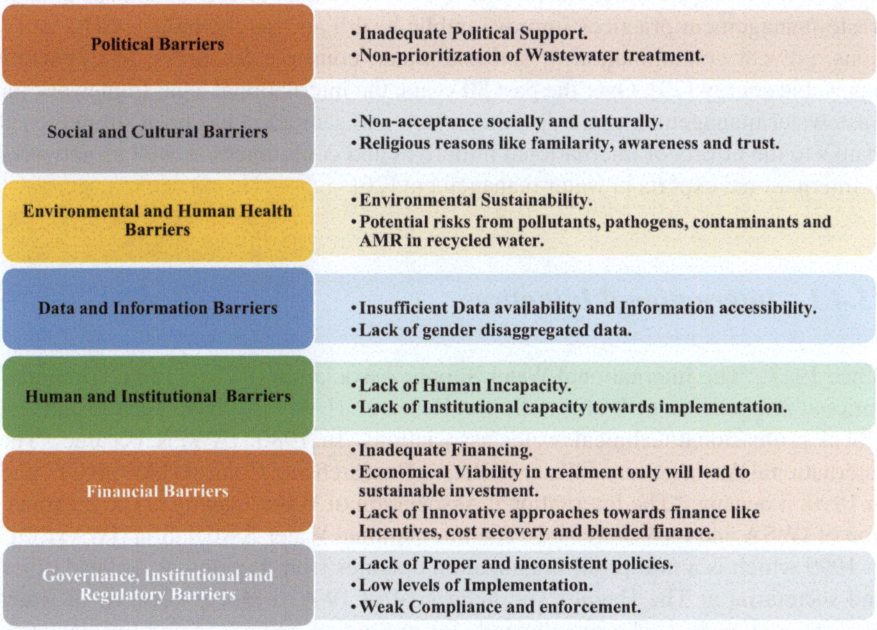

Fig. 15.4 Barriers and concerns of wastewater resource recovery and its safe reuse. *Source* UNEP Report [3]

management [26]. The Global Wastewater Initiative (GW^{2}I) was officially endorsed by the UN Global Programme for Action in 2013 as a means to foster knowledge exchange and promote awareness regarding the challenges and opportunities associated with wastewater management in the context of sustainable development. The GW^{2}I, led by UN-Habitat and UNEP, is a collaborative platform involving several stakeholders. Its primary objective is to facilitate coordinated efforts in wastewater management and foster partnerships that may drive the implementation of comprehensive, efficient, and long-lasting programs addressing wastewater-related issues. There is an anticipation for the enhancement of legislative and institutional frameworks, the creation of tools for the management and assessment of wastewater effects, and the exploration of integrated methods and solutions, such as collaborative demonstration projects [26].

Municipalities in developed and developing countries have increasing challenges in collecting, processing, and disposing of solid waste and wastewater as a result of rising population, income, urbanization, and industrialization [27]. Wastewater from agriculture, livestock, mining, industry, urban waste, and sewage presents operational issues. Financing, managing, and maintaining wastewater treatment infrastructure is difficult, especially in underdeveloped nations [28]. Despite its obstacles, wastewater treatment offers operational options for reuse following treatment. Use it for irrigation, biogas production, industrial use, and drinking water. Direct and indirect potable reuse are other wastewater uses. Affordable, effective, and sustainable waste-management practices improve public health and safety, reduce GHG emissions, prevent contamination of soil and water, conserve resources, and generate renewable energy [29]. Over the past 80 years, the international legal framework for wastewater management, safe drinking water, and sanitation has been strengthened thanks to the efforts of international initiatives and conferences as well as networks of international experts in water management [26].

15.4.1 International Initiatives

Since 1947, "The International Water Supply Association (IWSA)" and "The International Association of Water Pollution Research (IAWPR)" have been prominent global professional/technical water associations. In 1980, IAWPR became "The International Association of Water Pollution Research and Control (IAWPRC)", and in 1990, it became "The International Association of Water Quality". The combination of IWSA and IAWQ created "The International Water Association (IWA) [30]" in 1999 which is a non-profit organization operates from London as its head office and secretariat at The Hague. The main goal of IWA, a global network of water management experts who recommends for promoting and advancing the standards for sustainable water management best practices. World water and sanitation coordination altered in 1990 where, WHO and UNICEF created the Joint Monitoring Programme (JMP) to analyze and document water and sanitation access worldwide. This illuminated global water and sanitation service and enabled cross-country comparisons [31].

The UN General Assembly established "The Water Supply and Sanitation Collaborative Council (WSSCC) [32]" after the 1981–1990 International Drinking Water Supply and Sanitation Decade to promote sustainable sanitation, hygiene, and safe drinking water. WSSCC is to provide sustainable sanitation, hygiene, and clean drinking water to all. Good sanitation and hygiene boost economic and social growth, improving health, productivity, education, and the environment. WSSCC maintains the Global Sanitation Fund, coordinates nationally, regionally, and globally, supports professional development, and campaigns for 2.5 billion people without clean, safe toilets. UNOPS hosts WSSCC, which supports coalitions in over 30 countries and has global members [26]. WSSCC contributed to attaining SDG Goal 6, Target 6.2: "By 2030, achieve access to adequate and equitable sanitation and hygiene for all and end open defecation, paying special attention to women and girls and vulnerable situations." [26] (Fig. 15.5).

The above timeline showcases how wastewater management has been regulated through international initiative from time to time but there is an absence of uniform regulation. International conferences have helped promote pollution control and wastewater management worldwide. This includes:

1. The first United Nations Conference on Human Environment, 1972 in Stockholm, Sweden stressed the importance of UN agencies like FAO, WHO, UNESCO in waste management and water treatment infrastructure development, financing, international cooperation in research, control, and regulation [34]. After that UNEP was established exclusively to deal with environmental issues worldwide.
2. Four water and sustainable development principles were established in "The International Conference on Water and the Environment (ICWE), Dublin in

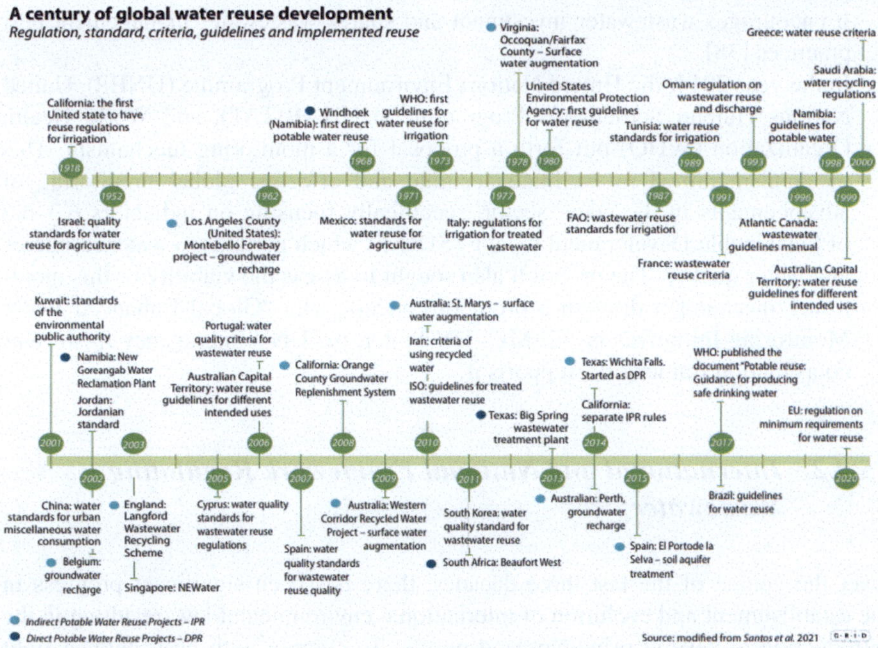

Fig. 15.5 Evolution of wastewater regulation and guidelines (1918–2020) *Source* UNEP [3], Santos et. al. [33]

1992". The action plan includes clean water, sanitation, and water conservation and reuse [35].

3. Agenda 21 and a sustainable development statement were adopted during Earth Summit 1992. Chapter 18 of Agenda 21 addressed freshwater quality and supply, including artificial groundwater recharge, marginal-quality water use, wastewater reuse and recycling, energy from wastes, and wastewater reuse in agriculture [36].
4. Millennium Development Goals (MDG) 2000- Goal 7C included "sustainable access to safe drinking water and basic sanitation as one of the goals to achieve".
5. The "Convention on the Prevention of Marine Pollution by Dumping of Wastes and Other Matter 1972," sometimes referred to as the "London Convention," represents an early international agreement aimed at safeguarding the marine environment against anthropogenic impacts. This convention has been in effect since 1975. The primary aim of this initiative is to facilitate the efficient regulation of various forms of marine pollution and to implement feasible measures to mitigate the contamination of the ocean through the disposal of waste materials and other substances. The parties later in 1996 established a protocol to replace the 1972 convention that forbids dumping with the possible exception of wastes on the so-called "reverse list" in the protocol's annex (IMO) [37].
6. UN Global Programme for Action (GPA) endorsed the Global Wastewater Initiative (GW^2I) to form partnerships and programs to address wastewater challenges and potential amongst various UN agencies, all Nation states, international and government organizations, scientists, the private sector, and other stakeholders. It encourages wastewater investment and shares wastewater management best practices [38].
7. In the year 2014, the United Nations Environment Programme (UNEP), United Nations Human Settlements Programme (UNHABITAT), and World Health Organization (WHO) put forth a proposal for a monitoring mechanism. This mechanism aimed to facilitate accurate and effective global monitoring of advancements in the water sector, specifically focusing on indicators 6.3–6.6 of Sustainable Development Goal 6 (SDG-6) which pertains to water resources and water quality. The proposal also sought to assess the viability of this monitoring mechanism through a proof of concept. The "Global Enhanced Water Monitoring Initiative" is "GEMI". UN-Water, the UN's interagency freshwater coordination framework, supports it.

15.4.2 International and National Framework Regulating Wastewater

Over the course of the last three decades, there has been significant progress in the establishment and evolution of international environmental law, resulting in the introduction of several principles and norms. To comply with basic human right principles, recognition of right to water and sanitation has stimulated the development of domestic laws and policies [26]. Wastewater management internationally is being regulated through set of Internationally recognized legal principles. These principles are applicable when these water sources either form international boundary lines or are intersected by such boundaries. The primary objective of these principles is to mitigate pollution and safeguard the quality of water sources that are shared among states [39]. The International and National Structures that govern the administration of wastewater are as follows:

I. **General Legal Principles**—The Precautionary, Polluter Pays, and Preventative principles are now international environmental law regulations that safeguard water before and after damage. The Environmental Impact Assessment and Liability clause monitors these principles and others [39]. Wastewater administration associated human rights include "the right to water and sanitation, which requires water to be safe and states party to the International Covenant on Economic Social and Cultural Rights to protect water resources from pollution by harmful chemicals and pathogenic microbes." The right to food requires water to be used sustainably to ensure food production's long-term sustainability.

Right to healthy environment has now been globally recognized environmental and human right. In 2012, 177 out of 193 UN member states had constitutional provisions, environmental laws, judicial rulings, or international agreements that recognized this right. The non- compliant countries includes "United States of America, Canada, Japan, Australia, New Zealand, China, Oman, Afghanistan, Kuwait, Brunei Darussalam, Lebanon, Laos, Myanmar, North Korea, Malaysia, and Cambodia". The environment right is recognized by a rising number of municipalities, as well as six states in the United States and five provinces or territories in Canada [40]. There are a number of human rights treaties that deal specifically with environmental rights; these include the ICCPR, ICESCR, ECHR, American CHR, and African CHPR [41]. The Aarhus Convention states in its preamble that "Every person has the right to live in an environment adequate to his or her health and well-being, and the duty, both individually and in association with others, to protect and improve the environment for the benefit of present and future generations."

II. **International Treaties**—When industrial or sewage effluents and agricultural runoff enter international rivers, lakes, or groundwater aquifers, states sharing transboundary waterways have legal obligations. Customary international water law imposes three primary duties on governments that share international watercourses: "(i) the obligation of due diligence not to cause significant harm to other riparian states (No Harm rule); (ii) the obligation of equitable utilization of an international watercourse; and (iii) the "duty to cooperate," which is necessary for the elaboration of procedural rules to achieve and maintain an equitable allocation" [42]. "The UN Convention on the Law of the Non-Navigational Uses of International Watercourses, 1997" [43] established these substantive and procedural regulations. Wastewater pollutes soil and waters, harming biodiversity and drinking water quality; environmental rules must be enforced. Below is a timeline of international wastewater conventions and legislation.

III. **National Framework**—Regulation of Domestic/National Wastewater vary from country to country. The country's structural laws utilize command and

> control, requiring polluters to reduce emissions. Enforcement of licensing regimes and requirements requires sufficient regulatory resources. Water quality and wastewater management operations are monitored and assessed against standards, which are based on national laws that provide the normative framework. National standards institutes from 166 nations form ISO. To analyze bacteria, cleanliness, and other qualities, ISO has produced over 260 water quality standards that provide uniform vocabulary, sampling techniques, reporting, and monitoring instructions. The State authorities, regulatory bodies and industries using water for processing and cooling, agriculture, monitoring laboratories and consultants, water and wastewater service providers, infrastructure manufacturers, planners, designers, contractors, and construction companies, and consumers benefit from ISO's water standards [44]. Public engagement, access to information and justice are crucial for assessing the efficacy and strengthening environmental law enforcement by the regulators.

National environmental, public health, and planning laws regulate wastewater management in addition to the Water Act or Law. "The National Environmental Management Act 107 of 1998 intends to provide for co-operative, environmental governance by establishing principles for decision-making on matters affecting the environment, institutions that will promote co-operative governance and procedures for co-ordinating environmental functions exercised by organs of state". The Integrated Coastal Management Act, 2008 protects the coastal ecosystem by controlling rubbish disposal. The EU Directive, 2000 aims "for 'good status' for all ground and surface waters (rivers, lakes, transitional waters, and coastal waters)". EU's Integrated Pollution Prevention and Control Directive from 1996 to 2010 aims at minimizing pollution from various industrial and agricultural source. The New Zealand-Code of Practice for Management of home Wastewater which is a best practice example for all as it includes the residential wastewater supply, wastewater treatment methods, and wastewater recycling procedures are all part of the system (Ministry of Environment, New Zealand) (Table 15.2; Fig. 15.6).

15.5 Regulation of AI Enabled Wastewater Treatment Plants

Climate change and water scarcity have influenced global water reuse regulations to protect the human health and the environment as a whole. The laws have been modified in order to accommodate technological improvements and new water demands. Scientific and technological advances led the state regulation to set more restrictive criteria's in the field of wastewater [45]. Artificial Intelligence used in Wastewater Treatment ultimately will be regulated through general principles regulating AI in other fields as the challenges and barriers will be almost same. Let us see how far we have moved globally for regulation of AI which has become an integral part each and every activity in today's world. In April 2021, the European Union (EU)

Table 15.2 Water quality management and environmental conventions, treaties and directives

S. No.	Convention/Treaty	Date of enforcement
1	The Basel Convention on the Control of Transboundary Movements of Hazardous Wastes and their Disposal	Adopted on 22nd March 1989 In Force from 5th May 1992
2	The Bamako Convention on the ban of the Import into Africa and Control of Transboundary Movement of Hazardous Waste within Africa	Adopted on 30th January 1991 In Force from 10th March 1999
3	Rotterdam Convention on the Prior Informed Consent (PIC) Procedure for Certain Hazardous Chemicals and Pesticides in International Trade	Adopted on 10th September 1998 In Force from 24th February 2004
4	Convention on Persistent Organic Pollutants	Adopted on 22nd May 2001 In Force from 17th May 2004
5	Convention on the Protection and Use of Transboundary Watercourses and International Lakes	Adopted on 17th March 1992 In Force from 6th Oct. 1996
6	Convention on Access to Information, Public Participation in Decision Making and Access to Justice in Environmental Matters	Adopted on 25th June 1998 In Force from 30th Oct 2001
7	Kiev Protocol on Pollutant Release and Transfer Registers under Convention on Access to Information, Public Participation in Decision Making and Access to Justice in Environmental Matters	Adopted on 21st May 2003 In Force from 8th Oct 2009
8	Convention on the Law of the Non-navigational Uses of International Watercourses (New York, 1997)	In Force from 17th August 2014

Source Good Practices for Regulating Wastewater Treatment—UNEP [26]

proposed "The Artificial Intelligence Act (AI Act)" to regulate AI through principles and rules for member states. Its objectives are to increase AI system safety and security, which must respect fundamental rights legislation and EU values; facilitate investment in automated systems; reinforce responsible AI governance through regulations and principles; and create a trustworthy and safe AI system development eco-system [46]. The 2018 "EU General Data Protection Regulation (GDPR)" requires enterprises worldwide to protect EU citizens' data (Table 15.3).

AI governance demands leaders to follow laws and regulations [47]. Additionally, they must observe ethical, moral and standard criteria's as well [48]. Practitioners should be trustworthy, conscientious, and accountable with intellectual capital and resources like IT, funds, and staff [49, 50]. AI technologies and ML/DL models need procedural governance to operate responsibly. Figure 15.7 shows some of the essentials for responsible AI governance. The following principles govern AI practitioners (including ML/DL developers) [46].

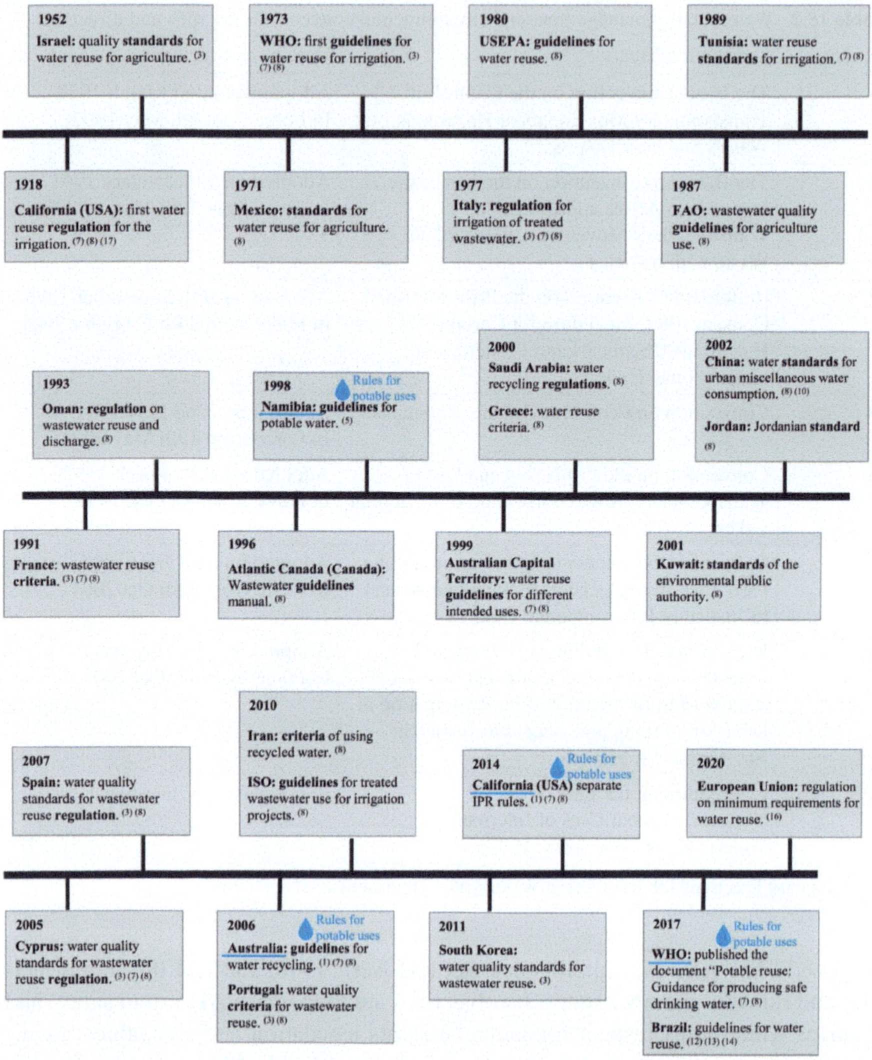

Fig. 15.6 Timeline of Evolution of Global Water Reuse (Regulation, Standard, Criteria or guidelines—RSCG). *Sources:* EU—European Union 2020 Regulation 2020/741 of the European Parliament and of the Council of 25 May 2020 on Minimum Requirements for Water Reuse [33]

A. Accountability and Transparency Denotes Compliance with Regulatory Frameworks Throughout the Lifecycle of Artificial Intelligence Systems (Design, Development, Testing, and Deployment). But in reality, it is difficult to understand the actual working of AI models.
B. Explainability and Interpretability of how AI models works will improve their transparency, trust and reduces bias and unfairness.

Table 15.3 Governance of AI: regulatory principles and guidelines

S. No.	Source of governance	Principle features
1	Singapore National AI Strategy 2019	Explainable; Fair; Reproducibility; Robustness; Transparent
2	Singapore National AI Strategy 2.0 (NAIS 2.0), 2023	"AI for the Public Good, for Singapore and the World", NAIS 2.0 focuses on two twin goals of Excellence and Empowerment. NAIS 2.0 will work with 3 Systems working through 10 Enablers 1. The system of "People & Community" enables through Talent, Capabilities and Placemaking 2. The system of "Infrastructure and Environment" enables through Compute, Data, Trusted Environment and Leader in thought and Action 3. The system of "Activity Drivers" through Industry, Government and Research Examples—Project Pensieve, SELENA+, OneService Chatbot
3	European Union (EU's) Artificial Intelligence Act, 2021	There are different rules for different level of risk like Unacceptable, High, Limited Risks through AI. Accuracy; Clear and adequate information; Detailed documentation; High quality datasets that reduce risks and discrimination; Human oversight measures; Logging of activities to trace any tampering of data; Robustness; Security
4	United States AI Bill of Rights (White House, 2022)	Based upon five principles—Algorithmic discrimination protection; Data privacy; Human alternatives consideration and fallback; Notices and explanation; Safe and effective systems
5	Institute of Electrical and Electronics Engineers (IEEE)'s AI Ethics and Governance Standards, 2023	Addressing ethical issues during design; Child friendly digital services framework; Ongoing evaluations on the impacts of automated systems on human well-being; Data privacy process; Ontological standards for ethically-driven automation systems and robotics; Transparency of autonomous systems; Transparent employer data governance
6	Organization for Economic Cooperation and Development (OECD)'s AI Principles, 2019	Five Principles for responsible stewardship of trustworthy AI-Inclusive growth, sustainable development and well-being, Human-centered values and fairness, Transparency and explainability, Robustness, security and safety Accountability
7	Microsoft's Responsible AI Principles (Microsoft, 2023)	Accountability and transparency; Fairness; Inclusiveness; Privacy; Safety and Security; Reliability
8	IBM's AIGovernance (IBM, 2022)	Explainability; Fairness; Privacy; Robustness; Transparency

Source Camilleri [46]

Fig. 15.7 Ideal framework of responsible AI governance. *Source* Camilleri [46]

C. Fairness and Inclusiveness in the AI governance will be helpful in correcting algorithmic biases created by the developers.
D. Personal data regulations must protect customers who give AI their data. Organizations must notify consumers that they are collecting and storing biometric data under many data privacy laws. Businesses using such technologies cannot utilize consumer data without approval.
E. Developers are responsible for ensuring the security and robustness of AI systems, reducing threats such as irregularities, leaks, privacy violations, and malicious activities [49].

15.6 Conclusion and Suggestions

It is crucial to establish comprehensive regulatory frameworks and policy recommendations for AI-enabled wastewater treatment in order to effectively manage the intersection of modern technology and critical environmental management. By considering concerns pertaining to data privacy, environmental impact, ethics, and international cooperation, it is possible to leverage the capabilities of artificial intelligence in order to optimize the effectiveness and sustainability of wastewater treatment procedures. This can be achieved while simultaneously ensuring the protection of our environment, public health, and data privacy. AI is being used in wastewater treatment now a days significantly. To ensure the responsible and efficient implementation of new technologies, it is imperative to establish proactive and well-designed rules. Ultimately, this will contribute to the advancement of a cleaner and more sustainable global environment.

Although we have plethora of laws, regulations internationally and nationally available but these laws are not uniform throughout as it varies country to country. Over the past century, wastewater reuse laws and practices have been tracked worldwide. Non-drinking uses, such as agriculture and irrigation, were the original inspiration for the criteria utilized in regulation. Because of population expansion and climate change, there has been a greater demand for potable water, and technological advancements have made it possible to produce higher-quality reclaimed water. As a result, areas hardest hit by water scarcity have begun implementing indirect and direct potable reuse programs tailored to their needs. Based on technology and scientific advancements, integrated water and wastewater management became an important consideration for decision makers to implement more effective strategies to solve future global water concerns. The present chapter aims to examine the governance of artificial intelligence (AI) in AI enabled Wastewater Treatment. It has provided insight into the significant legislation, as well as the reflexive principles and recommendations, aimed at practitioners engaged in the study, testing, development, and implementation of AI models and wastewater treatment. The chapter also elucidated the efforts made by institutions, non-governmental groups, and technological conglomerates to implement protocols, including self-regulations, in order to proactively mitigate the occurrence of contingencies resulting from inadequate AI governance.

Policymakers face pacing issues because innovation moves faster than lawmaking. Thus, innovation regulation by governments worldwide is reactive. Some global AI-enabled wastewater treatment regulation improvements: Policy makers are encountering a challenge in keeping pace with the rapid proliferation of innovation, which outpaces the process of lawmaking. Consequently, governments across the globe tend to exhibit a tendency towards reactive approaches when it comes to implementing regulatory measures pertaining to innovation. Several recommendations can be made to enhance the regulation of AI-enabled wastewater treatment on a worldwide scale -

1. To maintain data quality and quantity we need to use different AI models for different purposes so as to minimize the chances of biasness.
2. An accurate AI model in wastewater treatment that can predict precision in in complex processes is possible only when there is transparency, accountability on the part of AI developers about the disclosure of human understandable algorithms.
3. There is requirement of uniform regulation and legal framework to control AI activities which will ensure water quality and standards in the treatment plants.
4. There is requirement of secured and robust AI system that is updated from time to time so as to minimize the chances of being cyberattacked and crimes in wastewater treatment plants.
5. To tackle the issue of ethics related to privacy, data use and decision-making in the available critical infrastructure, we need to have robust and uniform data protection laws on the lines of EU GDPR.

6. To ensure the availability of trained and specialized operators who can work with AI systems effectively are required to be recruited. They are being appraised with innovative changes happening around in the technology and regulations from time to time.
7. In the era of technological transformation and rapid changes in the technology, to ensure long-term maintenance and reliability on AI systems we need to update the technology continuously.
8. Financial Mechanism—To ensure the long-term viability of sewage treatment facilities, it is important to set up reliable finance mechanisms (funds, taxes, tariffs) that allow for adequate cost recovery.
9. Institutional Monitoring—To make recycled materials more marketable, relevant organizations and stakeholders must be better supervised and engaged.
10. Target based Regulation—The development of goal-oriented rules, clearly articulated national reuse criteria for treated sewage, and robust enforcement mechanisms is urgently required [51–53].

References

1. World Health Organization (WHO), UN HABITAT Geneva (2018) Progress on safe treatment and use of wastewater: piloting the monitoring methodology and initial findings for SDG indicator 6.3.1.Licence: CC BY-NC-SA 3.0 IGO. Accessed 23 Oct 2023
2. United Nations DSDG (2022) The sustainable development goals report. https://unstats.un.org/sdgs/report/2022/The-Sustainable-Development-Goals-Report-2022.pdf. Accessed 30 Sep 2023
3. United Nations Environment Programme (2023). Wastewater—turning problem to solution. A UNEP rapid response assessment. Nairobi. https://doi.org/10.59117/20.500.11822/43142. Accessed 25 Oct 2023
4. Malviya A, Jaspal D (2021) Artificial intelligence as an upcoming technology in wastewater treatment: a comprehensive review. Environ Technol Review 10:177–187.
5. Chahal C, Akker VD et al (2016) Pathogen and particle associations in wastewater: significance and implications for treatment and disinfection processes. Adv Appl Microbiol 97:63–119
6. Jasim NA (2020) The design for wastewater treatment plant (WWTP) with GPS X modelling. Cogent Eng 7:1723782
7. Zaibel I, Arnon S et al (2022) Treated municipal wastewater as a water source for sustainable aquaculture: a review. Rev Aquaculture 14:362–377
8. Ren J, Shen W et al (2021) Applications of artificial intelligence in process systems engineering. Elsevier, Amsterdam, The Netherlands
9. Wang Y, Cheng Y, Liu H et al (2023) A review on applications of artificial intelligence in wastewater treatment. Sustainability 15:13557. https://doi.org/10.3390/su151813557
10. Safeer S, Pandey RP et al (2022) A review of artificial intelligence in water purification and wastewater treatment: recent advancements. J Water Process Eng 49:102974
11. Basheer IA, Hajmeer M (2000) Artificial neural networks: fundamentals, computing, design, and application. J Microbiol Method 43:3–31
12. Samuel AL (1959) Some studies in machine learning using the game of checkers. IBM J Res Dev 3:210–229
13. Dey A (2016) Machine learning algorithms: a review. Int J Comput Sci Inf Technol 7:1174–1179

14. Zhao L, Dai T et al (2020) Application of artificial intelligence to wastewater treatment: a bibliometric analysis and systematic review of technology, economy, management, and wastewater reuse. Process Saf Environ Prot 133:169–182. https://doi.org/10.1016/j.psep.2019.11.014
15. Krishnan SR, Nallakaruppan MK et al (2022) Smart water resource management using artificial intelligence—a review. Sustainability 14:13384
16. Abidli A, Huang Y et al (2022) Sustainable and efficient technologies for removal and recovery of toxic and valuable metals from wastewater: recent progress, challenges, and future perspectives. Chemosphere 292:133102
17. Altowayti WAH, Shahir S et al (2022) The role of conventional methods and artificial intelligence in the wastewater treatment: a comprehensive review. Processes 10:1832
18. Fan C, Lv C et al (2022) Influence of regular addition of ore on treatment efficiency and aquatic organisms in living machine system for black water treatment. J Clean Prod 341:130928
19. Alam G, Ihsanullah I et al (2022) Applications of artificial intelligence in water treatment for optimization and automation of adsorption processes: recent advances and prospects. Chem Eng J 427:130011
20. Singh NK, Yadav M et al (2023) Artificial intelligence and machine learning-based monitoring and design of biological wastewater treatment systems. Bioresource Technol 369:128486
21. Shi S, Xu G (2018) Novel performance prediction model of a biofilm system treating domestic wastewater based on stacked denoising auto-encoders deep learning network. Chem Eng J 347:280–290
22. Bein E, Zucker I et al (2021) Ozone membrane contactors for water and wastewater treatment: a critical review on materials selection, mass transfer and process design. Chem Eng J 413:127393
23. Gohil J, Patel J et al (2021) Advent of Big Data technology in environment and water management sector. Environ Sci Pollut Res 28:64084–64102. https://doi.org/10.1007/s11356-021-14017-y
24. Rao Mohd MM et al (2022) Artificial intelligence in waste management/wastewater treatment. In: Garg VK et al (ed) Omics for environmental engineering and microbiology systems. CRC Press. https://doi.org/10.1201/9781003247883
25. Singh BJ, Chakraborty A, Sehgal R (2023) A systematic review of industrial wastewater management: evaluating challenges and enablers. J Environ Manage 348:119230. https://doi.org/10.1016/j.jenvman.2023.119230
26. United Nations Environment Programme (UNEP) (2015) Good practices for regulating wastewater treatment: legislation, policies and standards copyright. Accessed 12 Oct 2023
27. Intergovernmental Panel on Climate Change (IPCC) (2007) Fourth assessment on climate change Chapter 10.1—Introduction. http://www.ipcc.ch/publications_and_data/ar4/wg3/en/ch10s10-1.html. Accessed 5 Oct 2023
28. Corcoran E, Nellemann C et al (2010) Sick Water? The central role of wastewater management in sustainable development. A Rapid Response Assessment. United Nations Environment Programme (UNEP), UN-HABITAT, GRID-Arendal, p. 31. www.grida.no
29. Meehl GA, Stocker TF et al (2007) Global climate projections. In: Climate change—the physical science basis. Contribution of Working Group I to the Fourth Assessment Report of the Intergovernmental Panel on Climate Change. https://archive.ipcc.ch/publications_and_data/ar4/wg1/en/ch10.html
30. International Water Association. http://www.iwa-network.org/. Accessed 20 Oct 2023
31. International Water Association. https://iwa-network.org/iwahistory/. Accessed 20 Oct 2023
32. The Water Supply and Sanitation Collaborative Council (WSSCC) https://wateractionhub.org/organizations/49/d/water-supply-and-sanitation-collaborative-council/. Accessed 15 Oct 2023
33. Santos A, Pachawo V et. al. (2022) Progress on legal and practical aspects on water reuse with emphasis on drinking water an overview. Water Supply 22(3):3000–3014. https://doi.org/10.2166/ws.2021.412
34. Report of the United Nations Conference on the Human Environment Stockholm 1972. http://www.unep.org/Documents.Multilingual/Default.asp?DocumentID=97. Accessed 25 Oct 2023
35. International Conference on Water and Environment (ICWE, Dublin 1992) https://www.wmo.int/pages/prog/hwrp/documents/english/icwedece.html. Accessed 15 Oct 2023

36. United Nations Conference on Environment & Development, Rio de Janeiro, Brazil (3 to 14 June 1992) Agenda 21. http://www.un.org/esa/sustdev/agenda21.htm. Accessed 18 Oct 2023
37. International Maritime Organization (IMO) http://www.imo.org/About/Conventions/ListOfConventions/Pages/Convention-on-the-Prevention-of-Marine-Pollution-by-Dumping-of-Wastes-and-Other-Matter.aspx. Accessed 14 Oct 2023
38. The United Nations World Water Development (UNWWD) Report (2014) Water and energy, united nations world water assessment programme, UNESCO 26. https://www.unwater.org/publications/un-world-water-development-report-2014. Accessed 17 Oct 2023
39. UNEP Training Manual for International Environmental Law (2006) 33. https://wedocs.unep.org/bitstream/handle/20.500.11822/20599/UNEP_Training_Manual_Int_Env_Law.pdf?sequence=1&amp%3BisAllowed. Accessed 10 Oct 2023
40. Boyd D (2012) The constitutional right to a healthy environment, environment magazine July–August
41. Boyle A (2012) The European Journal of International Law (Oxford University Press) 23(3)
42. McCaffrey SC (2001) The law of international watercourses, non-navigational uses. Oxford University Press 399
43. Convention on the Law of Non-Navigational Uses of International Watercourses (CLNUIW) (1997 May 21) reprinted in 36 I.L.M. 700. Entered into force on 17 August 2014
44. ISO and Water, Global solutions to global challenges, January 2012 http://www.iso.org/iso/iso_and_water.pdf. Accessed 12 Oct 2023
45. Olivieri A, Pecson W et al (2020) Chapter 2: California water reuse-Past, present and future perspectives. In: Advances in chemical pollution, environmental management and protection 5. https://doi.org/10.1016/bs.apmp.2020.07.002
46. Camilleri MA (2023) Artificial intelligence governance: ethical considerations and implications for social responsibility. Expert Syst:e13406. https://doi.org/10.1111/exsy.13406
47. Mantymaki M, Minkkinen M et al (2022) Defining organizational AI governance. AI Ethics 2(4):603–609
48. Koniakou V (2023) From the rush to ethics to the race for governance in artificial intelligence. Inf Syst Front 25(1):71–102
49. Agbese M, Alanen HK et al (2023) Governance in ethical and trustworthy AI systems: extension of the ECCOLA method for AI ethics governance using GARP. E-Informatica Softw Eng J 17(1):23010
50. Smuha NA (2019) The EU approach to ethics guidelines for trustworthy artificial intelligence. Computer Law Review Int 20(4):97–106
51. Jones ER, Van Vliet MTH et al (2021) Country-level and gridded estimates of wastewater production, collection, treatment and reuse. Earth Syst Sci Data 13(2):237–254. https://doi.org/10.5194/essd-13-237-2021
52. Malviya A, Jaspal D (2021) Artificial intelligence as an upcoming technology in wastewater treatment: a comprehensive review. Environ Technol Rev 10(1):177–187. https://doi.org/10.1080/21622515.2021.1913242
53. Santos ASP, Pachawo V et al (2022) Progress on legal and practical aspects on water reuse with emphasis on drinking water—an overview. Water Supply 22(3):3000–3014. https://doi.org/10.2166/ws.2021.412

SPRINGER NATURE

GPSR Compliance

The European Union's (EU) General Product Safety Regulation (GPSR) is a set of rules that requires consumer products to be safe and our obligations to ensure this.

If you have any concerns about our products, you can contact us on ProductSafety@springernature.com

In case Publisher is established outside the EU, the EU authorized representative is:

Springer Nature Customer Service Center GmbH
Europaplatz 3
69115 Heidelberg, Germany

The manufacturer's authorised representative in the EU is Springer Nature Customer Service Centre GmbH, Europaplatz 3, 69115 Heidelberg, Germany. If you have any concerns regarding our products, please contact ProductSafety@springernature.com

Printed and bound by CPI Group (UK) Ltd, Croydon, CR0 4YY
01/10/2026
02225904-0015